Understanding Spectroscopy

Understanding Spectroscopy

Georgia Hunt

MURPHY & MOORE
www.murphy-moorepublishing.com

Understanding Spectroscopy
Georgia Hunt
ISBN: 978-1-63987-559-7 (Hardback)

MURPHY & MOORE
Published by Murphy & Moore Publishing,
1 Rockefeller Plaza,
New York City, NY 10020, USA

Cataloging-in-Publication Data

Understanding spectroscopy / Georgia Hunt.
p. cm.
Includes bibliographical references and index.
ISBN 978-1-63987-559-7
1. Spectrum analysis. 2. Spectroscope. 3. Chemistry, Analytic--Qualitative.
I. Hunt, Georgia.
QD95 .U53 2022
543.085 8--dc23

For more information regarding Murphy & Moore Publishing and its products, please visit the publisher's website www.murphy-moorepublishing.com

Table of Contents

Preface

The branch of science which studies the interaction between various forms of matter and electromagnetic radiation is known as spectroscopy. It measures the radiation intensity as a function of its frequency or wavelength. Spectroscopy is used to explain various phenomena in the field of chemistry, physics and astronomy. It also allows the study of the physical and chemical composition of matter at atomic as well as astronomical distances. Spectroscopic techniques can be classified into acoustic spectroscopy, absorption spectroscopy, emission spectroscopy, atomic spectroscopy, nuclear spectroscopy, etc. It is used for various purposes such as measurement of toxic elements in blood, non-destructive elemental analysis, measuring different compounds present in food sample, etc. This book is a compilation of chapters that discuss the most vital concepts in the field of spectroscopy. It presents the complex subject of spectroscopy in the most comprehensible and easy to understand language. This book will provide comprehensive knowledge to the reader.

Given below is the chapter wise description of the book:

Chapter 1- The branch of science which deals with the study of the spectra produced during the interaction between matter and electromagnetic radiation is known as spectroscopy. There are various types of spectroscopy such as absorption spectroscopy, astronomical spectroscopy, atomic absorption spectroscopy, circular dichroism spectroscopy, etc. This is an introductory chapter which will briefly introduce all these types of spectroscopy.

Chapter 2- Physical spectroscopy is one of the main branches of spectroscopy. Some of its significant subfields of are photoelectron spectroscopy, vibrational spectroscopy and microwave spectroscopy. This chapter has been carefully written to provide an easy understanding of these varied facets of physical spectroscopy.

Chapter 3- Organic spectroscopy derives structural information by using various techniques. A few of these techniques are UV-Visible spectroscopy, infrared spectroscopy and nuclear magnetic resonance spectroscopy. The aim of this chapter is to explore these techniques of organic spectroscopy which are closely related to its progress.

Chapter 4- Atomic and molecular spectroscopy is a widely used set of analytical techniques. It involves the study of the structure of atoms and molecules. All the diverse principles and theories of atomic and molecular spectroscopy such as Pauli Exclusion Principle and molecular orbital theory have been carefully analyzed in this chapter.

Chapter 5- In laser spectroscopy, lasers are used as light sources. The invention and development of the laser have helped in advancing the field of spectroscopy. Science and technology have undergone rapid developments in the past decade which has resulted in the discovery of significant tools and techniques in the field of laser spectroscopy; which have been extensively detailed in this chapter.

At the end, I would like to thank all those who dedicated their time and efforts for the successful completion of this book. I also wish to convey my gratitude towards my friends and family who supported me at every step.

Georgia Hunt

1

Spectroscopy: An Introduction

The branch of science which deals with the study of the spectra produced during the interaction between matter and electromagnetic radiation is known as spectroscopy. There are various types of spectroscopy such as absorption spectroscopy, astronomical spectroscopy, atomic absorption spectroscopy, circular dichroism spectroscopy, etc. This is an introductory chapter which will briefly introduce all these types of spectroscopy.

Spectroscopy is the technique of splitting light (or more precisely electromagnetic radiation) into its constituent wavelengths (a spectrum), in much the same way as a prism splits light into a rainbow of colours. However, in general, a spectrum is generally more than a simple 'rainbow' of colours. The energy levels of electrons in atoms and molecules are quantised, and the absorption and emission of electromagnetic radiation only occurs at specific wavelengths. Consequently, spectra are not smooth but punctuated by 'lines' of absorption or emission.

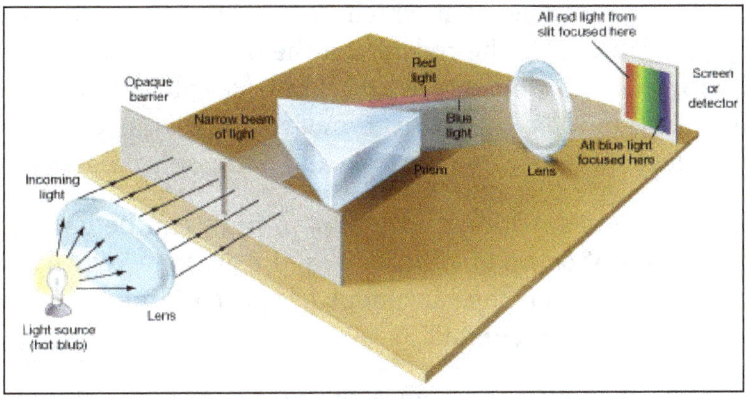

A spectrograph separates light into its constituent wavelengths. The spectroscopy of optical light produces the familiar rainbow.

Old style spectroscopy was carried out using a prism and photographic plates. These days, modern spectroscopy uses diffraction gratings to disperse the light, which is then projected onto CCDs (Charge Coupled Devices) similar to those used in digital cameras. The 2-dimensional spectra are easily extracted from this digital format and manipulated to produce 1-dimensional spectra like the galaxy spectrum shown below.

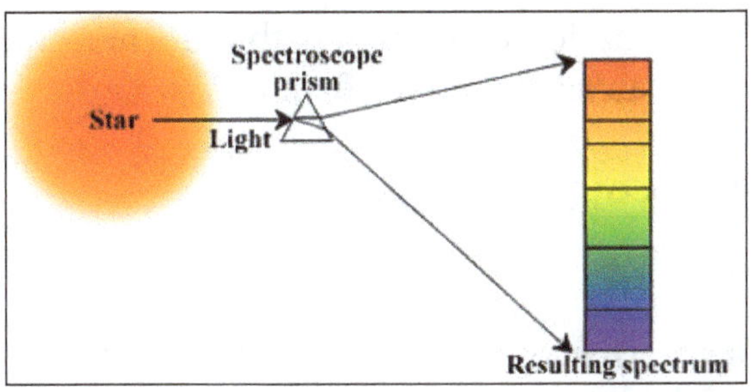

The spectra of stars are not simple smooth 'rainbows' of colors, rather they are punctuated by absorption and emission lines.

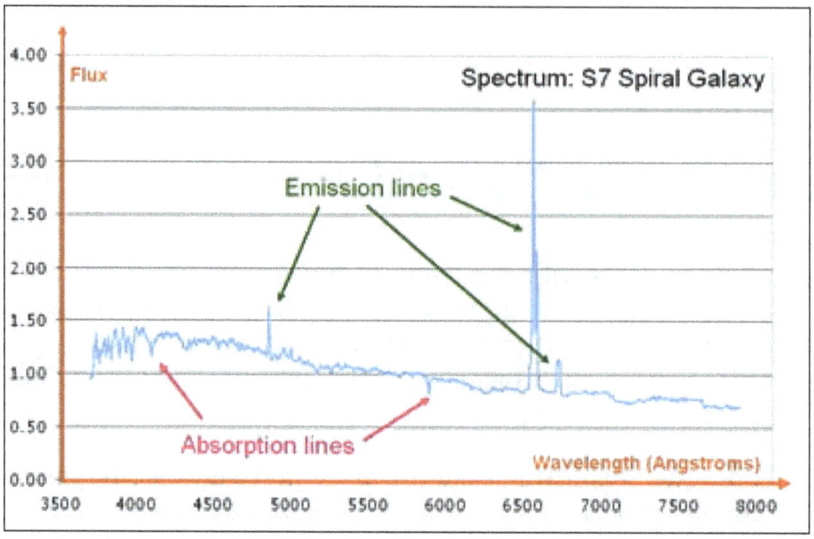

The spectrum of an S7 spiral galaxy showing both emission and absorption line features. Wavelength is measured in angstroms while the flux is in arbitrary units.

These 1-dimensional spectra contain an astonishing amount of useful data:

- The precise position (wavelength) at which known emission and absorption lines are detected can be used to measure the redshift of the observed object. For example, if the Hβ (486.2 nm) hydrogen line is detected at 487.8 nm, we calculate that the object has a recession velocity of 1,000 km/sec. Apart from measuring distances, this type of analysis can also be used to detect spectroscopic binaries and extra-solar planets.
- The widths of both the absorption and emission lines can be used to measure the internal velocity dispersion of complex objects (e.g. the average velocity of stars within galaxies). This is achieved through measurement of spectral line broadening from which the velocity broadening can be calculated.
- Many (indeed most) absorption and emission lines are produced by metals. The depths or heights of these lines can be used to estimate the abundances of the metals responsible. In stars, these also permit the measurement of the temperature and pressure of the stellar atmosphere. In objects such as galaxies, they permit an estimation of age, though the spectra are somewhat susceptible to the age-metallicity degeneracy making this a difficult procedure.
- Finally, the overall shape of the spectrum contains clues to many other aspects of the observed population including age and interstellar reddening (extinction).

The enormous quantities of information contained within a single spectrum makes spectroscopy one of the most powerful tools at an astronomer's disposal.

Various Types of Spectroscopy

Absorption Spectroscopy

Absorption spectroscopy is a spectroscopic technique that measures the frequency of wavelength of absorbed light as a result of the interaction between the light and a sample.

Principle of Absorption Spectroscopy

Absorption spectroscopy is based on the principle that materials have an absorption spectrum which is a range of radiation absorbed by the material at different frequencies. The absorption spectrum of materials depends on the atomic and molecular composition of that material. The frequency of light radiation absorbed by a material is dependent on the energy difference between the two energy states of the molecules. The absorption results in the formation of an absorption line, which, together with other lines, form an absorption spectrum.

Thus, when a photon with sufficient energy reaches an object, the energy is absorbed by the electrons causing them to bump into a higher energy state. The amount of photon (radiation) absorbed results in an absorption spectrum which can then be measured in terms of absorbance. The absorbance of a sample is dependent on the number of excited electrons which in turn is dependent on the concentration of molecules in the sample.

Steps of Absorption Spectroscopy

- Solvent liquid and the sample solution are taken in two transport vessels, also termed as cuvettes.
- The vessel with solvent liquid is then placed in the spectrometer to determine the light loss due to scattering and absorbance by the solvent. Any absorbance observed in this process is to be subtracted from the absorbance of the sample.
- The cuvette with the sample solution is then placed in the spectrometer.
- The absorbance of the sample is noted in different frequencies which usually ranges from 200-800 nm.
- A similar spectrum is formed from a different concentration of the samples.
- A graph of the absorbance measured against the concentration of the sample is plotted, which can then be used for the determination of the unknown concentration of the sample.

Uses of Absorption Spectroscopy

- Absorption spectroscopy is used to determine the presence of a particular substance in a sample and then to quantify the present substance.
- This technique has been applied in remote sensing, which allows the determination of the concentration of hazardous substances without the direct interaction of the instrument and the sample.
- Absorption spectroscopy is also used for the determination of the atomic and molecular structure of various substances.

Astronomical Spectroscopy

Astronomical spectroscopy is the study of astronomical structures by using the principle of spectroscopy for the measurement of the electromagnetic spectrum, which is radiated from stars or other celestial bodies.

Principle of Astronomical Spectroscopy

The spectrum produced by the light coming from celestial objects is not as smooth as the spectrum of white light. Both the absorption and emission spectrum of light is produced by passing the light through a diffraction grating. The light, when dispersed, forms absorption and emission lines which are dependent on the energy levels of electrons in atoms and molecules of the object present in the source. Based on the absorption of radiation at a particular wavelength, a graph is plotted between the wavelength and the flux of the absorption and emission lines. Most of these lines are formed due to the presence of metals. Thus, depending on the height and depth of these lines, the abundance of these metals can be quantified.

Steps of Astronomical Spectroscopy

- In astronomical spectroscopy, the incoming light from various stars is passed through a telescope into the spectroscope.
- As the light reaches the diffraction grating of the spectroscope, the light is dispersed into different wavelengths.
- The dispersed wavelengths land on the photo-detectors which analyze the nature of the wavelengths.
- The detectors form the flux scale of the spectrum depending on the wavelength by comparison with observations of standard stars.

Uses of Astronomical Spectroscopy

- The absorption of light from the stars reveals various properties of stars like chemical composition, temperature, density, mass, and relative motion.
- The study of the stellar spectrum helps in the study of galaxies and the composition of galaxies.
- Astronomical spectroscopy can also be used for the determination of the motion of stars and galaxies through the Doppler effect and redshift.

Atomic Absorption Spectroscopy

Atomic absorption is an analytical technique utilizing the principle of spectroscopy for the quantitative determination of chemical elements.

Principle of Atomic Absorption Spectroscopy

Atomic absorption spectroscopy utilizes the principle that free electrons generated in an atomizer absorb radiation of different wavelengths. The free electrons absorb UV or visible light, causing the electrons to transfer to higher energy orbits. During this process, the absorption spectrum is released, which is detected by the photo-detectors. The absorption spectrum formed allows the quantification of free electrons in the gaseous state of the matter. The amount of photon (radiation) absorbed results in an absorption spectrum which can then be measured in terms of absorbance. The absorbance of a sample is dependent on the concentration of molecules in the sample.

Steps of Atomic Absorption Spectroscopy

- The liquid sample is mixed with a particular volume of spirit which is added to a flask which is then vaporized into a gas by a fuel-rich acetylene-nitrous oxide flame.
- A lamp is set with the necessary wavelength as a light source.
- The gas formed from the liquid sample is then passed through a detector that detects the absorbance of the atoms in the gas.
- A similar process is performed for the detection of absorbance of solvent bank and standard solution.
- A graph is plotted for the absorbance against the concentration of the molecules in the sample.

Uses of Atomic Absorption Spectroscopy

- Atomic absorption spectroscopy can be used for the quantitative and qualitative determination of metallic elements in biological systems.
- This also helps in the detection of metals as an impurity in alloys and other mixtures.
- Atomic absorption spectroscopy has been utilized for the purification of environmental samples like water and soil.
- Detection of metals in pharmaceutical products and oil products can also be done by this method.

Circular Dichroism Spectroscopy

Circular dichroism spectroscopy is a type of light absorbance spectroscopy that measures the differences in the absorbance of right and left polarized light.

Principle of Circular Dichroism Spectroscopy

Left- and right-handed polarized components of the incident light are absorbed differently by the sample, which yields a difference in the absorption coefficients. This difference is termed circular dichroism. Circularly-polarized light rays will travel through an optically active medium with different velocities due to the different indices of refraction for right- and left-circularly polarized light. Optically active chiral molecules will preferentially absorb one direction of the circularly polarized light. The peptide bonds in protein act as chromophores. The peptide bonds are the optically active chiral molecules of protein, and the number of chromophores is proportional to the magnitude of absorption. Thus, the magnitude of absorption is then used for the verification of the adopted secondary structure of proteins.

Steps of Circular Dichroism Spectroscopy

- The sample is placed in a transport vessel with buffers which is then placed in the spectrometer.
- In the spectrometer, left and right circularly polarized light passes through the sample in an alternating fashion.

- The photomultiplier detector in the spectrometer produces a voltage proportional to the circular dichroism (the difference between the absorbance of left and right polarized light) of the resultant beam emerging from the sample.
- The circular dichroism of the sample is then compared with standard proteins to determine the differences in the secondary structure of the proteins.

Uses of Circular Dichroism Spectroscopy

- The primary application for CD spectroscopy is the verification of the assumed secondary structure of the protein. This method allows the detection of percentages of α-helix and β-sheet in proteins base on their circular dichroism.
- Circular dichroism spectroscopy can also be used to monitor changes of secondary structure within a sample over time.
- This technique can also be used to compare two macromolecules to detect the differences in the structure of the molecules.
- It can also be used for the analysis of pharmaceutical products to ensure that they are still present in the folded active conformations.

Electrochemical Impedance Spectroscopy (EIS)

Electrochemical impedance spectroscopy is an advanced electrochemical technique that measures the impedance of a system by applying different AC potential frequencies.

Principle of Electrochemical Impedance Spectroscopy (EIS)

- Electrochemical Impedance Spectroscopy is a technique that measures how a new material or device impedes the flow of electricity.
- This is done by applying an AC signal through the electrodes connected to the sample.
- AC voltage at different frequencies is applied to a sample, and the electrical current is measured.
- A Nyquist plot is generated from the response of frequency to the electrical impedance by plotting the impedance on the y-axis and the frequency on the x-axis.
- The instrument applies an alternating field voltage to the sample and measures the current response.
- The real and imaginary components of impedance are calculated by determining the phase shift and change in amplitude at different frequencies.

Steps of Electrochemical Impedance Spectroscopy (EIS)

- A test module is first hooked to the EIS that confirms that the wires connected to the system are hooked correctly, and all the parts are working correctly.

- To start the flow of current in the system, the Zplot software is operated on the computer. The required parameters are set, and the AC amplitude is set to 10mV. The initial frequency is set to 1× 106 and the final frequency to 100Hz.
- The Zview software is then operated to view the results.
- The electrodes are then removed from the test module.
- The sample is prepared and assembled into the test furnace located in the hood fume.
- The electrodes are attached to the assembly, and the EIS is operated as the same previous procedures.

Uses of Electrochemical Impedance Spectroscopy (EIS)

- The impedance of a sample can be used to determine the population of microorganisms when bacteria grow in a sample.
- EIC can also be used to screen for cancer tissues where the impedance of the electric current changes as the cell structure and size changes.
- This has also been used to study the application of layers of chemicals, polymers, or coatings to electrodes which provide useful enhancements in terms of electron transfer and sensitivity.

Electron Spin Resonance (ESR) Spectroscopy

Electron Spin Resonance (ESR) also known as Electron Magnetic Resonance (EMR) or Electron Paramagnetic Resonance (EPR) is a type of absorption spectroscopy in which radiations having a frequency in the microwave region (0.04 – 25 cm) are absorbed by paramagnetic substances resulting in transitions between magnetic energy levels of electrons having unpaired spins.

Principle of ESR Spectroscopy

Electron spin resonance spectroscopy is based on the fact that the atoms, molecules, or ions with unpaired electrons exhibit magnetic properties because of the electron spin. When a molecule or compound is placed in a magnetic field, the spin of the unpaired electrons can align in two different states. If the spin is aligned in the direction of the magnetic field, it results in a lower energy state. In contrast, the opposite alignment of the magnetic field results in a higher energy state. As a result, the unpaired electron can move between these two energy states by either absorbing or emitting photons of energy resulting in a resonance condition. The spectrum produced during this absorption is passed to the detectors for the analysis of the electronic structures of the sample.

Steps of ESR Spectroscopy

- The ESR instrument is turned on and warmed up for 30 minutes. The necessary parameters are set in the ESR.
- Calibration is performed by adding an empty ESR tube to ensure no background signals from the tube or the instrument is present.

- The sample is then placed in the ESR tube, and the ESR spectrum is measured.

Uses of ESR Spectroscopy

- Electron spin resonance spectroscopy is one of the essential methods for the study of transition metal-containing metalloproteins.
- This method also allows the study of the denaturation of proteins and protein folding.
- ESR also helps determine the rate of catalysis of different metals.
- It has also been widely applied in the field of biological research of quantitative and qualitative analyses of reactive nitrogen species (RNS) and reactive oxygen species (ROS).

Emission Spectroscopy

Emission spectrometry is a spectrometric technique that measures the wavelengths of photons emitted by atoms or molecules while transitioning from a high energy level to a lower energy state.

Principle of Emission Spectroscopy

When electrons or compounds are heated either on a flame or by an electric heater, they emit energy in the form of light. The light emitted from the compound is passed into a spectrometer then disperses the light into separate wavelengths. Each element forms a different atomic spectrum that indicates that an atom can radiate only a certain amount of energy. Each element emits a set of discrete wavelengths that is characteristic to it based on its electronic structure, and from these wavelengths, the elemental composition of the sample can be determined.

Steps of Emission Spectroscopy

- The solution containing the sample is heated either in a flame or with an electric heater.
- The solvent evaporates first, whereas the finely divided solid particle remains in the center of the flame along with other molecules and ions.
- This causes the excitation of electrons which produces the radiation of a specific wavelength.
- The radiation is passed through the spectrometer where the monochromator disperses the light into different wavelengths.
- Detectors detect the wavelengths in the spectroscope.
- A graph of wavelength against the concentration is plotted to determine the concentration of the sample.

Uses of Emission Spectroscopy

- Emission spectroscopy has wide applications in agricultural and environmental analysis along with industrial analysis for the detection of metals and alloys.

- It can also be used in the determination of lead in petrol.
- This has been applied for the determination of an equilibrium constant of ion exchange resins.

Energy Dispersive Spectroscopy

Energy dispersive spectroscopy, also known as Electron dispersive X-ray spectroscopy, is an analytical technique for the elemental or chemical characterization of the sample by using the X-ray interaction with the sample.

Principle of Energy Dispersive Spectroscopy

The principle of energy dispersive spectroscopy is similar to emission spectroscopy except that the excitation of electrons is brought about by X-ray beams. The compound in a sample has electrons in the ground state bound to the nucleus. When X-rays are focused on a compound, the electrons in the ground level get excited and reach an inner shell. This results in an electron-hole in the outer shell which is then filled by another electron from an outer high energy shell. The difference in the energy between the high-energy shell and the low-energy shell is released in the form of X-ray beams. The energy dispersive spectroscopy determines the energy and number of X-ray beams released from the sample. The difference in energy between two shells is characteristics of the atomic structure of the sample, which helps in the determination of the composition of the sample.

Steps of Energy Dispersive Spectroscopy

- Two samples of known and unknown concentrations are taken in a transport vessel, also termed as a cuvette.
- The vessels are then placed, one after the other, in the spectrophotometer that is provided with an excitation source which can either be an electron beam or an x-ray beam and detectors.
- The spectrophotometer is operated that passes an x-ray through the sample.
- The photosensitive detectors present in the spectrophotometer detect the x-ray passing through the sample, which is then converted into voltage signals.
- The voltage signals then enter the pulse processor that measures the signals and passes them to an analyzer for data display and further analysis.

Uses of Energy Dispersive Spectroscopy

- Energy-dispersive X-ray spectroscopy (EDS) as a part of electron microscopy has been widely used in many research areas as it provides precise information on the chemical composition of subcellular structures.
- EDS has been used in medical diagnostics as X-ray microanalysis may be useful in defining criteria for malignant tissue formation.

- The application of electron dispersive spectroscopy in environmental analysis is a useful tool to estimate the risk of polluted air in terms of human health.
- Droplets of pesticides (0.5 µl – 1 µl) at varying concentrations deposited on plant surfaces can also be analyzed by this method.

Fluorescence Spectroscopy

Fluorescence spectroscopy is a type of electromagnetic spectroscopy that utilizes the fluorescence produced by objects in a sample which is not necessarily in the visible range of the spectrum.

Principle of Fluorescence Spectroscopy

The principle of fluorescence spectroscopy is similar to emission spectroscopy, where the transition of electrons from one state to another causes the emission spectrum. Fluorescence is an emission phenomenon where a transition from a higher to a lower energy state is accompanied by radiation. Only molecules in their excited forms can emit fluorescence; thus, they have to be brought into a higher energy state prior to the emission phenomenon. The emitted radiation appears as a band spectrum because there are many closely related wavelength values dependent on the vibrational and rotational energy levels attained. The fluorescence spectrum of a molecule is independent of the wavelength of the exciting radiation and has a mirror image relationship with the absorption spectrum. The probability of the transition from the electronic excited to the ground state is proportional to the intensity of the emitted light. The fluorescence properties of a molecule are determined by features of the molecule itself and thus help in the determination of the composition of the molecules.

Steps of Fluorescence Spectroscopy

- Two samples of known and unknown concentrations are taken in a transport vessel, also termed as a cuvette.
- The vessels are then placed, one after the other, in the spectrofluorimeter that is provided with light source and detectors.
- The spectrofluorimeter is operated that passes light of a particular wavelength through the sample.
- The photosensitive detectors present in the spectrophotometer detect the light passing through the sample, which is then converted into digital values.
- A graph of the fluorescence measured against the concentration of the sample is plotted, which can then be used for the determination of the unknown concentration of the sample.

Uses of Fluorescence Spectroscopy

- Fluorescence spectroscopy is used in biomedical, medical, and chemical research for the analysis of organic compounds.

- This has also been used to differentiating malignant tumors from benign tumors.
- Atomic fluorescence spectroscopy can also be used for the detection of metals in various environmental samples like air, water, and soil.
- In analytical chemistry, fluorescence detectors are used along with HPLC.

Fourier-transform Infrared (FTIR) Spectroscopy

Fourier-transform infrared spectroscopy is a technique used for the detection of the infrared spectrum of absorption and emission of molecules in solid, liquid, or gaseous phase. The term Fourier-transform has been used because this process requires a mathematical process termed Fourier transform to convert the raw data into the actual spectrum.

Principle of FTIR Spectroscopy

Fourier-transform infrared spectroscopy uses the same principle of dispersive spectroscopy except that instead of focusing a beam with a single frequency, this technique uses a beam with multiple frequencies. This technique then measures how much of that beam is absorbed by the sample. Similarly, a second beam with different frequencies is used for sample processing, and the data is collected. This process is repeated a number of times over a short period of time. All the data obtained is then given to a computer that works backward to determine the absorption at each wavelength. The spectrometer has a special structure termed Michelson interferometer which consists of a certain number of mirrors, one of which moves towards the light source. With the movement of the mirror, each wavelength of light is periodically blocked and transmitted due to wave interference. As a result of this, the beam coming out of the interferometer has a different spectrum each time.

Steps of FTIR Spectroscopy

- The sample is placed in the FTIR spectrometer. The sample should be thin enough for the IR rays to pass through.
- The device directs the IR towards the sample and measures how much and which frequency of the beam is absorbed by the sample.
- With the help of the reference database, the sample is identified.

Uses of FTIR Spectroscopy

- FTIR spectroscopy is commonly used for the analysis of organic, polymeric, and even inorganic compounds.
- This technique allows the determination of the chemical composition of various samples.
- FTIR has been used with gas chromatography for the detection of substances separated from the chromatographic technique.
- FTIR can directly detect the polarity of the given site while investigating hydrophobic membrane environments.

Gamma-ray Spectroscopy

Gamma-ray spectroscopy is an analytical technique used for the study of the energy spectrum of gamma rays formed from radioactive objects in a sample.

Principle of Gamma-ray Spectroscopy

During radioactive decay, gamma decay is also seen in radioactive substances. In gamma decay, the nucleons in the nucleus found in discrete levels move towards the lower energy state, and the energy difference is released in the form of a gamma-ray. The gamma rays are then passed through detectors that convert the radiation into electric signals for the computer. The detection and quantification of gamma rays help to determine the nature of the source. By comparing the measured energy to the known energy of gamma-rays produced by radioisotopes, the identity of the emitter can be determined.

Steps of Gamma-ray Spectroscopy

- The counting chamber is opened using the handle. The interior of the counting chamber is usually shielded with lead and then lined with copper to reduce background radiation caused by x-rays emitted by the lead.
- The sample container is placed carefully on top of the detector, and the counting chamber is closed.
- Background counting is performed to determine if contamination or natural radiation is present in amounts that would interfere with sample data.
- Further analysis is done through computing systems.

Uses of Gamma-ray Spectroscopy

- Gamma-ray spectroscopy is exclusively used for the determination of nuclear structure, nuclear transitions, and nuclear reactions.
- This has also been employed for astronomical research like fort detection of water on other planets.
- Gamma-ray spectroscopy has been utilized for the elemental and isotopic analysis of airless bodies in the solar system, especially the moon and mars.

Infrared (IR) Spectroscopy/Vibrational Spectroscopy

Infrared spectroscopy, also termed vibrational spectroscopy, is a technique that utilizes the interaction between infrared and the sample.

Principle of IR Spectroscopy/Vibrational Spectroscopy

The wavelength utilized for the analysis of organic compounds ranges from 2,500 to 16,000 nm, with a corresponding frequency range from 1.9×10^{13} to 1.2×10^{14} Hz. These rays don't have enough energy to excite the electrons, but they do, however, cause the vibrational excitation of covalently bonded atoms or groups. The vibration observed in the atoms is characteristic of these atoms and thus helps in the detection of the molecules. The infrared spectrum is the fundamental measurement obtained in infrared spectroscopy. The spectrum is a plot of measured infrared intensity versus wavelength (or frequency) of light. IR Spectroscopy measures the vibrations of atoms, and based on this; it is possible to determine the functional groups.

Steps of IR Spectroscopy/Vibrational Spectroscopy

- The IR spectrometer is turned on and allowed to warm up for 30 minutes.
- The unknown sample is taken, and its appearance is recorded.
- The background spectrum is collected to remove the spectrum obtained from natural reasons.
- A small amount of sample is placed under the probe by using a metal spatula.
- The probe is set in place by twisting it.
- The IR spectrum of the unknown sample is obtained. The process is repeated, if necessary, to get a good quality spectrum.
- The absorption frequencies that indicate the functional groups present are recorded.
- The obtained spectrum is analyzed to determine the probable identification of the unknown sample.

Uses of IR Spectroscopy/Vibrational Spectroscopy

- Infrared spectroscopy has been widely used for the characterization of proteins and the analysis of various solid, liquid, and gaseous samples.
- IR spectroscopy can be used for the detection of functional groups which helps in the identification of molecules and their composition.
- Applications of IR spectroscopic techniques allow identifying molecular changes due to bodily changes, understanding of the molecular mechanism of various diseases, and identifying specific spectral biomarkers that can be used in diagnosis.

Magnetic Resonance Spectroscopy

Magnetic resonance spectroscopy is a diagnostic technique used alongside magnetic resonance imaging (MRI) which measures biochemical changes that take place in different parts of the body.

Principle of Magnetic Resonance Spectroscopy

Magnetic resonance spectroscopy analyzes the hydrogen atoms present in the sample. The nucleus of the atom is excited by applying a magnetic field that causes the spinning of the nucleus in a particular direction. The difference in the energy of the different directions of spin results in a spectrum characteristic to the nucleus of the atom. The frequencies of the spectrum are determined, which helps to analyze the concentration of various metabolites. By measuring the concentration of the metabolites, and comparing it to the normal concentration in various cells, determination of the type of tissue found in the body can be determined.

Steps of Magnetic Resonance Spectroscopy

- The MR instrument is turned on and warmed up for 30 minutes. The necessary parameters are set in the magnetic resonance spectrometer.
- Calibration is performed by adding an empty MR tube to ensure no background signals from the tube or the instrument is present.
- The sample is then placed in the MR tube, and the magnetic resonance spectrum is measured.

Uses of Magnetic Resonance Spectroscopy

- Magnetic resonance spectroscopy enables doctors and chemists to obtain biochemical information regarding tissue in the body which helps in the diagnostic process.
- MR spectroscopy is used to assist the MRI technique generally performed for diagnostic purposes in hospitals.
- It is increasingly used in research for medical projects.
- This technique has also been used in creating profile serum samples metabolically from patients diagnosed with hepatocellular carcinoma (HCC).

Mass Spectroscopy

Mass spectroscopy is a type of spectroscopic technique that helps to identify the amount and type of chemicals present in the sample by analyzing the mass to charge ratio of the ions.

Principle of Mass Spectroscopy

Mass spectroscopy is based on the principle that when a sample is bombarded with electrons, the molecules in the compounds are ionized into ions. The separation of ions is dependent on their mass to charge ratio. For most ions, the charge is one which means that the ratio is simply the molecular mass of the ion. The ions are then subjected to electric and magnetic fields which causes deflection of the ions. Ions with a similar charge to mass ratio show similar deflection. The relative abundance of each of such ions is then detected with the help of the detectors. The mass spectrum is formed by plotting the relative abundance of the ions against the ratio of mass to charge. The spectrum can then be used for the determination of the elemental configuration of the sample, the masses of the particle or molecules, and the chemical structure of the sample.

Steps of Mass spectroscopy

- 200 μl of the sample is mixed with 1.8 ml of 65% nitric acid.
- The mixture is then added to the water bath at 50 °C overnight.

- The tubes are then cooled down to room temperature, and the sample is diluted by adding 8 ml distilled water to obtain nitric acid concentration below 20%.
- The sample is then added to the spectroscope and run.
- The results are obtained through the software on the computer in the form of the mass spectrum.

Uses of Mass Spectroscopy

- Mass spectroscopy is a valuable tool to quantify known materials.
- It also allows the identification of unknown compounds and determination of the structure and chemical composition of various substances.

Molecular Spectroscopy

Molecular spectroscopy is a type of spectroscopy that utilizes the interaction between molecules and electromagnetic radiation to determine the structural composition of samples.

Principle of Molecular Spectroscopy

Molecular spectroscopy involves the interaction of materials with electromagnetic radiation in order to produce an absorption pattern (i.e. a spectrum) from which structural or compositional information can be deduced. The molecular spectrum is formed when the molecules move from one energy state to another as a result of interaction with different electromagnetic radiation. The mechanisms involved are similar to atoms, but the process is more complicated. The interaction between different nuclei and electrons takes place in molecules that are not observed in atoms. These interactions can be electronic, rotational, or vibrational. Based on the type of interactions, different spectrometric techniques can be applied for the analysis of the compounds. Similarly, either the absorption or emission spectrum can be utilized based on the type of interactions being exploited.

Steps of Molecular Spectroscopy

- Different types of spectrometer systems like emission spectrometer and Fourier-transform spectrometer can be used for molecular spectroscopy.
- All these systems include a source of radiation, a sample, and a system for detecting and analyzing the sample.
- The sample is added to the spectrometer where the radiation source focuses on it.
- Two samples of known and unknown concentrations are taken in a transport vessel, also termed as a cuvette.
- The vessels are then placed, one after the other, in the spectrophotometer that is provided with light source and detectors.
- The spectrophotometer is operated that passes light of a particular wavelength through the sample.

- The photosensitive detectors present in the spectrophotometer detect the light passing through the sample, which is then converted into digital values.
- A graph of the absorbance measured against the concentration of the sample is plotted, which can then be used for the determination of the unknown concentration of the sample.

Uses of Molecular Spectroscopy

- Molecular spectroscopy allows the analysis of the atomic and molecular structures of various compounds.
- It also helps in the determination of gas composition as well as the composition of other compounds.

Mössbauer Spectroscopy

Mössbauer spectroscopy is a technique based on the Mossbauer Effect discovered by Rudolf Mössbauer which utilizes the spectrum formed by the absorption or emission of nuclear gamma rays in solid particles.

Principle of Mössbauer Spectroscopy

Mössbauer Effect states that the recoil energy associated with absorption or emission of a photon can be described by the conservation of momentum. In solids, the nuclei of the atoms are not free but are bound to a particular lattice. When a nucleus in a solid particle emits or absorbs a gamma-ray, the energy always lost occurs in discrete packets called phonons which are the quantized vibration of the crystal lattice. In some cases, however, the no phonons might be released, and the conservation of momentum in such cases is maintained by the momentum of the lattice as a whole. Thus, gamma rays emitted by one nucleus can be completely absorbed by a sample containing nuclei of the same isotope, and this absorption can be measured. The measured Mossbauer spectrum can be used to detect the presence of radioactive samples.

Steps of Mössbauer Spectroscopy

- A sample is placed inside the spectrometer.
- A beam of gamma radiation is then passed through the sample. The atoms in the radiation source emitting the gamma rays must have the same isotope as the atoms in the sample absorbing them.
- The detector then measures the intensity of the beam transmitted through the sample.
- The spectrum formed is then analyzed for various purposes.

Uses of Mössbauer Spectroscopy

- Mössbauer spectroscopy is a sensitive technique that can detect even subtle changes in the nuclear environment.

- It provides precise information about the chemical, structural and magnetic properties of the various objects.
- It has also been used in geology to detect the composition of meteors and moon rocks.
- This technique has been highly used in bioorganic chemistry for the study of iron-containing proteins and enzymes.

Nuclear Magnetic Resonance (NMR) Spectroscopy

Nuclear magnetic resonance spectroscopy is a technique that utilizes the magnetic resonance of the nucleus to determine the structure of various compounds.

Principle of NMR Spectroscopy

Nuclear magnetic resonance spectroscopy is based on the fact that the nuclei in the atoms, molecules, or ions are charged and have spin. When a molecule or compound is placed in a magnetic field, the spin of the nucleus can align in two different states. If the spin is aligned in the direction of the magnetic field, it results in a lower energy state, whereas the opposite alignment of the magnetic field results in a higher energy state. As a result, the nucleus can move between these two energy states by either absorbing or emitting photons of energy resulting in a resonance condition. The spectrum produced during this absorption is passed to the detectors for the analysis of the nuclear structures of the sample.

Steps of NMR Spectroscopy

- The NMR instrument is turned on and warmed up for 30 minutes. The necessary parameters are set in the NMR.
- Calibration is performed by adding an empty NMR tube to ensure no background signals from the tube or the instrument is present.
- The sample is then placed in the NMR tube, and the NMR spectrum is measured.

Uses of NMR Spectroscopy

- NMR spectroscopy is used in quality control.
- It is also used in the determination of the content and purity of the sample while detecting the structure of the compounds.
- Solid-state NMR spectroscopy is performed to determine the molecular structure of solids.
- It can also be used for the determination of protein structure.

Photoelectron Spectroscopy

Photoelectron spectroscopy, also known as photoemission spectroscopy, is a spectrometric technique that measures the electrons released from a substance as a result of the photoelectric effect to determine the binding energies of the electrons.

Principle of Photoelectron Spectroscopy

- Photoelectron spectroscopy utilizes the principle of the photoelectric effect.
- The sample is exposed to UV rays that induce photoionization of the substances.
- The emitted photoelectrons have energies that are characteristic of their original energy states and the vibrational and rotational level of the electrons.
- The energy of these electrons is used to determine the binding energy of the electrons in the atoms by the given formula:

  BE= hv- KE where hv is the incident radiation, and KE is the energy of the emitted photo-electrons.
- A graph of BE against KE is then plotted.
- The peaks in the spectrum indicate the electrons in different subshells of an atom. The lowest peaks indicate the valence electrons, whereas the highest peaks correspond to the core electrons.

Steps of Photoelectron Spectroscopy

- The sample is placed in the spectrometer, which is exposed to the ionizing radiation from the radiation source.
- The emitted electrons strike the detectors, which then converts the energy into an electric signal.
- The signal is transferred to the analyzer to obtain the analog data from the signal.
- Then a graph is plotted of the kinetic energy of the emitted electrons and the binding energy.
- Based on the peaks formed in the graph, the unknown element can be detected.

Uses of Photoelectron Spectroscopy

- Electron spectroscopy is sensitive to every element except hydrogen and helium and thus can play an important role in all phases of chemistry.
- Since the electronic energy levels of atoms are sensitive to their environment, this technique is the best available for determining the oxidation states of atoms in molecules.
- From the photoelectron spectra, one can obtain the relative ratios of different elements and different atoms of the same element.
- This technique is sensitive to many essential areas of research such as corrosion studies, catalysis, semiconductor studies, adsorbed gas work, and electrode studies.

Raman Spectroscopy

Raman spectroscopy is an analytical technique that provides detailed information about the chemical structure, polymorphism, crystallinity, and molecular interactions within a compound.

Principle of Raman Spectroscopy

A Raman spectrum arises from the analysis of scattered light from the sample. The largest part of an incident light beam passes through the sample (transmission). A small part is scattered isotropically, i.e. uniformly in all directions, and possesses the same wavelength as the incident beam. The Raman spectrum arises from the fact that a tiny proportion of light scattered by the sample will have a different frequency than the incident light. As a result of the incident photon, the molecules move to an exciting rotational-vibrational-electronic state, and the photons are emitted from the sample.

In order to maintain the energy of the system after the molecule moves to a higher state, the scattered photons shift to a new state. The energy difference between the new state and the original state of the scattered photon is equal to the difference in energy between the two vibrational states of the molecules. The criterion for a peak to appear in the Raman spectrum is a change in polarizability of the molecule during the transition. The depolarization is calculated as the ratio of two intensities with perpendicular and parallel polarization with respect to the incident beam.

Steps of Raman Spectroscopy

- Mostly aqueous solutions are used the required laser is turned on after selecting the appropriate wavelength being used.
- The spectrometer is then calibrated by using a reference sample by using appropriate exposure energy and time.
- The sample is then placed under the microscope while focusing on the layer to be investigated.
- A range of wavenumbers is scanned by the monochromator, which results in the Raman spectrum.
- The data is then investigated by appropriate software, and the analysis is made based on the peaks formed on the spectrum.

Uses of Raman Spectroscopy

- Raman spectroscopy is employed over infrared spectroscopy as it allows the analysis of the sample in aqueous conditions.
- This technique can also be used to determine the chemical composition and structure of various samples.

UV Spectroscopy

UV spectroscopy is a type of absorption spectroscopy where UV lights are absorbed by the electrons that cause them to excite to a high energy state.

Principle of UV Spectroscopy

In UV spectroscopy, the UV rays passed to the sample are absorbed by the electrons, which increase the energy of the system. This causes the excitation of an electron from a lower energy state to a higher energy state. This excitation forms an absorption spectrum that can be detected by the detectors in the spectrometer. The amount of photon (radiation) absorbed results in an absorption spectrum which can then be measured in terms of absorbance. The absorbance of a sample is dependent on the number of excited electrons which in turn is dependent on the concentration of molecules in the sample.

Steps of UV Spectroscopy

- Two samples of known and unknown concentrations are taken in a transport vessel, also termed as a cuvette.
- The vessels are then placed, one after the other, in the spectrophotometer that is provided with light source and detectors.
- The spectrophotometer is operated that passes light of a particular wavelength through the sample.
- The photosensitive detectors present in the spectrophotometer detect the light passing through the sample, which is then converted into digital values.
- A graph of the absorbance measured against the concentration of the sample is plotted, which can then be used for the determination of the unknown concentration of the sample.

Uses of UV Spectroscopy

- UV spectroscopy is a technique used for the detection of impurities in organic substances.
- This can also be used for the quantitative determination of compounds that can absorb UV radiation.
- It can also be used for the study of the kinetics of a reaction where the UV rays are passed through the reaction cell, and the changes in absorbance are studied.

Ultraviolet and Visible (UV/Vis) Spectroscopy

Ultraviolet and visible spectroscopy is an absorption spectroscopy technique which uses the radiation in the UV range and the adjacent visible range of the electromagnetic radiation.

Principle of UV/Vis Spectroscopy

UV/Vis spectroscopy is based on the principle that materials produce an absorption spectrum which is a range of absorbance resulting from the radiation absorbed by the material at different frequencies. The absorption spectrum of materials depends on the atomic and molecular composition of that material. The frequency of light radiation absorbed by a material is dependent on the energy difference between the two energy states of the molecules. The absorption results in the formation of the absorption line, which, together with other lines, form an absorption spectrum. The incident light in this spectrometer is in the range of UV and visible spectrum of the electromagnetic spectrum. Thus, when a photon with sufficient energy reaches an object, the energy is absorbed by the electrons causing them to bump into a higher energy state. The amount of photon (radiation) absorbed results in an absorption spectrum which can then be measured in terms of absorbance. The absorbance of a sample is dependent on the number of excited electrons which in turn is dependent on the concentration of molecules in the sample.

Steps of UV/Vis Spectroscopy

- Solvent liquid and the sample solution are taken in two transport vessels, also termed as cuvettes.
- The vessel with solvent liquid is then placed in the spectrometer to determine the light loss due to scattering and absorbance by the solvent. Any absorbance observed in this process is to be subtracted from the absorbance of the sample.
- The cuvette with the sample solution is then placed in the spectrometer.
- The absorbance of the sample is noted in different frequencies which usually ranges from 200-800 nm.
- A similar spectrum is formed from the different concentrations of the samples.
- A graph of the absorbance measured against the concentration of the sample is plotted, which can then be used for the determination of the unknown concentration of the sample.

Uses of UV/Vis Spectroscopy

- Qualitative analysis may be performed in the UV/Vis regions to identify certain classes (proteins and nucleic acids) of compounds both in the pure state and in biological mixtures.
- This type of spectroscopy is used for the quantification of biological samples either directly or via colorimetric assays.

X-ray Photoelectron Spectroscopy

X-ray photoelectron spectroscopy is a sensitive, quantitative spectroscopic technique based on the photoelectric effect used for the identification of electrons inside a compound, on the surface as well as the chemical state and the electronic configuration of the compound.

Principle of X-ray Photoelectron Spectroscopy

Photoelectron spectroscopy utilizes the principle of the photoelectric effect. The sample is exposed to X-rays with specific wavelengths that induce photoionization of the substances. The emitted photoelectrons have energies that are characteristic of their original energy states and the vibrational and rotational level of the electrons. The energy of these electrons is used to determine the binding energy of the electrons in the atoms by the given formula:

$$BE = h\nu - KE$$

where hv is the incident radiation, and KE is the energy of the emitted photoelectrons.

- A graph of BE against KE is then plotted.

- The peaks in the spectrum indicate the electrons in different subshells of an atom. The lowest peaks indicate the valence electrons, whereas the highest peaks correspond to the core electrons.

Steps of X-ray Photoelectron Spectroscopy

- The sample is placed in the spectrometer, which is exposed to the ionizing radiation from the radiation source.
- The emitted electrons strike the detectors, which then converts the energy into an electric signal.
- The signal is transferred to the analyzer to obtain the analog data from the signal.
- Then a graph is plotted of the kinetic energy of the emitted electrons and the binding energy.
- Based on the peaks formed in the graph, the unknown element can be detected.

Uses of X-ray Photoelectron Spectroscopy

- X-ray photoelectron spectroscopy is commonly used for the measurement of composition, chemical state, and electronic configuration of various organic and inorganic substances.
- This also assists in the surface analysis of various compounds.

Electromagnetic Radiation

Electromagnetic radiation is the name given to a pair of electric and magnetic fields that propagate together at the speed of light. Examples of electromagnetic radiation are visible light, radio waves, X-rays and gamma rays – all parts of the electromagnetic spectrum.

Light is also described as a wave. A wave is defined as a disturbance which travels and spreads out through some medium. We are all familiar with ripples in a pond and ocean waves. Now when you throw a cork on the water in a pond, you will observe waves going outwards, but the cork only bobs up and down. This means that the water does not move outwards. Indeed if it were to do so, all the water would end up at the banks of the ocean. It only moves up and down, normal to the direction of propagation of the wave. Such waves are called transverse waves, and light waves are also transverse.

The other kinds of waves are longitudinal, in which the motion is parallel to the direction of motion. Examples are sound waves, and such waves require a medium for their propagation, and are thus also known as mechanical waves. Transverse waves can travel in vacuum: In space no one can hear you scream - but they can see you.

Characteristic Properties of Waves

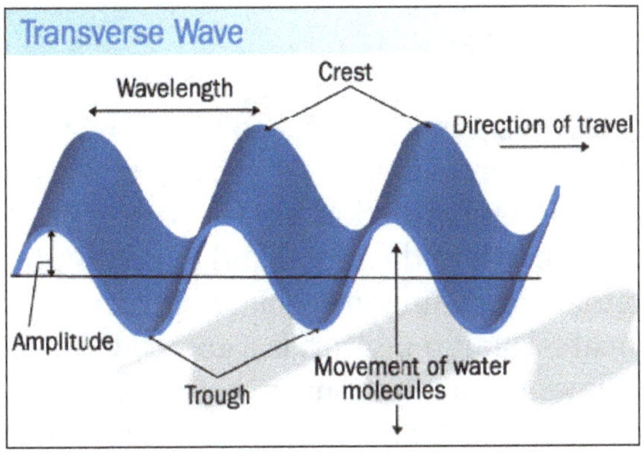

Consider the transverse wave shown above moving to the right. The water molecules move only up and down. The high points in the wave are called crests and the low points are called troughs. The distance between two consecutive peaks or troughs in a wave is called the wavelength and is symbolized by λ. The height of the crest is termed the amplitude of the wave, and is usually given the symbol A. Therefore, the height of the water molecules (y) varies periodically between $-A$ and $+A$ through zero. Such a wave can be conveniently described as a sine function since at the starting point of the wave, the height is zero, and sin(o) is also zero. We thus write,

$$y(t) = A \sin\theta(t)$$

where θ is expressed in radians. We can trace y(t) by imagining an object moving counter-clockwise on the circumference of a circle of radius A. When the object has traversed 2π radians, it completes one cycle, as shown in Figure. If we express the angular frequency as ω radians per second, then $\theta(t)$ is given by ωt, so we can equivalently write,

$$y(t) = A \sin \omega t$$

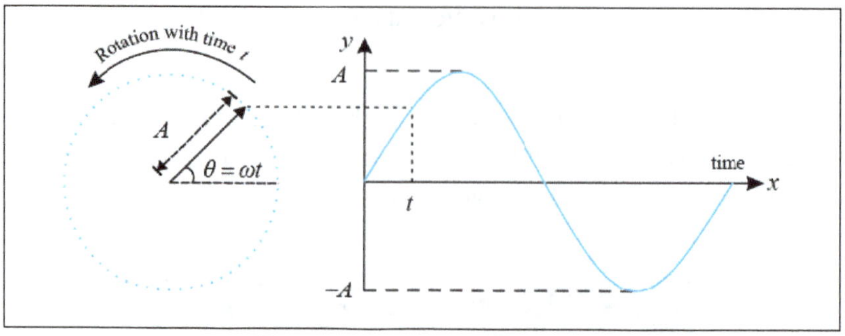

Oscillation of a ball in circular motion.

The time taken for a complete cycle is called the time period of the wave and is denoted by T. A more important quantity than the time period is the frequency, defined as the number of waves (cycles) per second that pass a given point in space, and symbolized by ν. Evidently, the two quantities are related as$\nu = 1/T$. As we shall soon see, the energy of a radiation is directly proportional to its frequency. Also, since a cycle is complete when θ changes by 2π, the number of cycles per second (the frequency) is given by $\varpi = \omega/2\pi$. We therefore write,

$$y(t) = A \sin 2\pi\nu t$$

The Speed of Light

The distance traversed in a cycle is λ; hence, the speed (v) of the wave is the distance travelled in a cycle divided by the time taken to complete a cycle, i.e. λ/T or λν. For light, the speed is given its own symbol c, since all electromagnetic radiation has the same speed, $\sim 3 \times 10^8\ ms^{-1}$. It takes 1.33 second for light to travel from the earth to the moon and 8.33 minutes for it to travel from the sun to the earth. Einstein showed that, according to special relativity, nothing can exceed the speed of light. Even at this speed, the nearest star, α Centuri is a good 9 years round trip away. We may thus write, $c = \lambda\nu$, a relation that allows us to interconvert between the wavelength (in which all spectroscopic measurements are made) and the frequency to which the energy is proportional. Thus, shorter wavelengths correspond to higher frequencies and larger energies and longer wavelengths to smaller energies.

Light travels in a straight line. Though the speed of light is constant in vacuum, it depends on the refractive index of the medium of propagation.

Equation $y(t) = A \sin 2\pi\nu t$ can be now rewritten as,

$$y(t) = A \sin(2\pi ct/\lambda)$$

Light waves are travelling waves, and, for these waves, we may combine equations $y(t) = A \sin \omega t$ and $y(t) = A \sin(2\pi ct/\lambda)$, and write:

$$y(t) = A \sin(\omega t - 2\pi ct/\lambda) = A \sin(\omega t - kz)$$

where $k = 2\pi/\lambda$ and we have substituted $z = ct$ for the distance travelled by a wave travelling in the z direction.

Classical Description of Electromagnetic Radiation

Up till now, we have not defined the quantity y(t) for light waves. James Clerk Maxwell was the first to realize more than a century ago the connection between optics and electricity and magnetism through his famous four equations, which govern all electromagnetic radiation. According to Maxwell's equations, there are oscillating magnetic ($\vec{B}$) and electric ($\vec{E}$) fields perpendicular to the direction of propagation of the wave. Thus, the y(t) of equation $y(t) = A \sin(\omega t - 2\pi ct/\lambda) = A \sin(\omega t - kz)$ could refer to either the sinusoidally oscillating electric or magnetic field.

Fields $\vec{E}$ and $\vec{B}$ are in phase, reaching their maximum and minimum values at the same time. The electric field oscillates in the xz plane and the magnetic field oscillates in the yz plane. This corresponds to a polarized wave. Conventionally, the plane in which the electric field oscillates is defined as the plane of polarization. In this case it is the xz plane. Figure shows a harmonic plane wave propagating in the z-direction. Note that $\vec{E}$, $\vec{B}$ and the direction of propagation $\hat{k}$ form a right handed triad.

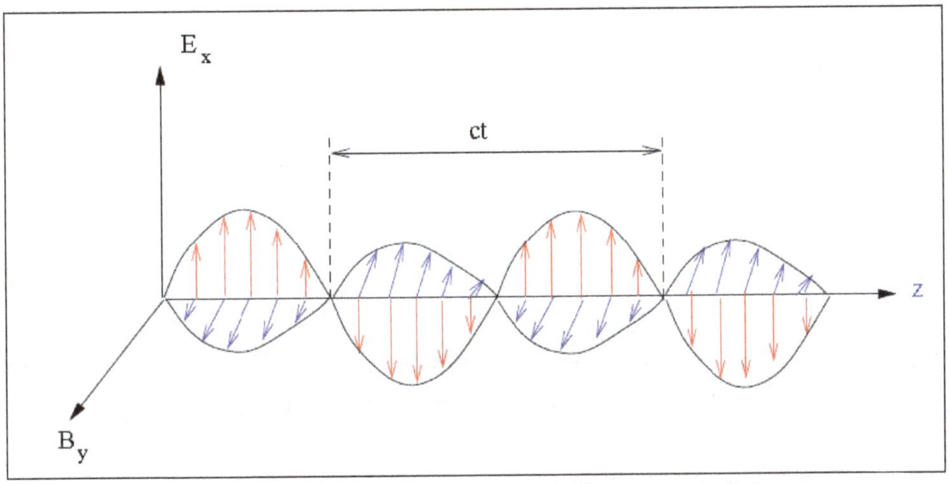

Sketch of the electric field and magnetic field vectors of a light wave.

Maxwell's Equation

Maxwell postulated the theoretical model of light as electromagnetic waves during the time in history when the wave and particle pictures of light were still duelling. Now let us see how Maxwell's equations can be manipulated to give classical wave equations for the space and time dependence of the electric and magnetic field. Maxwell's equations for vacuum in SI units are:

$$\vec{\nabla}.\vec{E} = 0$$

$$\vec{\nabla}.\vec{B} = 0$$

$$\vec{\nabla} \times \vec{E} = -\frac{\partial \vec{B}}{\partial t}$$

$$\vec{\nabla} \times \vec{B} = \varepsilon_0 \mu_0 \frac{\partial \vec{E}}{\partial t}$$

where $\mu_o = 4\pi \times 10^{-7}$ N s^2 C^{-2} is the permeability of free space and $\varepsilon_o = 8.85 \times 10^{-12}$ C^2 N^{-1} m^{-2} is the permittivity of free space.

The last two equations couple the electric and the magnetic fields. Equation $\vec{\nabla} \times \vec{E} = -\frac{\partial \vec{B}}{\partial t}$ implies that, if $\vec{B}$ is time dependent, $\vec{\nabla} \times \vec{E}$ is non-zero, i.e. $\vec{E}$ is a function of position. Further, if $\frac{\partial \vec{B}}{\partial t}$ itself changes with time, so does $\vec{\nabla} \times \vec{E}$. In such a case $\vec{E}$ also varies with time since the $\vec{\nabla}$ operator cannot cause time variation. Thus, in general, a time varying magnetic field gives rise to an electric field which varies both in space and time. It will be seen that these coupled fields propagate in space.

We will first examine whether the equations lead to transverse waves. For simplicity, let us assume that the electric field has only the x-component and the magnetic field only the y-component. Note that we are only making an assumption regarding their directions – the fields could still depend on all the space coordinates x, y, z, in addition to time t.

Gauss's law gives,

$$\vec{\nabla} \cdot \vec{E} = \frac{\partial E_x}{\partial x} + \frac{\partial E_y}{\partial y} + \frac{\partial E_z}{\partial z} = 0$$

Since only $E_x \neq 0$, this implies that,

$$\frac{\partial E_x}{\partial x} = 0$$

Thus E_x is independent of the x coordinate and can be written as $E_x(y, z, t)$. A similar analysis shows that B_y is independent of the y coordinate and can be written explicitly as $B_y(x, z, t)$.

Consider now the time dependent equations $\vec{\nabla} \times \vec{E} = -\frac{\partial \vec{B}}{\partial t}$ and $\vec{\nabla} \times \vec{B} = \varepsilon_o \mu_o \frac{\partial \vec{E}}{\partial t}$. The curl equation for $\vec{B}$ gives, taking the z-component,

$$\frac{\partial B_y}{\partial x} - \frac{\partial B_x}{\partial y} = \frac{\partial E_z}{\partial t} = 0$$

Since $B_x = 0$, this gives:

$$\frac{\partial B_y}{\partial x} = 0$$

Showing that B_y is independent of x and hence depends only on z and t. In a similar manner we can show that Ex also depends only on z and t. Thus the fields $\vec{E}$ and $\vec{B}$ do not vary in the plane containing them. Their only variation takes place along the z-axis which is perpendicular to both $\vec{E}$ and $\vec{B}$. The direction of propagation is thus the z-direction.

To see that the propagation is really a wave disturbance, take the y-component of equation $\vec{\nabla} \times \vec{E} = -\frac{\partial \vec{B}}{\partial t}$ and the x-component of equation $\vec{\nabla} \times \vec{B} = \varepsilon_o \mu_o \frac{\partial \vec{E}}{\partial t}$.

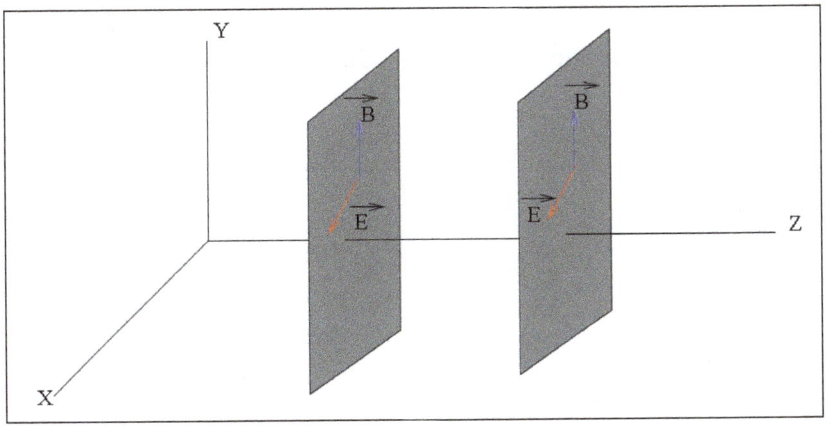

$$\frac{\partial E_x}{\partial x} = -\frac{\partial B_y}{\partial t}$$

$$-\frac{\partial B_y}{\partial z} = \mu_o \varepsilon_o \frac{\partial E_x}{\partial t}$$

To get the wave equation for E_x, take the derivative of equation $\frac{\partial E_x}{\partial x} = -\frac{\partial B_y}{\partial t}$ with respect to z and substitute in equation $-\frac{\partial B_y}{\partial z} = \mu_o \varepsilon_o \frac{\partial E_x}{\partial t}$ and interchange the space and time derivatives,

$$\frac{\partial^2 E_x}{\partial z^2} = \mu_o \varepsilon_o \frac{-\partial^2 B_y}{\partial z \partial t} = -\frac{\partial}{\partial t}\left(\frac{\partial B_y}{\partial z}\right) = \mu_o \varepsilon_o \frac{\partial^2 E_x}{\partial t^2}$$

Similarly, we can show,

$$\frac{\partial^2 B_y}{\partial z^2} = \mu_o \varepsilon_o \frac{\partial^2 B_y}{\partial t^2}$$

Each of the above equations represents a wave disturbance propagating in the z-direction with a speed,

$$c = \frac{1}{\sqrt{\mu_o \varepsilon_o}}$$

On substituting numerical values, the speed of electromagnetic waves in vacuum turns out to be 2.99792×10^8 m s^{-1}.

Consider plane harmonic waves of angular frequency ω and wavelength $\lambda = 2\pi/k$.

We can express the waves as,

$$E_x = E_o \sin(kz - \omega t)$$
$$B_y = B_o \sin(kz - \omega t)$$

The amplitudes E_o and B_o are not independent as they must satisfy equations $\frac{\partial E_x}{\partial x} = -\frac{\partial B_y}{\partial t}$ and $-\frac{\partial B_y}{\partial z} = \mu_0 \varepsilon_0 \frac{\partial E_x}{\partial t}$:

$$\frac{\partial E_x}{\partial z} = E_o k \cos(kz - \omega t)$$

$$\frac{\partial B_y}{\partial z} = B_o \omega \cos(kz - \omega t)$$

Using equation $\frac{\partial E_x}{\partial x} = -\frac{\partial B_y}{\partial t}$ we get,

$$E_o k = B_o \omega$$

The ratio of the electric field amplitude to the magnetic field amplitude is given by,

$$\frac{E_o}{B_o} = \frac{\omega}{k} = c$$

Example: Determine the state of polarization and the direction of propagation of the wave if the electric field of a plane electromagnetic wave in vacuum is given as,

$$E_y = 0.5 \cos\left[2\pi \times 10^8 (t - x/c)\right] \text{ V/m},$$

$$E_x = E_z = 0.$$

Also determine the magnetic field.

Comparing with the standard form for a harmonic wave,

$$\omega = 2\pi \times 10^8 \text{ rad/s}$$
$$k = 2\pi \times 10^8 / c$$

so that $\lambda = c/10^8 = 3$ m. The direction of propagation is the x-direction. Since the electric field oscillates in the xy plane, this is the plane of polarization. Since $\vec{B}$ must be perpendicular to both the electric field direction and the direction of propagation, $\vec{B}$ has only the z-component with an amplitude $B_o = E_o / c \simeq 1.66 \times 10^{-9}$ T. Thus,

$$B_z = 1.66 \times 10^{-9} \cos\left[2\pi \times 10^8 (t - x/c)\right] \text{T}$$

Maxwell's relations could explain most phenomena involving light:

- Law of Reflection – Angle of Incidence = Angle of reflection.
- Law of Refraction – A light beam is bent towards the normal when passing into a medium of higher Index of Refraction, and it is bent away from the normal when passing into a medium of lower Index of Refraction.

$$\text{Index of Refraction} - \frac{\text{Speed of light in vacuum}}{\text{Speed of light in a medium}}.$$

- Inverse square law – The light intensity diminishes with the square of distance from the source.

Hence, Maxwell's relations could provide explanations for all of light's interesting properties such as diffraction (rainbows) and reflection (mirrors), and light was firmly established as a wave.

The third property implies that the speed of light varies with the refractive index of the medium, and only its speed in vacuum can be considered a fundamental constant. The frequency remains constant in any medium, and is thus a more fundamental property. This means that the wavelength changes on changing the medium of propagation.

Maxwell's relations show that the electric and magnetic fields recreate each other as the wave propagates. Thus, the oscillating electric field creates magnetism, as in an electromagnet, and similarly, the oscillating magnetic field creates electricity as in electric generators, and this goes on until the electromagnetic wave meets an object.

Regions of Electromagnetic Spectrum

Visible light ranges through seven major colours from short wavelengths (high frequency - violet) to long wavelengths (low frequency - red) – Violet, indigo, blue, green, yellow, orange, red (VIBGYOR) – the colours of the rainbow. Our eyes recognize ν (or λ) as colour.

However, at the beginning of the nineteenth century, it became apparent that there is radiation beyond the visible and the electromagnetic spectrum is not restricted to the visible region. It encompasses a wide range of frequencies. This continuous range of frequencies is known as the electromagnetic spectrum. The entire range of the spectrum is often sub-divided into specific regions. The subdivision of the entire spectrum into smaller regions is done mostly on the basis of how each region of electromagnetic waves interacts with matter. In the various kinds of spectroscopy, depending upon the region of the electromagnetic spectrum that we are scanning, different kinds of energy changes occur. For example, in the microwave region, it is mainly the rotational energy that changes. In the next higher energy range, the infrared region, another degree of freedom, the vibrational motion is also affected. Similarly, electronic energy changes occur in a still higher energy range, i.e. the ultraviolet and visible regions. Other kinds of spectroscopy are classified according to the techniques adopted for their study. For example, the resonance techniques, the electron spin resonance and nuclear spin resonance techniques, involve smaller energy changes, and are studied in the microwave and radiofrequency regions, respectively. Photoelectron spectroscopy involves higher energy photons, while Raman spectroscopy is based on the scattering of light using high energy laser beams. The figure below shows the electromagnetic spectrum.

We see that the wavelength has the dimension of length and is expressed in various units, depending upon the region of the electromagnetic spectrum. In the visible region, the units are nanometre (nm) and the range is 400 nm (violet) to 700 nm (red). White light is a mixture of the seven colours of the rainbow. No object has its own color. When white light strikes the object, it may absorb a particular colour from the white light and reflect the remaining colour. Our eye detects the reflected colour, which is complementary to the absorbed colour.

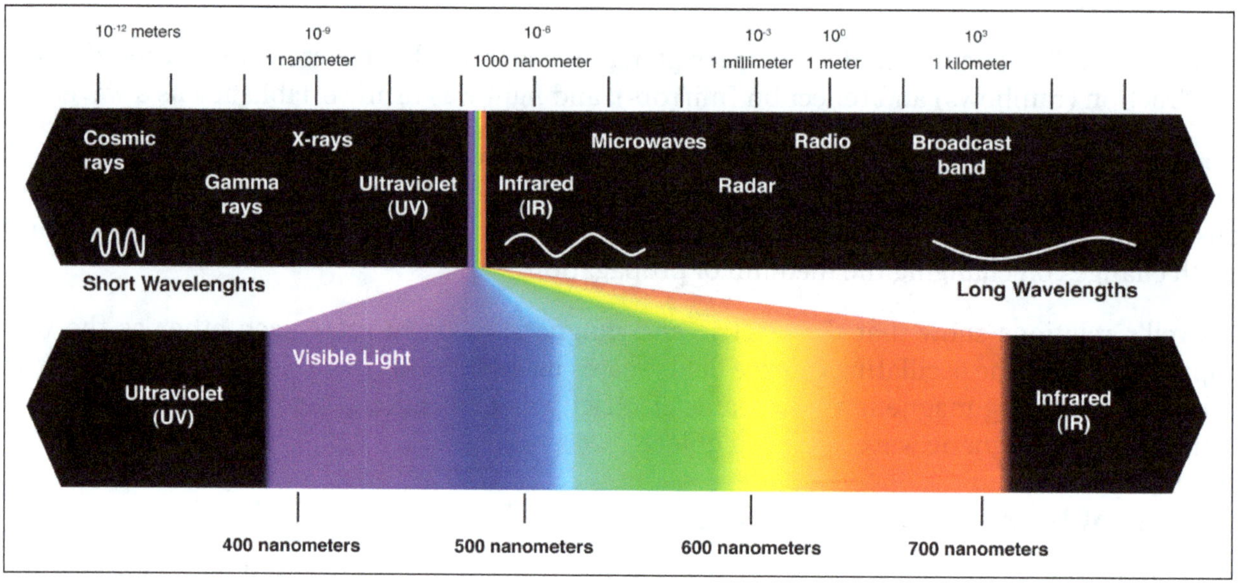

Variables and Units

Wavelength units: length.

Nanometre (nm): 1 nm = 1×10^{-9} m.

Micrometre (μm): 1 μm = $1{\times}10^{-6}$ m.

Frequency units: Cycles per second 1/s (or s^{-1}) is called hertz (abbreviated Hz) after the physicist Heinrich Hertz, who discovered radiowaves in 1887, 22 years after Maxwell's equations. To keep the numbers manageable, the prefixes kilo (1 kHz = 10^3 Hz), mega (1 MHz = 10^6 Hz) and giga (1 GHz = 10^9 Hz) are often used.

Wavenumber: In spectroscopy, the wavenumber is often used, particularly in infrared and UV/Vis spectroscopy. It is defined as a count of the number of wave crests (or troughs) in a given unit of length (symbolized by $\tilde{v}$). It has the dimensions of inverse length and is given by $v/c = 1/\lambda$. Though not an SI unit, cm^{-1} is the preferred unit. This means that in the expression $\tilde{v} = 1/\lambda$, the wavelength must be expressed in cm. Alternatively in the expression, $\tilde{v} = v/c$, c must be expressed in cm s^{-1} and v in s^{-1}.

Quantity	Symbol	Unit
Wavelength	l	m
Frequency	v	s^{-1} or Hz
Speed	c	m s^{-1}
Wavenumber	$\tilde{v}$	m^{-1}

Planck's Blackbody Radiation Law

Planck's radiation law is a mathematical relationship formulated in 1900 by German physicist Max Planck to explain the spectral-energy distribution of radiation emitted by a blackbody (a hypothetical body that completely absorbs all radiant energy falling upon it, reaches some equilibrium temperature, and then reemits that energy as quickly as it absorbs it). Planck assumed that the sources of radiation are atoms in a state of oscillation and that the vibrational energy of each oscillator may have any of a series of discrete values but never any value between. Planck further assumed that when an oscillator changes from a state of energy E_1 to a state of lower energy E_2, the discrete amount of energy $E_1 - E_2$, or quantum of radiation, is equal to the product of the frequency of the radiation, symbolized by the Greek letter ν and a constant h, now called Planck's constant, that he determined from blackbody radiation data; i.e., $E_1 - E_2 = h\nu$.

Planck's law for the energy E_λ radiated per unit volume by a cavity of a blackbody in the wavelength interval λ to λ + Δλ (Δλ denotes an increment of wavelength) can be written in terms of Planck's constant (h), the speed of light (c), the Boltzmann constant (k), and the absolute temperature (T):

$$E\lambda = \frac{8\pi hc}{\lambda^5} \times \frac{1}{\exp(hc / kT\lambda) - 1},$$

The wavelength of the emitted radiation is inversely proportional to its frequency, or $\lambda = c/\nu$. The value of Planck's constant is defined as $6.62607015 \times 10^{-34}$ joule second.

For a blackbody at temperatures up to several hundred degrees, the majority of the radiation is in the infrared radiation region of the electromagnetic spectrum. At higher temperatures, the total radiated energy increases, and the intensity peak of the emitted spectrum shifts to shorter wavelengths so that a significant portion is radiated as visible light.

Spectrometer

A spectrometer is any instrument used to view and analyze a range (or a spectrum) of a given characteristic for a substance (for example, a range of mass-to-charge values as in mass spectrometry), or a range of wavelengths as in absorption spectrometry like nuclear magnetic radiation spectroscopy or infrared spectroscopy). A spectrophotometer is a spectrometer that only measures the intensity of electromagnetic radiation (light) and is distinct from other spectrometers such as mass spectrometers.

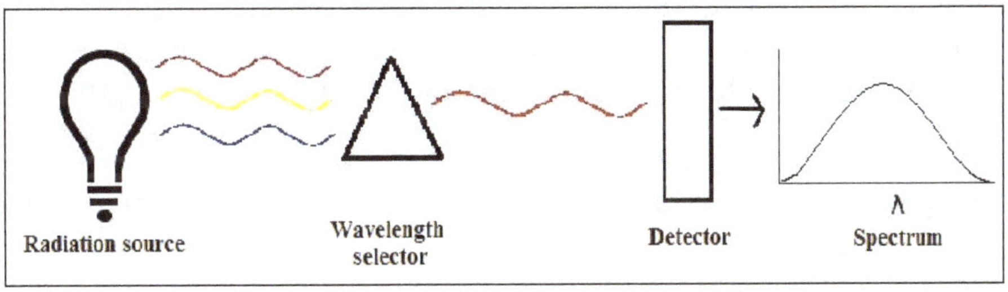

A spectrometer is typically used to measure wavelengths of electromagnetic radiation (light) that has interacted with a sample. Incident light can be reflected off, absorbed by, or transmitted through a sample; the way the incident light changes during the interaction with the sample is characteristic of the sample. A spectrometer measures this change over a range of incident wavelengths (or at a specific wavelength).

There are three main components in all spectrometers; these components can vary widely between instruments for specific applications and levels of resolution. Very generally, these components produce the electromagnetic radiation, somehow narrows the electromagnetic radiation to a specified range, and then detect the resulting electromagnetic radiation after is has interacted with the sample.

Sources of Radiation

There are two classes of radiation sources used in spectrometry: continuum sources and line sources. The former are usually lamps or heated solid materials that emit a wide range of wavelengths that must be narrowed greatly using a wavelength selection element to isolate the wavelength of interest. The latter sources include lasers and specialized lamps that are designed to emit discrete wavelengths specific to the lamp's material.

Electrode lamps are constructed of a sealed, gas-filled chamber that has one or more electrodes inside. Electrical current is passed through the electrode, which causes excitation of the gas. This excitation produces radiation at a wavelength or a range of wavelengths, specific to the gas. Examples include argon, xenon, hydrogen or deuterium, and tungsten lamps, which emit radiation in the following ranges.

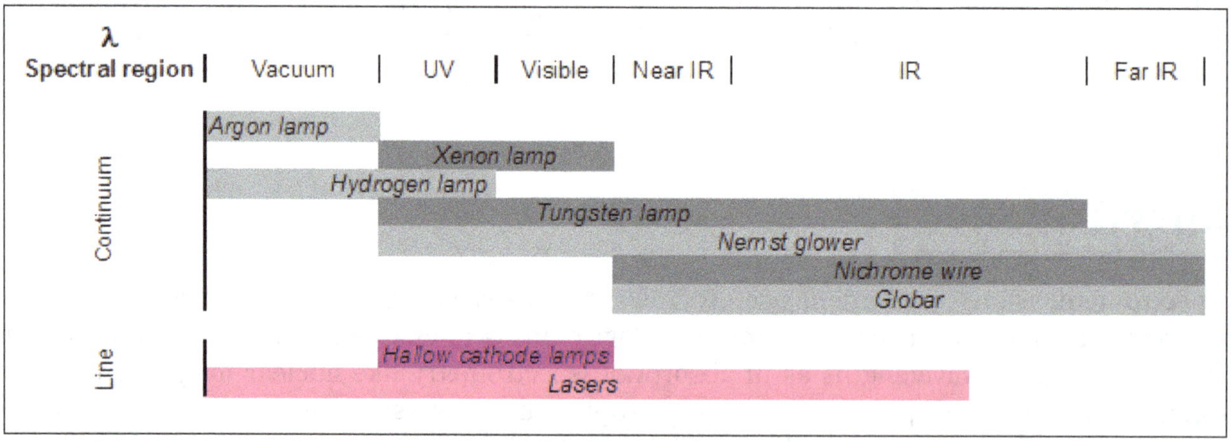

There are also non-electrode lamps used as line sources that contain a gas and a piece of metal that will emit narrow radiation at the desired wavelength. Ionization of the gas occurs from radiation (usually in the radio or microwave frequencies). The metal atoms are then excited by a transfer of energy from the gas, thereby producing radiation at a very specific wavelength.

Laser (an acronym for light amplification by stimulated emission of radiation) sources work by externally activating a lasing material so that photons of a specific energy are produced and aimed at the material. This triggers photon production within the material, with more and more photons being produced as they reflect inside the material. Because all the photons are of equal energy they are all in phase with each other so that energy (and wavelength) is isolated and enhanced. The photons are eventually focused into a narrow beam and then directed at the sample.

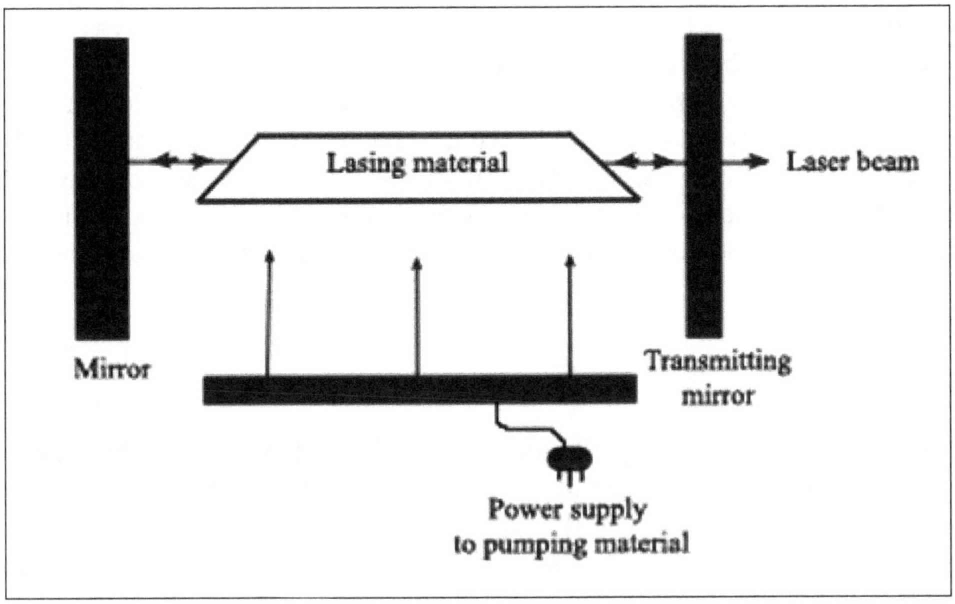

Mass Spectrometry

Mass spectrometry, also called mass spectroscopy is the analytic technique by which chemical substances are identified by the sorting of gaseous ions in electric and magnetic fields according to their mass-to-charge ratios. The instruments used in such studies are called mass spectrometers and mass spectrographs, and they operate on the principle that moving ions may be deflected by electric and magnetic fields. The two instruments differ only in the way in which the sorted charged particles are detected. In the mass spectrometer they are detected electrically, in the mass spectrograph by photographic or other nonelectrical means; the term mass spectroscope is used to include both kinds of devices. Since electrical detectors are now most commonly used, the field is typically referred to as mass spectrometry.

Mass spectroscopes consist of five basic parts: a high vacuum system; a sample handling system, through which the sample to be investigated can be introduced; an ion source, in which a beam of charged particles characteristic of the sample can be produced; an analyzer, in which the beam can be separated into its components; and a detector or receiver by means of which the separated ion beams can be observed or collected.

Many investigations have been conducted with the help of mass spectrometry. These include the identification of the isotopes of the chemical elements and determination of their precise masses and relative abundances, the dating of geologic samples, the analysis of inorganic and organic chemicals especially for small amounts of impurities, structural formula determination of complex organic substances, the strengths of chemical bonds and energies necessary to produce particular ions, the identification of products of ion decomposition, and the analysis of unknown materials, such as lunar samples, for their chemical and isotopic constituents. Mass spectroscopes also are employed to separate isotopes and to measure the abundance of concentrated isotopes when used as tracers in chemistry, biology, and medicine.

General Principles of Mass Spectromatry

The evolution of mass spectrometry has been marked by an ever-increasing number of applications in science and technology. New applications and new developments have gone hand in hand to create a complex array of instruments, but all may be understood by tracing the ions through three basic elements: an ion source, a method of analyzing the ion beams according to their mass-to-charge ratio, and detectors capable of measuring or recording the currents of the beams. These elements exist in many forms and are combined to produce spectrometers with specialized characteristics. The needs of users vary, as do the chemical form and the amount of sample available for analysis, which may be in sub-microgram quantities. The result is a great variety of design.

Ion Sources

Direct-current Arc

Historically this was the first way of producing a beam of ions and came quite naturally out of the 19th-century experiments for observing the passage of electricity in gases at low pressure. Two planar electrodes oriented perpendicular to the axis of the electric field can, with a few hundred-volt potential differences, form a plasma discharge. (Plasma refers to an ionized gas containing an approximately equal number of positive ions and electrons.) Electrons attracted to the anode collide with molecules of the gas to form ions and free more electrons; the positive ions contribute in turn to further ionization by their collisions. A hole in the cathode allows positive ions to emerge collimated into a beam. Such sources are found with many electrode configurations, including electron-emitting filaments, and operate with wide ranges of pressures and voltages. Sources with magnetic fields parallel to the electric fields can yield beams greater than one milli-ampere. Direct-current sources were widely used during the first decades of mass spectrography. They served well for gases and liquids introduced as vapors and for many solids as well, because these could be transformed into gaseous atoms and incorporated into the plasma through impact by the ions, a process called sputtering. One disadvantage of this kind of ionization is the wide band of energies attained by the ions, ranging from the maximum electrode potential to almost zero. Such a distribution of energies was the cause of Thomson's parabolas, but accurate work requires a narrow energy range, which in this case must be achieved in the analyzer section of the instrument.

Electron Bombardment

Electrons extracted from a glowing filament may be used to ionize gases. This is the basis for the electron bombardment ion source. A satisfactory electrode arrangement enables the production of a beam of ions much more nearly homogeneous in energy than with the arc, greatly simplifying the ensuing analyzing method. Electron impact has remained the most widely used method of ionization in mass spectrometry. It is subject to problems common to the arc: an almost total lack of selectivity as to the chemical element ionized and, to a lesser extent, the production of ions with degrees of ionization greater than one. Electron impact is utilized extensively in fields of study in which the sample is gaseous or prepared in gaseous form. Isotopic studies of carbon, nitrogen, oxygen, sulfur, and the noble gases make up a large field of endeavor. Electron impact is useful for studying organic compounds. Organic molecules are ionized not only as ions of the whole molecule but in a range of fragments as well.

This property, which may at first seem disadvantageous, is actually quite valuable in organic identifications because the resulting mass spectrum allows the identification of the source molecule as uniquely as fingerprints are used in human identification. This forms the basis of a powerful method of organic analysis. If the fragmentation of the molecule is harmful to the objectives of the experiment, another method of ionization can be employed that produces few fragments. In this technique a reagent such as methane (CH_4) is mixed with the sample gas and subjected to electron bombardment. The ionized methane (CH^{4}_{+}) reacts to form CH^{5}_{+}, which in turn reacts to ionize the sample gas by proton or charge transfer. This process is called chemical ionization, and in some cases it increases the mass of the ion formed by one unit.

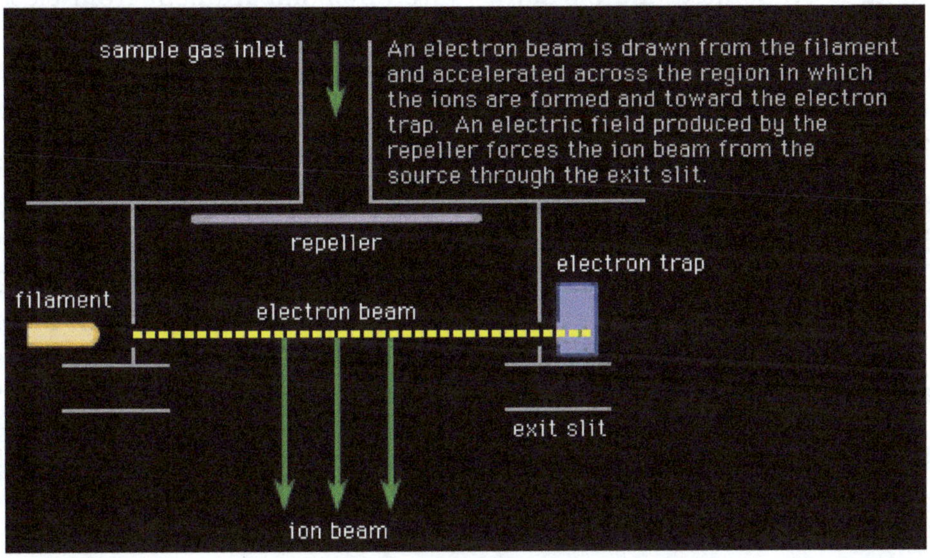

An electron bombardment ion source in cross section. An electron beam is drawn from the filament and accelerated across the region in which the ions are formed and toward the electron trap. An electric field produced by the repeller forces the ion beam from the source through the exit slit.

Thermal Ionization

Atoms with low ionization potentials can be ionized by contact with the heated surface of a metal, generally a filament, having a high work function (the energy required to remove an electron from its surface) in a process called thermal, or surface, ionization. This can be a highly efficient method and has the experimental advantage of producing ions with a small energy spread characteristic of the filament temperature, typically a few tenths of an electron volt, as compared with beam energies of thousands of electron volts. The filaments, generally made of platinum, rhenium, tungsten, or tantalum, are heated by current. Surface ionization requires a nearby source of atoms, often another filament operating at lower temperatures. Samples can also be loaded directly on the filament, a widely used and successful technique and one that has resulted in many interesting chemical treatments of the sample when it is deposited on the filament.

One such application changed lead from a difficult to an easy element to analyze, enabling important geo-chronological and environmental measurements. A disadvantage of thermal ionization is the possible change in isotopic composition during the measurement. This effect is caused by Rayleigh distillation, wherein light isotopes evaporate faster than heavy ones. Studies done on isotopes that come from radioactive decay, such as those used in determining the ages of rocks, encounter this problem, but it is correctable using the measured values of the isotopes that are not radiogenic. With few exceptions the use of a thermal source requires the chemical separation of the sample. Useful data are commonly obtained on extremely small (e.g., nanogram) samples.

Spark Discharge

In the vacuum spark source, a pulsed, high-frequency potential of about 50 kilovolts is built up between two electrodes until electrical breakdown occurs. Hot spots appear on the electrodes, and electrode material is evaporated and partially ionized by bombardment from electrons present between the electrodes. The principal merit of the vacuum spark source is its ability to produce copious quantities of ions of all elements present in the electrodes.

Secondary-ion Emission

Direct analysis of solids can be accomplished by bombarding the surface with an ion beam, the impact of which creates additional ions from the solid surface. The bombarding ions transfer substantial momentum to the target atoms, knocking them loose from the crystal lattice of the solid. The process is, generally speaking, not selective, although there are significant differences by element in the efficiency of ionization. The bombarding ions can be given a fine focus, with beam diameters of a few micrometres attainable. This allows the observer to select specific regions of the solid surface for analysis through the use of an auxiliary microscope and micrometre values for sample motion. Ion bombardment eats away the surface with time, allowing the solid to be analyzed for depth as well. This method is the basis for the ion microprobe.

Field Ionization

Intense fields, of the order of 10^8 volts per centimeter, can be generated in the neighborhood of sharp points and edges of electrodes, and these have been used as field ionization, or field emission, sources. This source is becoming popular in the study of organic compounds, which can be introduced as vapours and ionized in the intense fields. The ions are formed with very little excitation energy, so that there is little fragmentation of the molecular ions, making molecular formulas easier to determine.

High-frequency-produced Plasma

An oscillator can create an electrode-less discharge in gas at low pressure within a glass tube. The plasma so produced is now a commonly used source for mass spectrometers but was first used in plasma-emission spectrometry (optical and near optical). Samples are introduced by means of a carrier gas, typically argon, and ions result as from the direct-current arc but with very few molecular ions and with the absence of impurities introduced by source electrodes. Such discharges are generally coupled by a coil to an oscillator having a frequency of about 20 megahertz and are called inductively coupled. Discharges can also be produced for specialized experiments in a device called a waveguide that is connected to a cavity magnetron, which has a frequency more than 100 times higher and significantly greater power. This is the basis for the inductively coupled mass spectrometer.

Photoionization

Instead of electrons, photons in the far ultraviolet region may be used, as they have sufficient energy to produce positive ions in a sample gas or vapor to be analyzed. A discharge in a capillary tube through which is passed a suitable gas, such as helium, is a good source for such radiation. Photoionization sources usually produce fewer ions than electron-bombardment sources but have advantages when the ionization chamber must be held at low temperature.

Resonance Photoionization

All of the methods of ionization suffer from a lack of selectivity as to which element is ionized and depend either on the mass spectrometer for differentiation or on careful sample chemistry. A technique that achieves higher elemental selectivity is resonance ionization. In this scheme, a laser with adjustable wavelength irradiates the volume of gas from which the ions are to be extracted, exciting a transition from an atom's ground state to one of its excited (high-energy) states. This strong excitation enables an equilibrium to be established between the two states, while at the same time other radiation—or sometimes the same radiation—takes atoms from the well-populated excited state to ionization. A slight change in the irradiating wavelength stops the equilibration and leaves the excited state unpopulated, which cuts off the ionization. The intense levels of radiation required are produced by pulsed lasers with very short duty cycles, however, making efficient sample use difficult. (The duty cycle is the ratio of the number of atoms irradiated in a given volume to the total number of atoms entering that volume.)

Negative Ions

Negative ions are formed by many of these same methods as well and can be useful in mass spectrometry. The accelerating voltages of the source and the direction of analyzing fields must be reversed, but the detectors respond equally well, with the exception of the Daly detector. Arc discharges and electron impact produce negative ions, although at rates varying widely according to the construction and mode of operation. Negative ions can be formed in a two-stage process wherein positive ions are accelerated into a gas from which they capture two electrons, a technique infrequently used in mass spectrometry. Negative ions can result from thermal ionization, with those of the halogens easily formed. The elements rhenium, iridium, platinum, and gold are efficiently ionized as molecular negative ions for important applications in geochemistry. The cesium sputter source produces copious quantities of negative ions and is used exclusively for accelerator mass spectrometry. In this source the low ionization potential of cesium is utilized in two ways: (1) surface ionization provides a beam of positive cesium ions that bombard a sample having a thin layer of cesium condensed on it; and (2) the atoms or molecules dislodged by the bombarding beam capture an electron from this layer. In addition to providing beams for accelerator mass spectrometry, this source completely changed the manner in which tandem Van de Graaff accelerators are employed in nuclear physics.

The wide use of mass spectrometers as analytical instruments is accompanied by a correspondingly wide range of forms that the sample can take. Gases for which electron impact is a suitable ionization method are introduced into the vacuum of the source through a fine valve from the sample reservoir, although in some cases the gas may be devolved from a solid by heating in the source. Liquids invariably have vapour pressures high enough for them to be handled as gases. Organic chemistry often furnishes mixtures of gas and liquid in need of analysis. Electron bombardment not only ionizes these molecules but fragments them as well with distributions by which they can be identified. By comparison with a catalog of mass spectra, one can even identify limited mixtures. In 1952 the invention of the gas chromatograph by A.T. James and A.J.P. Martin provided chemists with a method of separating mixtures of volatile substances into their component fractions. In this technique the substance to be analyzed is introduced into a stream of gas, usually helium or nitrogen, and carried by it through a capillary containing or coated with an absorbing substance. The various fractions move with different speeds, and the arrival of each at the end of the column is signaled by a suitable detector. In 1957 a mass spectrometer was first employed as the detector, and an important instrument for organic analysis found its place in the modern laboratory, the gas chromatograph–mass spectrometer. The chromatograph causes the fractions of the sample mixture to arrive at the ion source in succession. Mass analyses of the fractions then allow determinations of high reliability. Liquid chromatography may be combined with mass spectrometry as well.

Ion-beam Analysis

The separation of ions according to their mass is accomplished with static magnetic fields, time-varying electric fields, or methods that clock the speeds of ions having the same energies—the time-of-flight method. Static electric fields cannot separate ions by their mass but do separate them by their energy and so provide an important design element by functioning as an energy filter.

Magnetic Field Analysis

Ions of mass m and charge z moving in vacuo with a velocity v in a direction perpendicular to a magnetic field B will follow a circular path with radius r given by,

$$r = \frac{mv}{Bz}.$$

Therefore, all ions with the same charge and momentum entering the magnetic field from a common point will move in the same radius r and will come to a first-order focus after 180°, as shown in, regardless of their masses. Hence, the mass spectrometer used by Dempster can be referred to as a "momentum spectrometer." If all ions of charge z enter the magnetic field with an identical kinetic energy zV, owing to their acceleration through a voltage drop V, a definite velocity v will be associated with each mass, and the radius will depend on the mass. Since $zV = {}^1/_2 mv^2$, substitution in the previous equation will give $m/z = B^2r^2/2V$. This formula shows that the radius of curvature r for ions in this spectrometer depends only on the ratio of the ions' mass to charge, as long as their kinetic energy is the same. Thus, a magnetic field can be used to separate a mono-energetic ion beam into its various mass components. A magnetic field will also exert a focusing action on a mono-energetic beam of ions of mass m as is shown in Figure. In this figure an ion beam emerges from a point A with a spread in direction 2α and comes to an approximate focus at B after traversing 180°. When a molecular ion of mass m_1 carries a single positive charge, it may decompose in front of the magnetic sector to form a fragment ion of mass m_2 and a neutral fragment.

If there is no kinetic energy of separation of the fragments, the ion m_2, and also the neutral fragment, will continue along the direction of motion of m1 with unchanged velocity. The equation of motion for the ion m2 entering the magnetic sector can now be written from a previous relationship, $r = m_2v/Bz$. In this equation v is the initial velocity appropriate to m_1 and given by $\sqrt{2zV/m_1}$.

Multiplying both sides of the equation $v=\sqrt{2zV/m_1}$ by m_2, one obtains $m_2\ v=\sqrt{2zV(m_2^2/m_1)}$.

Since the general momentum equation for any mass m can be written $mv=\sqrt{2zVm}$, it is apparent from the former equation that the momentum m_2v is appropriate to an ion of mass m_2^2/m_1. Thus, the decomposition of the metastable ion will give rise to a peak at an apparent mass $m^* = m_2^2/m_1$, not necessarily an integral number. This peak is known as a metastable peak. Generally, metastable peaks occur at non-integral mass numbers, and, because there usually is a kinetic energy of separation during fragmentation of the polyatomic ion, they tend to be more diffuse than the normal mass peaks and thus are recognized easily. For any value of m* a pair of integers m_1 and m_2 can be found such that $m^* = m_2^2/m_1$. Thus, the action of the magnetic field on the charged metastable-ion decomposition product can be used to give information on the individual fragmentation processes taking place in a mass spectrometer.

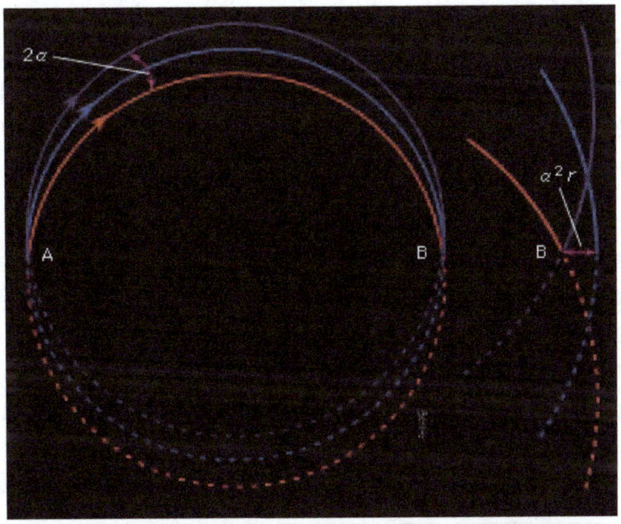

Paths of mono-energetic ions moving in a plane perpendicular to a magnetic field, passing through focal point B after originating at point A.

Electrostatic Field Analysis

An electrostatic field that attracts ions toward a common center—i.e., a radial field—will also exert a focusing action on a divergent beam of ions as shown in Figure. The radial force on the ions due to the electrostatic field will be Ez, the product of the field E and the ionic charge z, and is equal to the centripetal force mv^2r, of mass m moving with velocity v about a radius r. Thus, one may write the equation,

$$Ez = \frac{mv^2}{r} \text{ or } r = \frac{1/2mv^2}{1/2\ Ez}.$$

The radius of the arc traversed by the ions will be proportional to their kinetic energy, and an electric sector will thus produce an energy spectrum of the ions passing through it. Alternatively, if narrow collimating slits are placed at either end of the sector, a mono-energetic beam of ions can be selected.

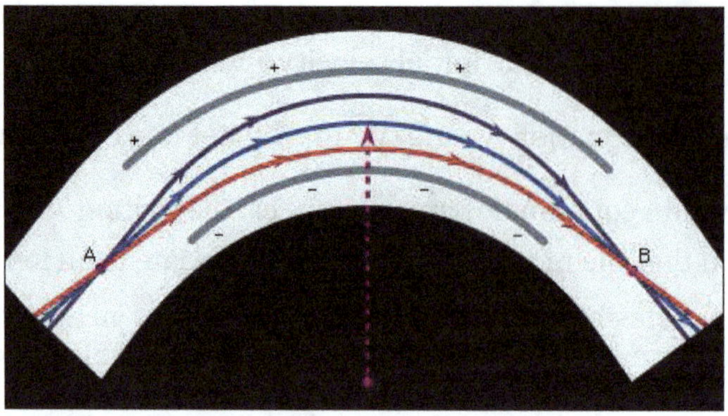

Focusing action of a radial electrostatic field produced by two semi-cylindrical charged electrodes on a mono-energetic beam of ions. The ions diverging from point A are brought to a focus at point B.

Combined Electric and Magnetic Field Analysis

Use of an electric wedge or sector to obtain a monoenergetic beam of ions, which is then separated for mass analysis by a magnetic sector, is another possible technique. General equations for combined electric and magnetic sectors have been developed; they show that a suitable combination of fields will give direction focusing for an ion beam of given mass-to-charge ratio, even though the beam may be heterogeneous in energy. The term double focusing is used for those combinations in which the angular and velocity aberrations effectively cancel.

Two of the best examples of double-focusing mass spectroscopes, both of which have been used in a variety of commercial instruments, were built by Mattauch and Richard Herzog in West Germany and by the American physicist Alfred O. Nier and his collaborators. The Mattauch-Herzog geometry is shown in Figure. Ions of all masses focus along a line that coincides with the second magnetic field boundary. Many versions of this design have been used when high resolution (up to 10^5) is desired for accurate mass and abundance measurements and general analytical work. It can give good spectra, even for a spread of 5 percent in energy of the ion beam, and can be used with virtually any ion source. The resolved ion beams can be recorded electrically or with a photographic plate.

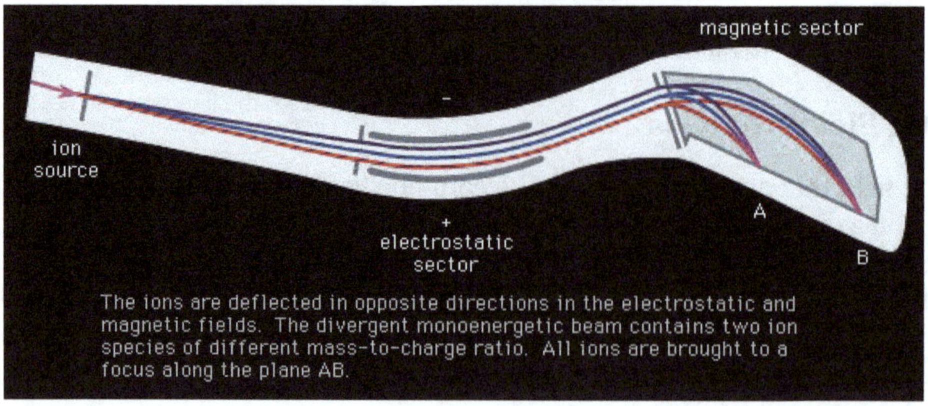

Arrangement of the electrostatic and magnetic sectors in the Mattauch-Herzog double-focusing mass spectrometer. The ions are deflected in opposite directions in the electrostatic and magnetic fields. The divergent monoenergetic beam contains two ion species of different mass-to-charge ratio. All ions are brought to a focus along the plane AB.

Nier's design is illustrated in Figure. In this instrument, Nier was able to achieve high sensitivity as well as high resolution. Using an electron-bombardment ion source, a resolving power of 2×10^5 has been attained in a commercial instrument. The design has yielded mass measurements of the highest precision for a very large number of the isotopes, and it has also been the most popular design for high-resolution work in organic chemistry.

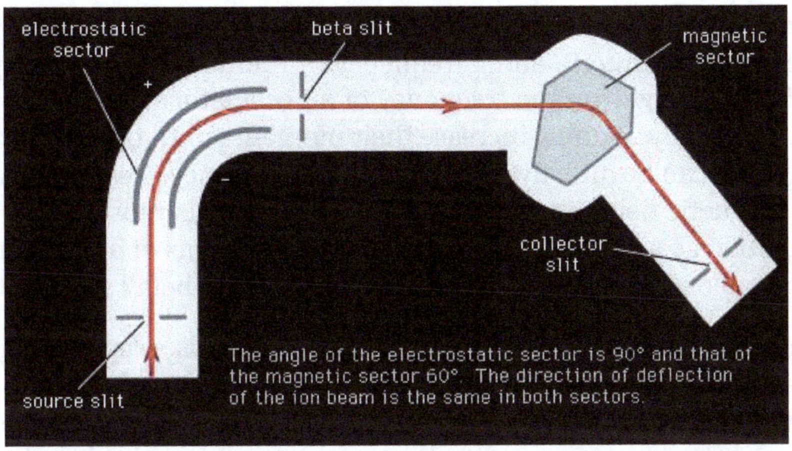

Arrangement of the electrostatic and magnetic sectors in the Nier double-focusing mass spectrometer. The angle of the electrostatic sector is 90° and that of the magnetic sector 60°. The direction of deflection of the ion beam is the same in both sectors.

z-axis Focusing

The form of focusing in the analyzers has assumed that the forces acting upon an ion lie entirely in the same plane, generally referred to as the x-y plane, with the y axis defining the direction of the beam. This is adequate for most applications in magnetic sector machines where the beam is ribbon-shaped and where slight deviations in velocity in the z direction, which is perpendicular to the principal plane, cause negligible beam loss. At the entrance and exit of a trajectory passing through a field produced by a magnet, the field is not parallel but rather is bowed out. As a consequence, an ion encounters a small, but not negligible, field component parallel to the x-y plane, which generates a correspondingly small force in the z direction. Magnets have been constructed that take advantage of such forces to focus the beam in the z direction. It is possible to make the x focus and z focus coincident. Focusing in the z direction is often accomplished electrostatically by a suitably arranged pair of electrodes in the ion source.

Time of Flight

The simplest form of mass analysis that does not use magnetic fields depends on the differing speeds of ions with the same energy but different masses. The ion source is generally of the electron-impact type and has one or more electrodes modulated so as to extract ions for a time that is brief compared with the time it takes them to reach the detector. The ion velocities, as given above by $v = \sqrt{2zV/m}$, and the distance between source and detector allow the mass to be calculated directly. In practice, the response of the detector is displayed on a cathode-ray oscillograph and recorded by a computer. This method has two advantages: it is fast and, if desired, can display the entire mass spectrum. Its deficiencies are poor resolution, poor accuracy of signal, and poor efficiency, due to the short period during which ions are extracted from the source.

Ion-trap Methods

It is possible to configure electric and magnetic fields so that ions can be held in stable orbits for a period of time long enough to perform useful measurements on them. Two forms of mass spectrometers are derived from this idea, the omegatron and the Fourier-transform spectrometer. Both make use of the cyclotron principle, in which positive ions produced by a beam of electrons flowing along the axis of a uniform magnetic field follow circular trajectories with a radius proportional to momentum, $r = mv/zB$, and a frequency of rotation inversely proportional to mass, $\omega = v / r = zB / m$. In the omegatron the frequency of an oscillator is varied so as to bring ions of various masses in tune and by so doing increase their momenta until they reach a radius at which a detector is located. Mass can be directly calculated from frequency. Resolution can be remarkably high if a sufficient magnetic field is provided, but this analyzer is most frequently operated with less than ideal resolution as a device for analyzing the residual gas of a vacuum, information that can be extremely valuable in diagnosing the problems that often befall such systems.

In the Fourier-transform method, the frequency of the oscillator is swept through the range corresponding to the mass range of interest. Each ion is placed into a circular orbit of approximately constant radius but well-defined frequency. The oscillator is turned off, and an electrode picks up radio-frequency radiation from the moving ions. The amplified output can be recorded either directly or after having been mixed with the frequency of a local oscillator, a standard radio technique. This yields a complex time-varying signal that follows the amplitude of the various ion radiators. The signal is converted to digital form and stored in a computer memory. The computer converts this periodic signal to its frequency spectrum by the mathematical technique known as the Fourier transform, with mass being inversely proportional to frequency. The process is repeated many times in order to enhance accuracy. These devices are capable of resolutions exceeding one million. In order to have orbital radii of convenient size, very high magnetic fields are required, generally provided by superconductors.

Tandem Spectrometry

The combination of two analytical techniques, such as resulted in the gas chromatograph–mass spectrometer, has been followed by the combination of two mass spectrometers, which has proved helpful in determining the structure of complicated molecules. A beam from the first spectrometer is passed into a gas cell (maintained in the vacuum system by differential pumping), where it is dissociated by collisions, and these fragments are passed on to a second mass analyzer, which generally discloses a more easily identified spectrum.

Quadrupole Spectrometer

Positive ions incident along an axis parallel to four cylindrical electrodes, as shown in Figure, experience for the static potentials indicated a focusing force along the x axis and a defocusing one in the z direction. If one superimposes a radio frequency voltage onto the static voltage, oscillatory ion trajectories can be found that allow ions of a given mass to pass through the quadrupole with other masses being defocused and lost from the beam. Knowledge of the potentials and frequency specifies the mass. This device is widely used where speed of data acquisition and high transmission are important. It is compact and lightweight and can trade off sensitivity for resolution by simple adjustment of electrical parameters. It is the common analyzing element of gas chromatograph–mass spectrometers.

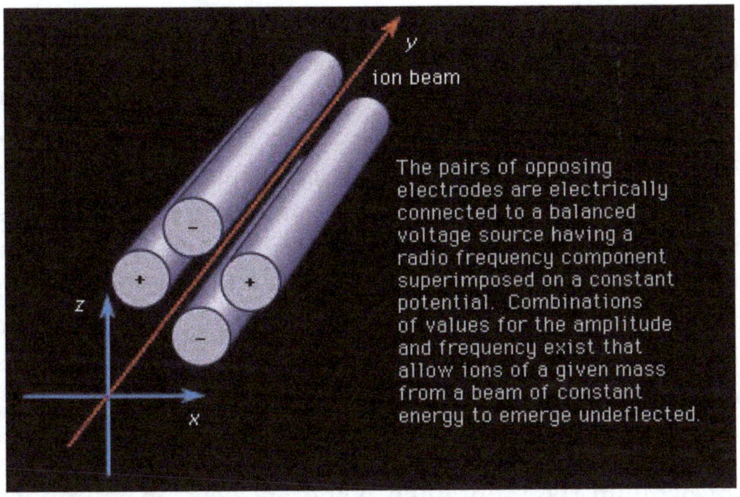

Schematic diagram of a quadrupole mass spectrometer. The pairs of opposing electrodes are electrically connected to a balanced voltage source having a radio frequency component superimposed on a constant potential. Combinations of values for the amplitude and frequency exist that allow ions of a given mass from a beam of constant energy to emerge undeflected.

Ion Beam Detection

Photographic Plates

Especially sensitive photographic plates are employed to compensate for the low penetrating power of the ions. It has proved possible with these to detect an element over a sensitivity range of one part in one billion. In addition to the sensitivity, a major advantage of the photographic plate arises when it is used in a double-focusing mass spectroscope in which the whole or a major part of the mass spectrum is focused in a plane. In this case, one can make use of the integrating action of the plate and compare the densities of lines due to different elements.

Faraday Cup

The direct measurement of ion currents collected by a shielded electrode, called a Faraday cup, became possible in the 1930s with the introduction of electrometer tubes capable of measuring currents below a nano-ampere, although sensitive galvanometers had been used for larger currents. The introduction of feedback led to greater stability and accuracy and faster response time, but it was the introduction of the vibrating-reed electrometer that allowed isotopic ratios to be routinely measured to a few parts in a hundred thousand. For more than three decades, these electrometers functioned unsurpassed as laboratory workhorses and were only slightly modified in design. They can now be equalled and in some respects surpassed in performance by the feedback electrometer, which uses a metal-oxide silicon field-effect transistor instead of a tube to measure extremely small currents.

Electron Multipliers

The development of electronic techniques for television during the 1930s yielded a device of extraordinary sensitivity for measuring small electron beams—namely, the secondary electron multiplier. Although originally invented for the amplification of the tiny currents from a photocathode, it soon proved to be an excellent detector for ion beams with sensitivity sufficient to record the arrival of single ions. The fundamental principle of the multiplier is, as the name suggests, a multiplication of the number of electrons emerging from an electrode as compared with the number incident upon it. Electrodes, called dynodes, are so arranged that each succeeding generation of electrons is attracted to the next dynode. For example, if 4 electrons are released at the first dynode, then 16 will emerge from the second and so forth. Gains of as much as one million are easily attained; the noise is limited to the currents originating from the few electrons leaving the first dynode as a result of thermal electron emission. Multipliers were originally constructed with discrete dynodes, a form still in wide use. Continuous dynode multipliers, which use a semiconducting glass to provide the distribution of electrostatic potential, are smaller and perform equally well in most applications. A multiplier can be employed in an analog mode, in which the output current is measured with an electrometer as is any small current, or in a pulse-counting mode, in which individual ions are counted.

The mass spectrum of osmium (Os), obtained using an electron multiplier detector, is shown in Figure. It is a recorder trace of the electrometer output from an electron multiplier detecting OsO_3^- taken as the field of the analyzing magnet was steadily increased. Owing to their small sizes, the leftmost and two rightmost peaks were recorded with an electrometer gain 100 times what was used for the other peaks; the change in gain is marked by a change in the position of the baseline. The osmium isotopes observed, from left to right, are 184, 186, 187, 188, 189, 190, and 192. The oxygen (O) in the ions results in very small satellite peaks caused by the low abundance isotopes ^{17}O and ^{18}O; the satellite peaks of ^{192}Os are at the right. This trace is typical of machines used in geochronology, where flat-topped peaks are desired rather than high resolution. The irregular signal of the three weak isotopes results from the low rate at which these ions are detected.

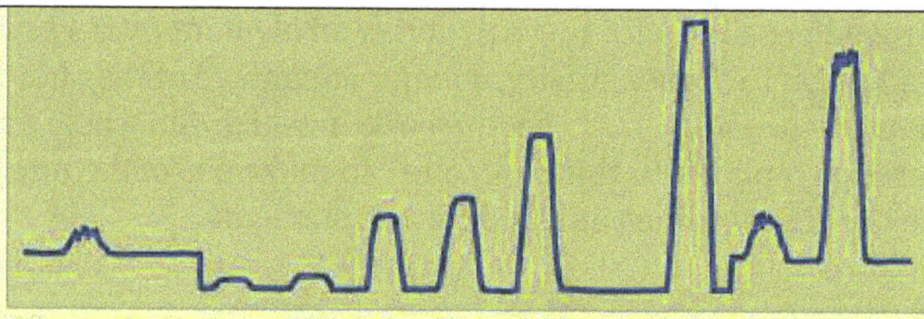

This recorder trace was obtained with an electron multiplier detecting OsO_3^-. The leftmost and rightmost peaks were recorded with the detector gain set at a value 100 times that used for the rest of the spectrum; this change is marked by the change in the baseline position. The small satellite peaks to the left are those of the low abundance oxygen isotopes ^{17}O and ^{18}O; the osmium isotopes are, from left to right, ^{184}Os, ^{186}Os, ^{187}Os, ^{188}Os, ^{189}Os, ^{190}Os, and the satellite peaks of ^{192}Os.

The mass spectrum of osmium.

This recorder trace was obtained with an electron multiplier detecting OsO_3^-. The leftmost and rightmost peaks were recorded with the detector gain set at a value 100 times that used for the rest of the spectrum; this change is marked by the change in the baseline position. The small satellite peaks to the left are those of the low abundance oxygen isotopes ^{17}O and ^{18}O; the osmium isotopes are, from left to right, ^{184}Os, ^{186}Os, ^{188}Os, ^{189}Os, ^{190}Os, and the satellite peaks of ^{192}Os.

Daly Detector

In 1960 N.R. Daly introduced a form of detector with properties superior to the electron multipliers.

In this design the incident ions are attracted to a rounded electrode of a few centimetres in dimension that is held at 10,000 to 20,000 volts negative. The ions strike the "door knob" and release a few secondary electrons for each incident ion; these electrons are then accelerated from the high negative potential to a scintillation crystal mounted on a photomultiplier at ground potential. The electrons generate a light signal in the scintillation crystal that is amplified by the photomultiplier. The output is then treated just as the output of an electron multiplier. The advantage of this more complicated device is an almost complete independence of the signal size with the position of the ion beam in the defining slit that precedes the detector. This is an important property for accurate measurement of isotopic ratios because the invariable instability of the analyzing magnet and the ion-source voltage cause the beam to drift within the limits set by the slit, and so a non-uniform response gives rise to errors in the measured ratios. Electron multipliers require various adjustments, not always satisfactory, to produce a signal size independent of beam position. The Daly detector cannot be used with negative ions.

Multiple Detectors

In instances in which the ratio of two or more isotopes is the experimental goal with a magnetic sector analyzer, there are advantages in measuring the currents of the beams simultaneously in multiple detectors rather than switching the magnetic field. This is relatively simple for light elements that have widely spaced beams at the detector location and is often employed in such cases. For heavy elements the small spacing necessitates more difficult techniques that came into routine use only in the 1980s.

Important Technical Adjuncts

Vacuum

In the devices heretofore described, the presence of a good vacuum system has been assumed. Mass spectroscopy originated at about the time that high vacuum was first attained in the laboratory. High vacuum refers to a pressure low enough that the mean free path (the distance travelled between collisions) of molecules in the residual gas is greater than the dimensions of the vacuum vessel. Mass spectroscopists invariably seek conditions of improved vacuum. The properties that render low pressures desirable include a reduction in the scattering of the beam in the analyzer, which causes interfering background effects and a reduction in the production of spurious beams out of the residual gases, particularly from the organic compounds that are present. The history of vacuum techniques is varied and great and has provided present mass spectrometrists with pressures that are routinely four to five orders of magnitude lower than those first used by Thomson, Aston, and Dempster. The invention of the diffusion pump by the German physicist Wolfgang Gaede in 1915, with important improvements by the American chemist Irving Langmuir shortly thereafter, freed mass spectroscopy from the severe limitations of poor vacuum. During the 1960s diffusion pumps began to be replaced by ion-getter pumps, with turbomolecular pumps becoming common in the 1980s.

Electronics

The operation of a mass spectrometer depends on elaborate electronic equipment: ion sources require extremely stable power supplies, magnets need instruments for measuring the magnetic field and controlling the current supply for the coils, detectors use a variety of power supplies and amplifiers, and general operation requires electronic auxiliary equipment. The rapid increase in the use of mass spectrometers following World War II can likely be attributed in part to the large number of physicists who had gained electronic training during the war, many of whom had utilized mass spectroscopy during that conflict to monitor uranium isotope separation and to analyze aviation gasoline.

Computers

The introduction of small computers for laboratory work during the 1960s altered entirely the manner in which mass spectrometry was performed and widened its applications to an extraordinary degree. Computers were interfaced with spectrometers, making it possible to repeat a measurement schedule on a steady basis and record the data acquired. In organic analysis the computer was programmed to store the spectra of thousands of compounds, allowing rapid identification of the substance under study. Users soon devised ways by which the answers to their questions came within minutes after the conclusion of the analysis.

Applications

Atomic Masses

The discovery of isotopes with the first mass spectrograph answered the question about the integer value of atoms only to a crude level of accuracy and made all too clear the need for more accurate mass determinations. These were first undertaken by Aston and repeated with increasing precision by succeeding generations. The first data showed slight deviations from an integer law but also showed a quasi-systematic variation as a function of atomic number. In the early 1930s these data explained the energies of nuclear reactions then being observed through the mass-energy relation that had been given two decades earlier by the special theory of relativity. Since that time mass spectroscopy and nuclear physics have combined to determine isotopic masses to a high degree of accuracy. The mass unit now used is defined so that the mass of the carbon-12 isotope is exactly 12 atomic mass units (amu). Nuclear theory is continually tested in its ability to reproduce the observed values. Indeed, these masses provide an early and critical test of nuclear models. Mass spectrometry has been closely associated with nuclear physics since its beginning. A mass spectrometer is frequently found in some form as a part of nuclear experiments to identify reaction products. Large mass spectrometers are employed as isotope separators and are capable of producing weighable amounts of selected stable isotopes that have valuable analytical applications. They are used for labeling compounds so that they can be traced through various chemical, physical, and environmental processes without the problems created by radioisotope tracers. Analysis is, of course, performed by mass spectrometer. In addition, nuclear reactions often produce extremely small amounts of radioactive products, which cannot easily be manipulated by normal chemical procedures. In such cases, stable isotopes of the product are added to the radioactive element, thereby increasing the concentration so that it falls within the range of the analytical method. These enriched isotopes are indispensable for research in nuclear physics.

Geochronology and Geochemistry

The early studies of the radioactive decay of uranium and thorium into lead caused the British physicist Ernest Rutherford to suggest that this process could be used to determine the age of rocks and consequently of the Earth by observing the amount of helium retained by a rock relative to its uranium and thorium contents. Mass spectrometers capable of measuring isotopic ratios allow the composition of elements to be determined in which one or more isotopes result from radioactive decay. The age of the rock from which the element has been obtained can be determined if the amount of the parent element can be measured and certain requirements on the environmental history of the rock are met.

The Earth's crust is generally richer in oxygen-18 (^{18}O) than is the mantle, as a result of the reaction of these upper-layer rocks with the hydrosphere and atmosphere. This fact allows oxygen-18 to be used to assess the degree to which ascending magmas have incorporated crustal rocks as they rise to the surface. The use of isotopes has proved especially valuable in understanding the origin and nature of the solar system. A great body of evidence now suggests that meteorites are objects that solidified very early in the history of the solar system. Extinct radioactivities of elements with various half-lives have been identified that set limits on the time between the synthesis of the elements and their condensation. (Extinct radioactivities are nuclides that have nearly completely decayed into their daughter elements.) An example is the excess of magnesium-26 (^{26}Mg) found in primitive meteorites that resulted from the decay of aluminum-26 (^{26}Al), which has a 720,000-year half-life.

Hydrogen, Carbon, Nitrogen, Oxygen and Sulfur in Nature

These elements, each of which has two or more stable isotopes, are vital to life. All show measurable variation in isotope composition as a result of natural and, in particular, metabolic processes. It was observed as early as 1939 that living matter preferentially incorporates the light isotope of carbon at rates differing according to species and environment. Knowledge of this is valuable in understanding the early biochemical evolution of the Earth. The evaporation of seawater causes a lower ratio of ^{18}O to ^{16}O in fresh water during times of high average temperature than in times of low temperatures. Examination of the oxygen preserved in polar ice and calcareous fossil shells has revealed the climatic evolution of the recent geologic past.

Trace Element Analysis

Mass spectrometry may be used to measure with high sensitivity trace amounts of an element through the technique of isotope dilution. A small, measured amount of an isotopically enriched sample, called a spike, is added to the original material, thoroughly mixed with it, and extracted with that element. The mass spectrum of this mixture will be a combination of the natural spectrum of the element plus the unnatural one of the spike. By knowing the amount of the spike added, one may calculate the amount of the unknown. Rigorous measures must be undertaken to ensure that the reagents and vessels used do not themselves supply the element, but this is easily controlled by chemically processing a known amount of the spike alone and determining the amount of the element picked up.

Molecular

Ion-molecule Reactions

Owing to the poor vacuums available prior to the contributions of Gaede and Langmuir, this subject was forced on the attention of early experimenters. They observed masses of 3 and 19, which could not have been produced by simple ionization and which arise from the following reactions, respectively:

$$H_2^+ + H_2 \rightarrow H_3^+ + H$$

$$H_2^+ + H_2O \rightarrow H_3O^+H.$$

These "problems" disappeared with improved vacuum. Ion-molecule reactions are important in understanding the chemistry of flames, of electrical discharges, of the upper atmosphere, and of samples subject to radiation. The mass spectrometer is the instrument of choice for such investigations.

Organic Chemistry

Mass spectrometry has a critical role in organic chemistry. The same techniques can be used in determining the structure of complicated molecules, but perhaps of even greater value for such work are high-resolution measurements.

With a high-resolution mass spectrometer it is possible to carry out mass measurements on the molecular ion (or any other ion in the spectrum) to an accuracy of approximately one part in one million. This mass provides the best index for determining ionic formulas. The accurate masses of the ions $C_6H_{12}^+$ and $C_4H_4O_2^+$ are, for example, 84.0939 and 84.0211, respectively, and these ions can easily be distinguished solely on the basis of their masses. Once the molecular formula is known it is possible to deduce the total of rings and double bonds making up the molecular structure and to begin to speculate on possible structural formulas. In order to deduce structural formulas from molecular formulas, it is essential to study the fragment ions in the mass spectrum. It is still not possible to predict definitively the fragmentation patterns for organic molecules, but many semi-empirical rules of fragmentation are known, and it is usually possible to pick out peaks in the spectrum that are characteristic of particular chemical groups. The technique is valuable in that it is generally not necessary to know any details of the composition of the unknown compound in order to deduce a complete or partial structure. Only a small quantity of compound, a hundred micrograms or less is necessary for an analysis.

Using a computer coupled to a high-resolution mass spectrometer, about 1,000 mass peaks per minute can be plotted at a resolving power of up to 20,000, accurate measurements can be made on each peak, and peak heights and ion compositions can be printed out in the form of an "element map" to aid in the interpretation of the spectrum. It is also possible for the computer to carry out many of the logical steps in reducing the data that lead to structural elucidation.

Continuous sampling of the materials contained in a reaction vessel, followed by analysis with a mass spectrometer, has been used to identify and measure the quantity of intermediate species formed during a reaction as a function of time. This kind of analysis is important, both in suggesting the mechanism by which the overall reaction takes place and in enabling the detailed kinetics of reactions to be resolved.

Isotopic labeling is widely used in such studies. It can indicate which particular atoms are involved in the reaction; in rearrangement reactions it can show whether an intra-molecular or intermolecular process is involved; in exchange reactions it can show that particular atoms of, for example, hydrogen are exchanging between the reacting species. Labeling is also widely used in mass-spectrometric research to give information about the fragmentation reactions occurring in the mass spectrometer.

Fields of investigation that employ mass spectrometry include studies of protein structure, drug metabolism, flavor and smell, petroleum and petrochemicals, organic fossils, inherited metabolic diseases, atmospheres and respiratory gases, and many other highly specialized subjects.

Space Probes

Future space exploration, addressing the question of whether life exists elsewhere in the solar system, will rely on the mass spectrometer to produce spectra of those molecules characteristic of life. An unmanned spacecraft equipped with a mass spectrometer has already revealed much about the surface and atmosphere of Mars and set limits on the amount of organic matter present.

Leak Detection

A widely used commercial device designed to locate leaks in vacuum systems consists of a small mass spectrometer with an electron-bombardment ion source that is connected to the troubled system. The mass spectrometer is set to detect helium, and the gas is played onto suspected parts through a capillary. A signal develops when the helium enters through the leak, and the exact location can be determined by adept manipulation of the capillary.

Accelerator Mass Spectrometry

Development of Accelerator Mass Spectrometry

The particle accelerators used in nuclear physics can be viewed as mass spectrometers of rather distorted forms, but the three principal elements—the ion source, analyzer, and detector—are always present. L.W. Alvarez and Robert Cornog of the United States first used an accelerator as a mass spectrometer in 1939 when they employed a cyclotron to demonstrate that helium-3 (^{3}He) was stable rather than hydrogen-3 (^{3}H), an important question in nuclear physics at the time. They also showed that helium-3 was a constituent of natural helium. Their method was the same as that for the omegatron except that a full-sized cyclotron was used, and it easily distinguished the two isotopes. The method was not employed again for nearly 40 years; however, it has found application in measuring cosmogenic isotopes, the radioisotopes produced by cosmic rays incident on the Earth or planetary objects. These isotopes are exceedingly rare, having abundances on the order of one million millionth of the corresponding terrestrial element, which is an isotopic ratio far beyond the capabilities of normal mass spectrometers.

If the half-life of a cosmogenic isotope is relatively short, such as beryllium-7 (^{7}Be; 53 days) or carbon-14 (^{14}C; 5,730 years), its concentration in a sample can be determined by radioactive counting; but if the half-life is long, such as beryllium-10 (^{10}Be; 1.5 million years) or chlorine-36 (^{36}Cl; 0.3 million years), such a course is ineffective. The advantage of the large, high-energy accelerator mass spectrometer is the great detector selectivity that results from ions having 1,000 times more energy than any previously available machine could provide. Conventional mass spectrometers have difficulty measuring abundances less than one hundred-thousandth of the reference isotope, because interfering ions are scattered into the analyzer location where the low-abundance isotope is to be sought. Extremes of high vacuum and anti-scattering precautions can improve this by a factor of 10 but not the factor of 100 million that is required. An accelerator suffers from this defect to an even greater degree, and large quantities of "trash" ions are found at the expected analyzer location of the cosmogenic isotope. The ability of certain kinds of nuclear particle detectors to identify the relevant ion unambiguously enables the accelerator mass spectrometer to overcome this shortcoming and function as a powerful analytical tool.

Operation of the Tandem Electrostatic Accelerator

The tandem electrostatic accelerator quickly displaced all other machines for this purpose, primarily because its ion source, the cesium sputter source, is located near ground potential and is easily accessible for changing samples. The ions must be negative, but this does not prove to be a handicap as they are easily and efficiently produced. Before entering the high-voltage tube, the ions are mass-analyzed so that only the beam emerging at the mass location of the cosmogenic isotope enters the accelerator; the intense reference isotope beam is often measured at this location without entering the accelerator at all. The cosmogenic isotope beam is attracted to the high-voltage terminal of the machine where collisions with gas or a thin carbon foil or both strip various numbers of electrons, thereby leaving the subject isotope with a distribution of multiple positive charge states that are repelled by the positively charged terminal. All molecular ions are broken up. The emerging beam then passes through analyzing fields of which a high dispersion magnet is the principal part. Upon leaving the analyzer, the beam enters the detector. Each ion is examined individually in a manner that allows its identity to be established. The most common way of doing this is by using a combination of two particle detectors: one detector measures the rate at which the particle loses energy when passing a given length of matter, while the other simultaneously measures the total energy of the particle. The counts are stored in the bins of a two-dimensional computer array, the coordinates of which are given by the amplitudes of the signals from the two detectors. The numerous "trash" ions take on values from the two detectors that fill regions of the data array but generally do not overlap the well-defined region occupied by the subject ion. Each kind of isotope requires a specially designed detector system with various additional analyzing fields and, in some cases, even the use of time-of-flight techniques. A schematic diagram of an accelerator mass spectrometer is shown in figure below.

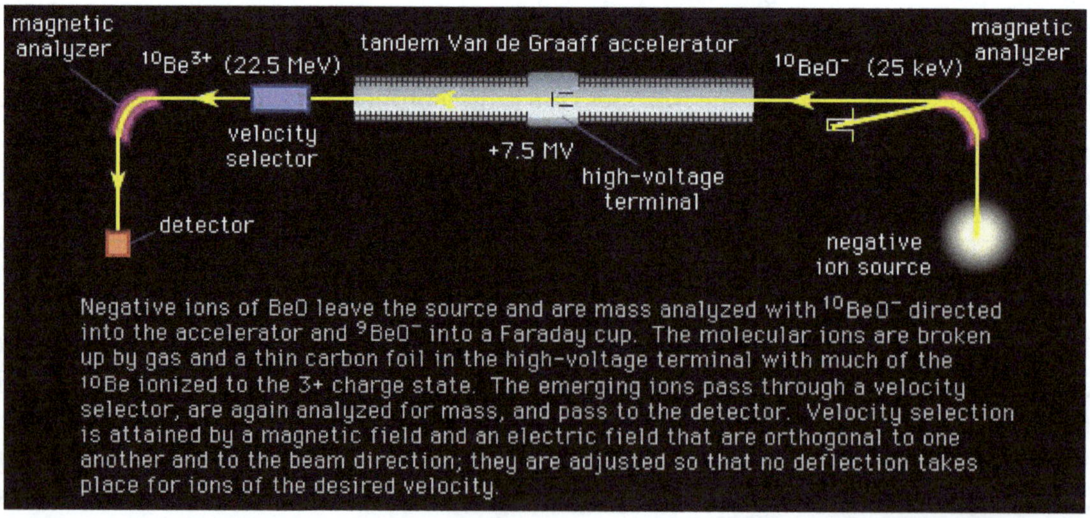

Negative ions of BeO leave the source and are mass analyzed with $^{10}BeO^{-}$ directed into the accelerator and $^{9}BeO^{-}$ into a Faraday cup. The molecular ions are broken up by gas and a thin carbon foil in the high-voltage terminal with much of the ^{10}Be ionized to the 3+ charge state. The emerging ions pass through a velocity selector, are again analyzed for mass, and pass to the detector. Velocity selection is attained by a magnetic field and an electric field that are orthogonal to one another and to the beam direction; they are adjusted so that no deflection takes place for ions of the desired velocity.

Schematic diagram showing the ion trajectories in an accelerator mass spectrometer with application to ^{10}Be.

Negative ions of BeO leave the source and are mass analyzed with $^{10}BeO^-$ directed into the accelerator and $^{9}BeO^-$ into a Faraday cup. The molecular ions are broken up by gas and a thin carbon foil in the high-voltage terminal with much of the ^{10}Be ionized to the 3+ charge state. The emerging ions pass through a velocity selector, are again analyzed for mass, and pass to the detector. Velocity selection is attained by a magnetic field and an electric field that are orthogonal to one another and to the beam direction; they are adjusted so that no deflection takes place for ions of the desired velocity.

Applications

The accelerator method has opened lines of investigation that had previously been inaccessible. A strong motivation for the inventors was the improvement of radiocarbon dating. Scientists are now able to make age determinations from much smaller samples and to make them much more rapidly than by radioactive counting, but carbon-14 proved to be a considerably more difficult problem for instrumental development than the other cosmogenic isotopes. The method was applied almost immediately to analyses involving beryllium-10 and chlorine-36, with aluminum-26 (^{26}Al), calcium-41 (^{41}Ca), and iodine-129 (^{129}I) following soon after; notable achievements resulted from all five. Cosmic rays striking the atmosphere are a strong source of beryllium-10, carbon-14, and chlorine-36, which are deposited in rain and snow, whence their migration may be followed. A question concerning the origin of the lavas of island-arc volcanoes, which had been disputed since the general acceptance of the plate tectonic theory of the Earth's structure, was settled from the observation of beryllium-10 in these lavas.

The presence of beryllium-10 proved that deep-ocean sediment, rich in the isotope, had been subducted (i.e., carried on the surface of a descending tectonic plate beneath another such plate) and some of the sediment incorporated into the magma. The first application of chlorine-36 was the study of the migration of ancient groundwater. Later improvements in instrumental techniques added iodine-129 as a needed tracer for this challenging problem. Nuclear bomb tests at oceanic sites produced huge amounts of chlorine-36 that were injected into the atmosphere. For a few years rain contained this isotope at a level up to 1,000 times higher than the cosmogenic level. This yielded a tracer with a well-defined time of origin that will be useful long into the future for following the course of such water in soils and aquifers (water-bearing layers of rock). The four lightest of these isotopes have proved useful in determining the ages and irradiation histories of meteorites and lunar samples. There have been extensive studies of beryllium-10 in cores of polar ice and ocean sediments that give unique information about the intensity of cosmic rays over the past few million years.

Magnetic Spectrometer

Magnetic spectrometer is a device which is useful in measuring the momentum of charged particles or their distribution of intensity versus momentum. Similarly, it happens by passing the particles through a magnetic field. It bends their paths in proportion to their momentum.

Time of Flight Mass Spectrometer

Time-of-flight mass spectrometers (TOF MS) do exactly what their name suggests: determine an ion's mass-to-charge ratio (m/z) based on how long it takes for that ion to travel down the instrument's "flight tube" and reach the detector. Frequently, though not exclusively coupled to matrix-assisted laser desorption ionization (MALDI) sources, TOF MS come in a variety of configurations, including a single-stage time-of-flight mass spectrometer, a multi-stage TOF/TOF, and even a hybrid quadrupole-TOF (Q-TOF) instrument; the latter two enable MS/MS applications. Often, time-of-flight mass spectrometers contain two detectors: a linear detector at the end of the flight tube, and a "reflector" detector, which captures ions that were redirected back towards the source by a "reflectron." The reflectron serves two purposes: correcting for discrepancies in initial kinetic energy of identically sized molecules (which would cause them to reach the detector at different times and thus, have different apparent m/z values) and lengthening the flight path, improving resolution.)

Optical Spectrometer

Optical spectrometry refers to the analysis of a light spectrum separated by wavelengths. It can be of two types – absorption or emission. An Atomic Optical/Emission Spectrometer (AES / OES) is one that analyses an optical (light) spectrum emitted by an excited sample. The excitation could be by a number of means, such as application of a spark, plasma, flame etc. That said, the term "OES" is now almost ubiquitously used by people to refer to the arc-spark OES technique.

Working Principle of Arc/Spark Optical Emission Spectrometer

The key principles of spectrometry used in an Arc/Spark Optical Emission Spectrometer (OES) are:

- Electrons in atoms absorb energy (get "excited") and move into higher energy states (also called orbits) when energy is applied. When this energy source is removed, the electrons fall into the ground state and release the absorbed energy in the form of photons.
- No two different element's atoms can emit photons at the same wavelength. Consequently, every wavelength is unique to a single element alone.

This means that once we know the wavelength of the photon emitted, we know which element is emitting it.

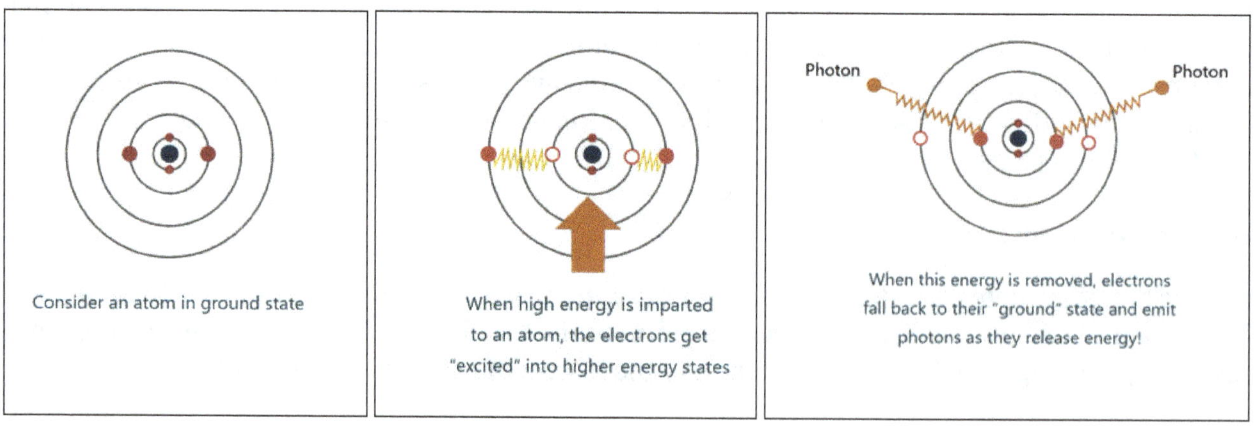

In an arc/spark OES, the principles outlined above are leveraged to analyze metallic samples to assess exactly which elements are present in it – and in what proportion. The output of the OES is a detailed assessment of the elemental composition of the sample in weight percentages.

First up, there is a need to "spark" the sample. The sample is therefore first prepared, i.e. one face of the sample is made absolutely uniform, clean, flat and as free from surface flaws as possible. Suitable methods of sample preparation must be used for this. The prepared sample is then placed on the sample stand as shown below. The sample stand has a hole in it which the sample must cover. Below this, there is an electrode at a fixed distance from the sample's exposed surface. This entire spark enclosure is filled with Argon when analysis is to be done. Then, a high current is applied to the sample.

The extremely high levels of DC current create a plasma in the Argon-purged atmosphere of the spark chamber, and a rapid series of high-energy sparks is therefore created between the electrode and the sample. Application of these sparks causes a part of the sample to vaporise. The vaporised atoms in the plasma absorb energy and their electrons move to higher energy-states with each spark. With each removal, the electrons move back into ground state and emit photons. Given the large number of elements simultaneously emitting photons, a composite emission is generated. This composite light is made to fall upon a diffraction grating. The diffraction grating separates each individual wavelength and creates a spectrum inside what is called the "optical chamber".

The spectrum can now clearly be analysed! The basis for analysis is of course, simplicity itself. We know the wavelengths that characterize each element. Further, the stronger the element, the higher its concentration. Were we to therefore measure the intensity of the emission at each wavelength of interest to us, we could compare with a database of knowledge and say with conviction what the concentration of individual elements is.

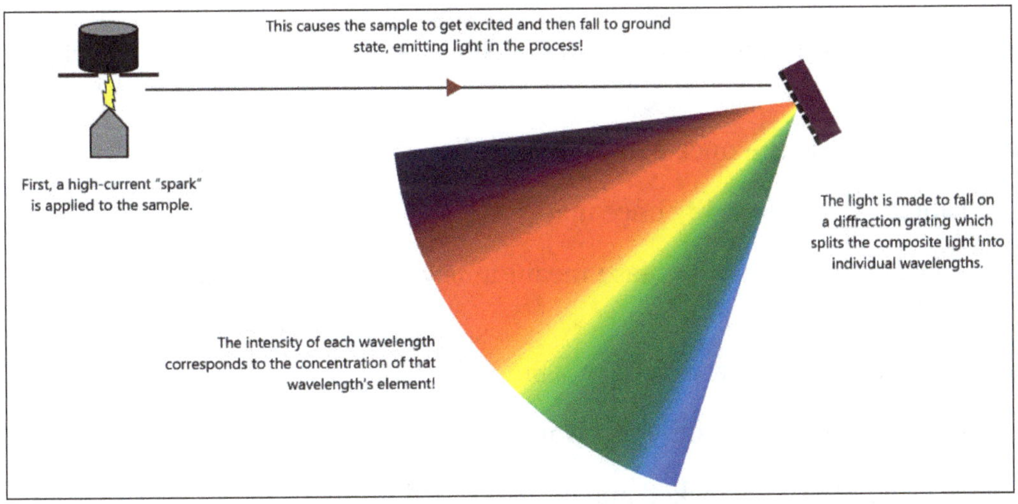

How has have Optical Emission Spectrometers (OES) evolved? What are PMTs and what are CCDs, for instance?

How the First Spectrometers Worked

The first instruments (very early) had to work without photo-emitters. The earliest researchers therefore had to rely on more mundane analog methods! They simply placed a photographic plate upon which the diffracted spectrum would fall. This plate was then developed and studied to arrive at the required results.

The First Move to Automation – The Introduction of PMT Detectors

In the 1930s however, there emerged the photomultiplier tube (PMT), a vacuum tube that emits electrons when light is incident upon it. Spectrometers therefore rapidly moved to using PMTs. A PMT was therefore placed inside the optical chamber in precise position for each wavelength that the user wished to analyze. Along with this, there was also a computer connected to the spectrometer. The computer stored the database against which the PMTs' outputs were compared to arrive at the elemental composition required. This automated the process and not only made it far more rapid and convenient, but also far more accurate and error-free.

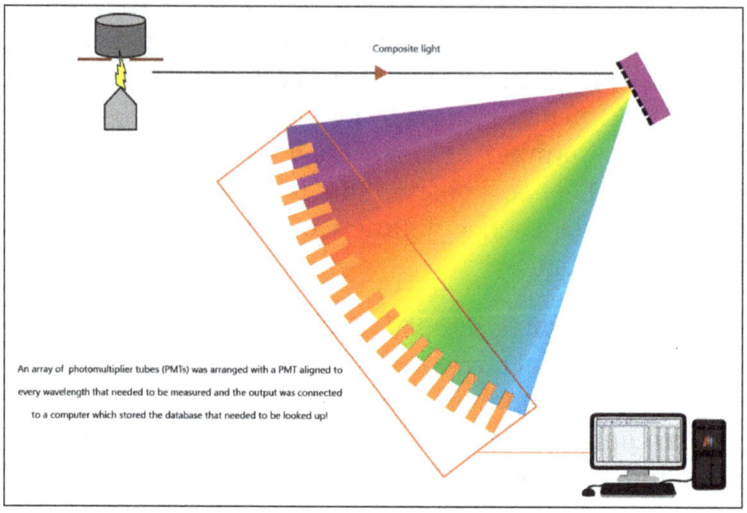

This worked very well for decades – but, as ever, technology moved on. PMTs clearly had a load of drawbacks:

- Flexibility was absent – once bought, that's that:
    - No ability at all to modify once made.
    - Even a single element increase meant a new optical emission spectrometer needed to be purchased.
- The number of elements analysed was limited:
    - Each PMT can analyze only one wavelength.
    - So wavelengths were picked and chosen – and limited.
- Cost and tedium was still very high:
    - Detectors, cards etc. were extremely expensive.
    - Regular profiling needed, vacuum pumps etc.

The Revolution

The introduction of CCD (Charged Coupled Device) and now CMOS (Complementary metal-oxide semiconductor) detectors solved literally every issue that the PMT devices posed and also offered several more upsides to spectrometer makers and users. Just a few of these are:

- Unmatched flexibility:
  - Every wavelength is captured – so can be analysed.
  - No limitations of space as CCDs are small.
- Instruments became smaller and less expensive:
  - High-res grating & CCDs results in shorter focal lengths.
  - Fewer detectors mean fewer cards and lower cost.
- Low tedium and low running costs:
  - No need for profiling etc. – as all this is automated.
  - No vacuum and efficient electronic = lower running costs.

Spectrometers therefore rapidly shifted towards using these devices and today, the modern OES exclusively comprises of optics with these devices.

Types of Optical Spectrometer

Now that the key component of a spectrometer has been identified, the different types of spectrometer, their role, and basic design can be discussed. Three of the most common optical spectrometers: spectrophotometers, spectrofluorometers and Raman spectrometers are introduced.

Spectrophotometer

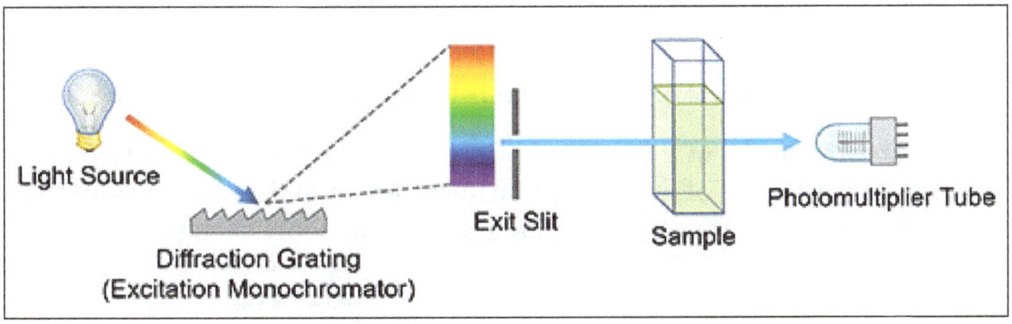

Simplified diagram of a single beam spectrophotometer.

The term spectrophotometer can refer to quite a variety of instruments that measure light, with the exact definition depending on the area of science or industry. In all cases the term 'photo' is used to indicate that the spectrometer is for the quantitative measurement of light intensity with wavelength. Within academic research (particularly chemistry and biology laboratories) the term spectrophotometer is used specifically to refer to a spectrometer which measures the absorption of light by a sample and that definition will be used here.

A stylised layout of a basic single beam spectrophotometer is shown in Figure. It comprises a white light source which is usually either a combination of a deuterium arc lamp to cover the UV range and a tungsten halogen lamp to cover the visible; or a single Xenon arc lamp to cover the entire range. This is followed by an excitation monochromator which selects the wavelength of light that reaches the sample. This light is then either transmitted through the sample (as shown in Figure) for transparent samples such as solutions or reflected off the surface for opaque samples. The intensity of the transmitted or reflected light is then monitored using a detector which is typically a photomultiplier tube or silicon photodiode.

In addition to stand alone spectrophotometers, the functionality of a spectrophotometer can also be incorporated as a secondary feature into other spectrometers. An example of this is the FS5 Spectrofluorometer which comes with a transmission detector as standard and has all the functionality of a single beam spectrophotometer in addition to being a spectrofluorometer.

The most common measurement undertaken in a spectrophotometer is measuring the absorption spectrum of a sample. The excitation monochromator is scanned and the change in light intensity transmitted through the sample recorded on the detector. This is then repeated with a reference sample and the absorption spectrum calculated as shown in Figure for a solution of fluorescein in phosphate buffered saline.

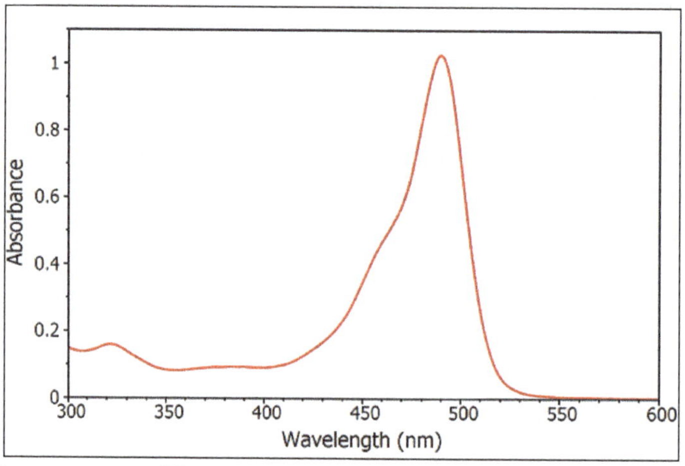

Absorption spectrum of fluorescein measured using the FS5 Spectrofluorometer.

A much more advanced version of a spectrophotometer is a transient absorption spectrometer, which can measure the evolution of the absorption spectrum with time and is essential for looking at temporary species generated through chemical reactions or short lived photo-excited states.

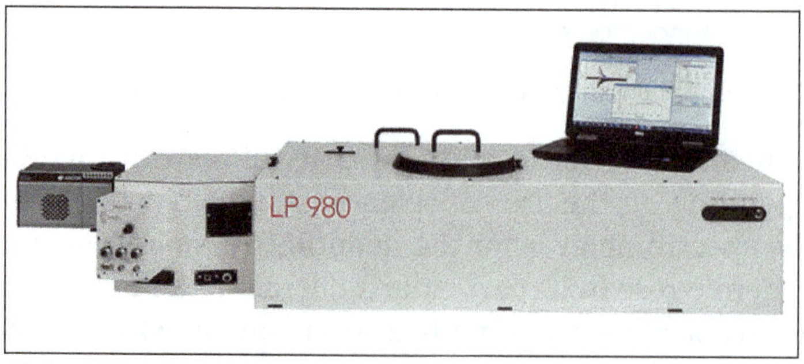

Edinburgh Instruments LP980 Transient Absorption Spectrometer.

Spectrofluorometer

A spectrofluorometer is used to measure the fluorescence emission (or more generally, the photoluminescence) from a sample. The terms spectrofluorometer and fluorescence/photoluminescence spectrometer are interchangeable and different manufacturers call them by different names. A general convention is that spectrofluorometer refers to a compact bench-top instrument that is similar in size to a spectrophotometer, such as the FS5 Spectrofluorometer, while the term fluorescence/photoluminescence spectrometer is used for larger spectrometers with superior performance and more diverse functionality such as the FLS1000 Photoluminescence Spectrometer.

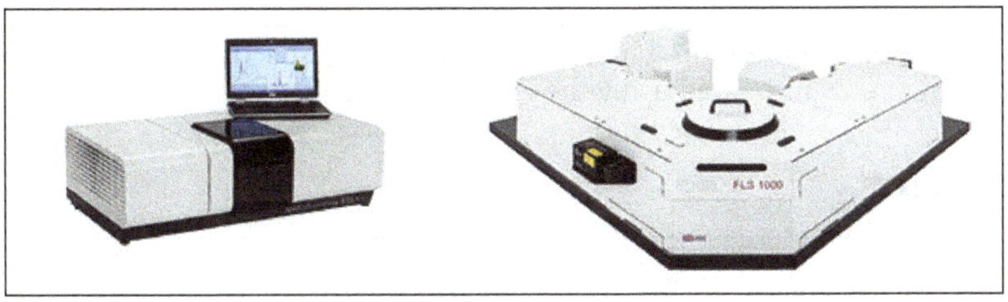

Example of a compact spectrofluorometer (FS5) and a more advanced photoluminescence spectrometer (FLS1000).

The layout of a typical steady state spectrofluorometer is shown in Figure. The excitation side of a spectrofluorometer is equivalent to the spectrophotometer: a white light source and an excitation monochromator. Xenon arc lamps are used as the light source as their high brightness is essential to measure the weak fluorescence emission. The sample is illuminated by the chosen excitation wavelength which causes it to fluoresce. The fluorescence emission is collected by the emission monochromator which is orientated at 90 degrees to the excitation monochromator and the selected wavelength reaches the detector; typically a photomultiplier tube.

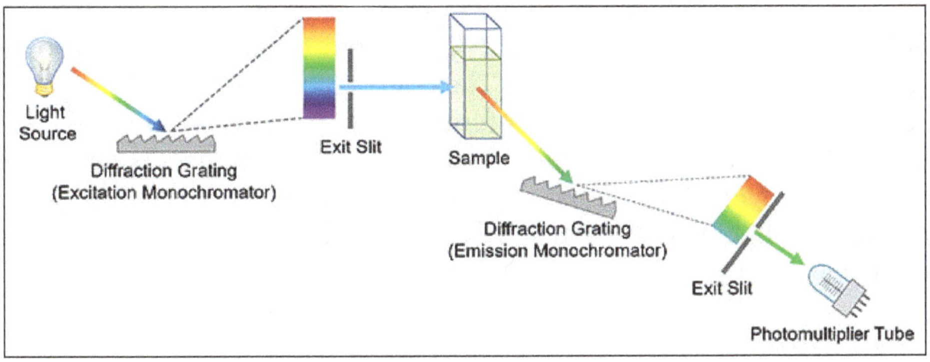

Simplified diagram of a spectrofluorometer.

The two most common spectral measurements undertaken in a spectrofluorometer are excitation and emission spectra. To measure an excitation spectrum the emission monochromator is set at a wavelength of strong fluorescence emission, the excitation monochromator is scanned across the wavelength region of interest and the change in fluorescence intensity monitored on the detector (Figure blue curve) which reveals the absorption features of sample. Emission spectra are the inverse of this; the wavelength of the excitation monochromator is set where there is strong absorption by the sample, the emission monochromator is swept across the wavelength region of interest and the change in fluorescence detected (Figure red curve) revealing the fluorescence features of the sample.

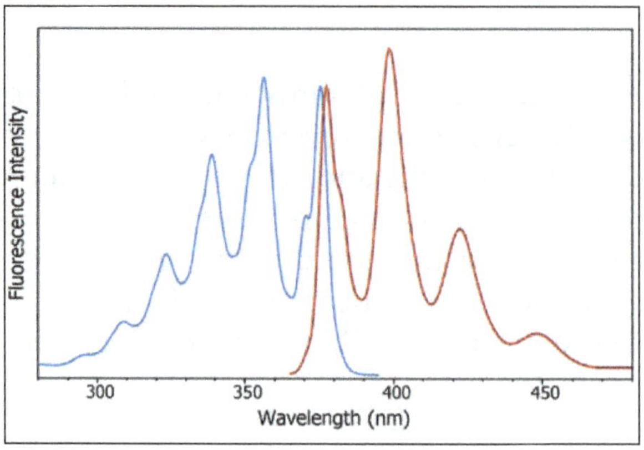

Fluorescence excitation (blue) and emission (red) spectra of anthracene measured using the FS5 Spectrofluorometer.

Raman Spectrometer

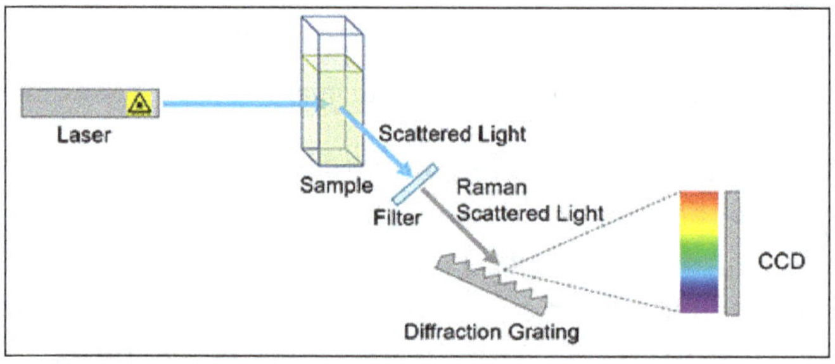

Figure: Simplified diagram of a Raman spectrometer.

Raman spectrometers are used to measure the Raman scattering of light from a sample. The design of a typical Raman spectrometer is shown in Figure and is similar to a spectrofluorometer but with a few key differences. The white light source and excitation monochromator found in spectrofluorometers are replaced with a laser. The reason for this is twofold. The first is that 'Raman' is a scattering effect and the light is therefore not absorbed by the sample. This means that a broadband tuneable light source for matching to the absorption features is not required. The second reason is that the Raman Effect is much weaker than fluorescence (ratio of Rayleigh scattered light to Raman scattered light is ~10^6) and sources with a high photon flux are therefore essential to maximise the signal.

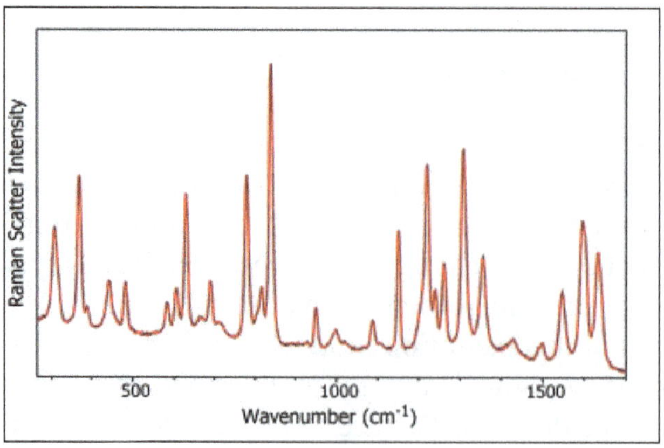

Raman spectrum of paracetamol.

The laser light is scattered off the sample and passed through a filter to remove the much more intense Rayleigh scattered component. The remaining Raman scattered light is then passed into a spectrograph and captured using a CCD detector to give a Raman spectrum like that shown in Figure of paracetamol.

Fluorescence Spectrometer

A fluorescence spectrometer or fluorescence spectroscopy is a device to detect and analyze organic compounds. Used in biology, chemistry, and environmental science laboratories, the fluorescence spectrometer measures the fluorescence of a molecule to gather information about its components and chemical environment. The two main types of fluorescence spectrometers are filter fluorometers, which use filters to isolate the different types of light, and spectrofluorometers, which isolate the types of light with diffraction grating monochromators.

The term fluorescence refers to one type of luminescence. Luminescence, broadly defined, is light emission from a molecule. There are several types of luminescence.

- Photoluminescence is when light energy, or photons, stimulates the emission of a photon.
- Chemiluminescence is defined as when chemical energy stimulates the emission of a photon, and this includes bioluminescence, as seen in fire flies and many forms of sea life.
- Electroluminescence is when electrical energy or a strong electric field stimulates the emission of a photon, such as in some lighting applications.
- Fluorescence, specifically, is a type of photoluminescence where light raises an electron to an excited state. The excited state undergoes rapid thermal energy loss to the environment through vibrations, and then a photon is emitted from the lowest-lying singlet excited state. This process of photon emission competes for other non-radiative processes including energy transfer and heat loss.

When the term "fluorescence" is used, the same methods of measurement can typically be applied to any of the above categories of luminescence.

What is a Fluorescence Spectrum?

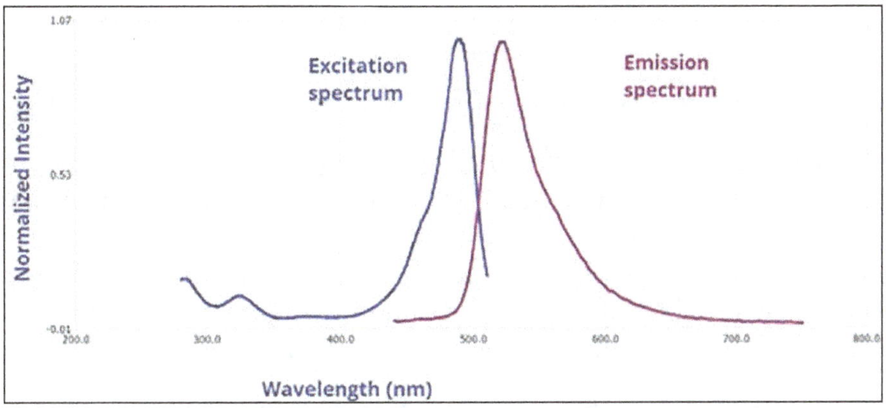

A fluorescence excitation spectrum (blue) and an emission spectrum (purple) are mirror images of each other.

Steady state fluorescence spectra are when molecules, excited by a constant source of light, emit fluorescence, and the emitted photons, or intensity, are detected as a function of wavelength. A fluorescence emission spectrum is when the excitation wavelength is fixed and the emission wavelength is scanned to get a plot of intensity vs. emission wavelength.

A fluorescence excitation spectrum is when the emission wavelength is fixed and the excitation monochromator wavelength is scanned. In this way, the spectrum gives information about the wavelengths at which a sample will absorb so as to emit at the single emission wavelength chosen for observation. It is analogous to absorbance spectrum, but is a much more sensitive technique in terms of limits of detection and molecular specificity. Excitation spectra are specific to a single emitting wavelength/species as opposed to an absorbance spectrum, which measures all absorbing species in a solution or sample. The emission and excitation spectra for a given fluorophore are mirror images of each other. Typically, the emission spectrum occurs at higher wavelengths (lower energy) than the excitation or absorbance spectrum.

These two spectral types (emission and excitation) are used to see how a sample is changing. The spectral intensity and or peak wavelength may change with variants such as temperature, concentration, or interactions with other molecules around it. This includes quencher molecules and molecules or materials that involve energy transfer. Some fluorophores are also sensitive to solvent environment properties such as pH, polarity, and certain ion concentrations.

Types of Molecules and Materials Exhibit Fluorescence

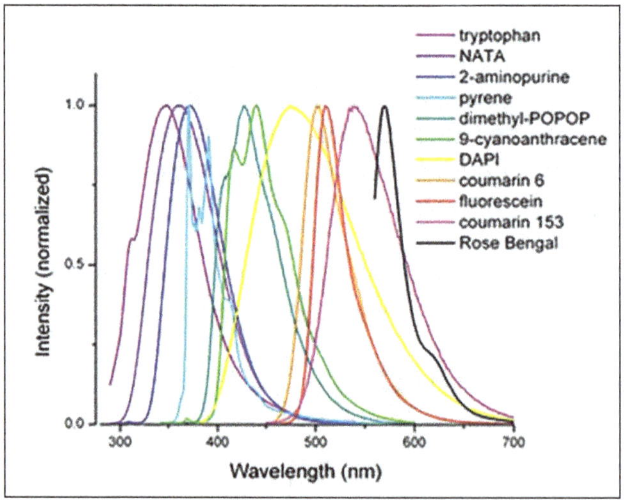

Fluorescence emission spectra of some common fluorophores across the UV and visible spectrum.

Fluorescent molecules and materials come in all shapes and sizes. Some are intrinsically fluorescent, such as chlorophyll and the amino acid residue tryptophan (Trp), phenylalanine (Phe) and tyrosine (Tyr). Others are molecules synthesized specifically as stable organic dyes or tags to be added to otherwise non-fluorescent systems. There are entire catalogs of these available. Typically, organic fluorescent molecules have aromatic rings and pi-conjugated electrons in them. Depending on their size and structure, organic dyes can emit from the UV out into the near-IR.

Here are a random sampling of a few common fluorophores that span the UV and Visible range. Some rare earth elements, or lanthanides, have higher electronic orbitals filled, where electrons transition due to metal ligand charge transfers happen between 4f-5d and even 4f-4f orbitals. There are many molecules that are luminescent in nature such as a few of the amino acids, chlorophylls, and natural pigments. Others are highly engineered for very specific uses of fluorescence spectroscopy. A few of the categories of fluorescent molecules and materials are:

- Amino acids (Trp, Phe, Tyr).
- Base pair derivatives (2-AP, 3-MI, 6-MI, 6-MAP, pyrrolo-C, tC).
- Chlorophylls.
- Fluorescent Proteins (FPs).
- Organic dyes (fluorescein, rhodamine, N-aminocoumarins and derivatives of these).
- Rare earth elements (lanthanides).
- Semiconductors.
- Quantum dots.
- Single Walled Carbon Nanotubes (SWCNTs).
- Solar cells.
- Pigments, brighteners.
- Phosphors.

Other molecules and materials such as fluorescent proteins, semiconductors, phosphors, and rare earth elements are among the commonly used fluorescent samples. Polymers with conjugated aromatics or dienes also commonly have fluorescent properties. Of course, new materials are being created all the time.

Electron Spectroscopy

Electron spectroscopy is a method of determining the energy with which electrons are bound in chemical species by measuring the kinetic energies of the electrons emitted upon bombardment of the species with X-ray or ultraviolet radiation. Details of the structure may be inferred from the results because differences in the arrangements of the atoms affect the amount of energy required to eject electrons.

The most common type is photoelectron spectroscopy, in particular X-ray photoelectron spectroscopy (XPS) but also the UV photoelectron spectroscopy (UPS) is widely used. In most XPS spectra also Auger precesses are seen. Even the electron energy loss spectroscopy (EELS) can give valuable information. The instruments for photoelectron spectroscopy are often equipped to show also other kinds of spectra such as secondary ion mass spectra (SIMS). A summary of some the surface analytical spectroscopies where the incident radiation is photons, electrons, or ions is given below.

	Incident radiation		
emission	hv	e^-	jon
hv	SEXAFS		
e^-	UPS	AES	INS
	XPS	LEED	
		ESD	SIMS
			ISS
			RBS

- SEXAFS: Surface extended X-ray Absorption Fine-structure Spectroscopy.
- UPS: Ultra-violet Photoelectron Spectroscopy.
- XPS: X-ray Photoelectron Spectroscopy (a.k.a. ESCA, Electron Spectroscopy for Chemical Analysis).
- AES: Auger Electron Spectroscopy.
- LEED: Low Energy Electron Diffraction.
- ESD: Electron Stimulated Desorption.
- INS: Ion Neutralization Spectroscopy.
- SIMS: Secondary Ion Mass Spectroscopy.
- ISS: Ion Scattering Spectroscopy.
- RBS: Rutherford Back-scattering Spectroscopy.

References

- Spectroscopy, cosmos: astronomy.swin.edu.au, Retrieved 07, July 2020
- Types-of-spectroscopy: microbenotes.com, Retrieved 15, February 2020
- Plancks-radiation-law, science: britannica.com, Retrieved 21, August 2020
- Spectrometer, Instrumental-Analysis, Supplemental-Modules-(Analytical-Chemistry): chem.libretexts.org, Retrieved 09, March 2020
- Mass-spectrometry, science: britannica.com, Retrieved 19, January 2020
- What-is-a-spectrometer: edin.granite-web.com, Retrieved 28, May 2020
- What-is-fluorescence-spectroscopy, fluorescence-spectroscopy, spectroscopy, measurement-and-control-techniques, technology: horiba.com, Retrieved 03, July 2020
- Electron-spectroscopy, science: britannica.com, Retrieved 16, April 2020

2

Physical Spectroscopy

Physical spectroscopy is one of the main branches of spectroscopy. Some of its significant subfields of are photoelectron spectroscopy, vibrational spectroscopy and microwave spectroscopy. This chapter has been carefully written to provide an easy understanding of these varied facets of physical spectroscopy.

Photoelectron Spectroscopy

Photoelectron spectroscopy (PES) is based on the photoelectric effect, the fact that matter irradiated by photons of sufficiently high energy emits electrons. Information about the sample is determined from the intensity, the angular distribution, and the spin of the emitted electrons. An additional attribute of electrons is that they can only travel a very short path in a solid before they collide with an atom and lose energy. This means that only electrons from the very top layers of a solid sample will emerge without energy loss, and energy-resolved electron spectroscopy is therefore a standard tool in surface science.

The kinetic-energy measurement of the electrons is often done by deflection in an electric field, most commonly in hemispherical mirror analyzers (as at both permanent end stations at FlexPES), or by measurement of the time of flight of the electrons (as will be used with the mobile ICE end station).

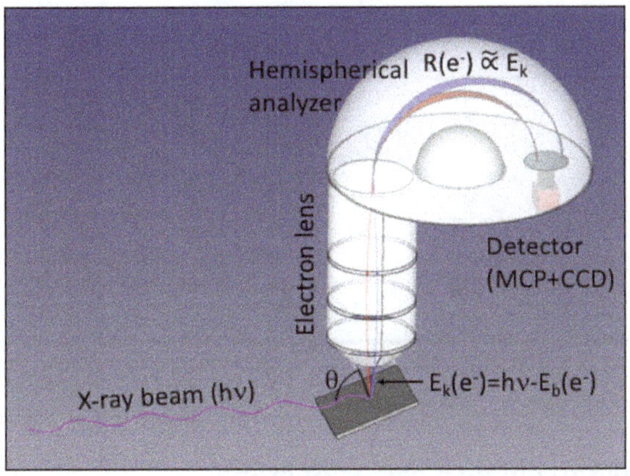

X-ray photons with a well-defined energy (hv) impinge on a sample, and the emitted electrons have a kinetic energy given by the difference between the photon energy and the electron binding energy. At the FlexPES beamline, the photon energy range is between 40-1500 eV, and there are provisions allowing a wide range of experimental situations for solid, liquid, or gas-phase samples. The emitted electrons are collected and transported by an electron lens to the hemispherical analyzer, where the electrons are deflected by the electrostatic field between two charged hemispheres.

The radius of curvature of the electron trajectory depends (nearly linearly) on the kinetic energy, and in the figure the red and blue-colored trajectories are for electrons with lower and higher kinetic energies, respectively. After traveling through the hemisphere, the electron is detected with a position sensitive detector (often a micro-channel plate in combination with a phosphor screen + fast camera or any other read-out).

Photoelectron Spectroscopy is useful in order to gain insight regarding:

- The abundance and/or chemical state of a particular element in a sample.
- Charge transfer processes.
- Nuclear motion (ex. ultra-fast dissociation).
- The mechanical, electrical and magnetic properties of a condensed matter sample.

Photoelectron Spectroscopy of Atoms

The Photoelectron Experiment

In photoelectron spectroscopy (PES), the atom is bombarded with high energy radiation from a He-I source. The He-I emission line is observed at 58.4 nm. The emitted photons have energy $hc/\lambda = (6.626 \times 10^{-34}\ \text{J s} \times 2.998 \times 10^{8}\ \text{m s}^{-1})/58.4 \times 10^{-9}\ \text{m} = 3.40 \times 10^{-19}\ \text{J}$. This is equivalent to 21.2 eV, the unit usually used to describe ionization energies.

When photons of this energy hit an atom, they transfer their energy to its electrons. The energized electrons overcome their attraction to the nucleus and escape. If the energy of the photon is higher than the minimum energy required to escape, the remaining energy is converted to kinetic energy of the electrons. Hence, the energy of the photon, hν, is split into the ionization energy of the atom and kinetic energy.

$$h\nu = IE + KE$$

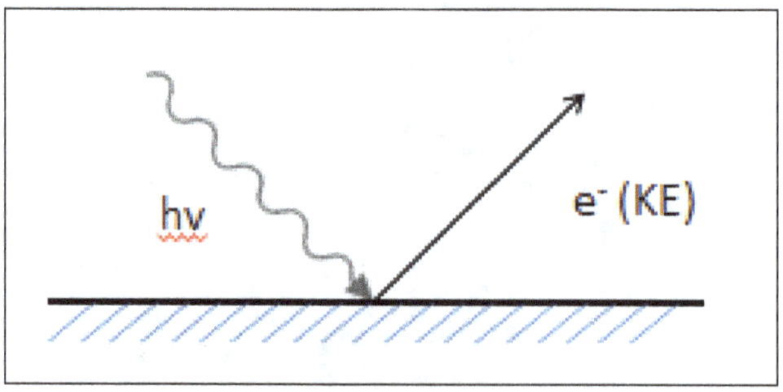

The entire process may be visualized as a series of steps,

$$A + h\nu \rightarrow A^{+} + e^{-}$$

Conservation of energy requires that the energies on both sides should be the same, i.e.,

$$E(A) + h\nu = E(A^{+}) + E(e^{-})$$

The last term, the energy of the electron, is entirely kinetic, so previous equation can be rearranged to,

$$KE = h\nu - \left[E\left(A^{+}\right) - E\left(A\right)\right]$$

The final term in brackets is the difference in energy between the ionized atom and the neutral atom, and this is frequently called the binding (or ionization) energy of the electron in the atom. We finally have,

$$KE = h\nu - BE,$$

The same as equation $h\nu = IE + KE$.

A photoelectron spectrum, such as the one shown in Figure, displays the kinetic energy distribution of the emitted photoelectrons. It is a plot of the number of emitted photoelectrons as a function of their kinetic energy. This can be measured using an electron energy analyzer.

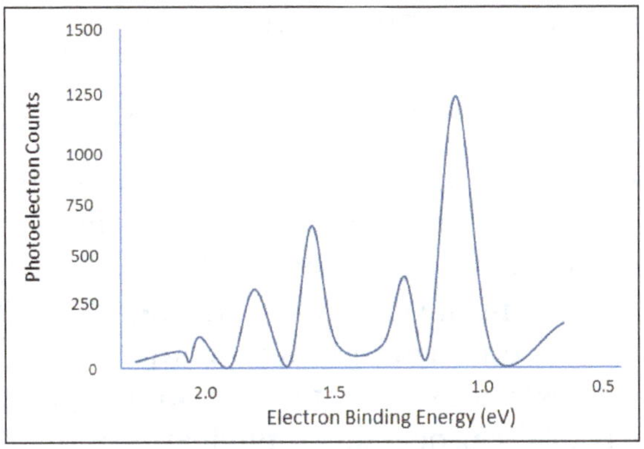

A typical photoelectron spectrum.

The He-I radiation has sufficient energy (21.2 eV) to cause emission of only the valence electrons. There is another helium emission He-II at 40.8 eV, and photoelectron spectroscopy studied using helium lamps with this emission falls in the domain of ultraviolet photoelectron spectroscopy (UPS). For analyzing the core electrons, another technique, X-ray Photoelectron Spectroscopy (XPS) is used. The PES techniques that are possible using the FlexPES hemispherical analyzer setups include:

- Core-level photoelectron spectroscopy.
- Auger electron spectroscopy.
- Resonant Photoemission/Resonant Auger Spectroscopy (RPES/RAES).
- Angle-Resolved Photoelectron Spectroscopy (ARPES).
- X-ray Photoelectron Diffraction (XPD).

Core-level Photoelectron Spectroscopy

Perhaps the most straightforward application of PES is core-level photoelectron spectroscopy, which involves the study of the deeper lying electron levels. The energies of the core levels are distinct for each element, which makes it easy to distinguish between for instance nitrogen and oxygen. The core levels do not contribute to the bonding in matter, but they are indirectly influenced by the chemical surroundings in such a way that the binding energy is changed, and this is referred to as a "chemical shift". Core-level PES is therefore often used to investigate the abundance and chemical state of a particular element in the sample, for instance the oxidation state of a catalyst, or the protonation state of an acid or a base.

Auger Electron Spectroscopy (AES)

After the core ionization event, the atom is in an unstable electron configuration, and decays rapidly to a lower energetic state. This can happen via the emission of an X-ray photon, or by emission of an electron, a so-called Auger decay, and this is the most common process for light elements and shallow core levels. In the same way as core-level photoelectrons, Auger electrons are sensitive to the chemical surrounding of the ionized atom, but the spectra are often more difficult to interpret (due to overlap of many possible final states with open-shell character). An advantage of Auger electron spectroscopy is that it is independent of the excitation, meaning that the resolution in the electron spectrum does not depend on the photon energy bandwidth.

Resonant Photoemission/Resonant Auger Spectroscopy (RPES, RAES)

In conventional photoelectron spectroscopy, the electron is ejected in an almost instantaneous process, using a photon energy that is higher than the threshold of ionization. By tuning the photon energy, one can instead excite a core electron to an unoccupied level near the threshold. This excited state, with a hole in the core level, will rapidly decay by emission of an electron, just as in the Auger decay after a core photoionization event, or a photon. The decay is fast, of the order of a few femtoseconds or faster, but some dynamic processes occur on the same time scale, e.g. charge transfer processes or nuclear motion of light atoms, and the electron spectra recorded using resonant excitation can thus give insight into such processes. The resonant process selects a particular element or even a particular chemical state, which means that the information will also be particular to the situation close to the selected atom. Since electron emission is induced by resonant photo-excitation, and the process is similar to an Auger decay filling a core hole, it is referred to as both Resonant Photoemission and as Resonant Auger Spectroscopy. This method has been used to; for instance, investigate how quickly charge moves in a light-harvesting molecule in a solar cell, or how the dissociation speed of a molecule is affected by condensation.

Angle-resolved Photoelectron Spectroscopy (ARPES)

In condensed matter, the valence electron orbitals overlap and form bands. By studying the photoelectron emission at different angles, it is possible to investigate the distribution of electrons of these bands (in the so-called reciprocal space). From this technique, often called band mapping, information about mechanical, electrical, and magnetic properties of the sample can be obtained. The Surface & Material Science (SMS) branch of the FlexPES beamline will offer possibilities to perform ARPES experiments.

X-ray Photoelectron Diffraction (XPD)

Electron diffraction is based on the wave character of electrons – just like light (electromagnetic waves), electron waves can be scattered and result in interference patterns. This technique is commonly used as a means to study crystal structure in connection to electron microscopy (SEM and TEM). The electrons emitted in a photoionization event will also be diffracted, and by utilizing this effect, it is possible to at the same time obtain structural information from the diffraction pattern, and chemical information from the spectrum.

X-ray Photoelectron Spectroscopy

Each element has a characteristic binding energy associated with each of its core atomic orbitals. Hence, each element gives rise to a characteristic photoelectron spectrum, which can be used to identify it. Moreover, there is a quantitative aspect, too. The intensity of the peaks is related to the concentration of the element in the sample. Thus, XPS is an important technique, providing both qualitative and quantitative information about the composition of a sample. For this reason, it is also known by the alternative acronym, ESCA (Electron Spectroscopy for Chemical Analysis). The following X-ray sources are commonly employed:

X-ray source	hv (eV)
Mg K_α radiation	1253.6
Al K_α radiation	1486.6
Cu K_α radiation	5414.7

The emitted photoelectrons will therefore have kinetic energies in the range of ca. 0 - 1250 eV, 0 - 1480 eV or 0 – 5410 eV and the Cu K_α radiation can probe the deepest electrons.

Photoelectron Spectrum of Ar

As an example, let us take up the interpretation of the photoelectron spectrum of a noble gas, argon. Its electronic configuration is $1s^2\ 2s^2\ 2p^6\ 3s^2\ 3p^6$ and the ground state is therefore a closed shelll, ${}^1S_0$. The 3s and 3p electrons are the valence electrons, and He-I radiation can only provide the energies of these electrons. Figure is a composite of UPS and ESCA for obtaining information about the core electrons.

Ionization of an electron results in a positively charged ion, $Ar^{+\bullet}$, having one electron less than the neutral atom. The easiest electron to remove is the 3p electron, resulting in the electron configuration $1s^2\ 2s^2\ 2p^6\ 3s^2\ 3p^5$. The binding energy of this electron is ~24 eV. The next easily ionizable electron is the 3s electron with a binding energy of 31.5 eV. These are the only two peaks observable in the He-I photoelectron spectrum. The peaks observed in the ESCA are at ~250 eV (2p) and 326 eV (2s).

Effect of Spin-orbit Coupling

Closer look at the 2p peak at ~250 eV shows that it is actually a doublet with peaks at 248.4 and 250.6 eV. This is due to spin-orbit coupling because of which the $2p^5$ configuration gives rise to two states $2^2P_{3/2}$ and $2^2P_{1/2}$, of which the former is higher in energy because the 2p orbital is more than half-filled. The same splitting should also be observed for the 3p peak, but the two states $3_2P_{3/2}$ and $3^2P_{1/2}$ are close in binding energy (15.8 and 15.9 eV), and the compound peak merges to one.

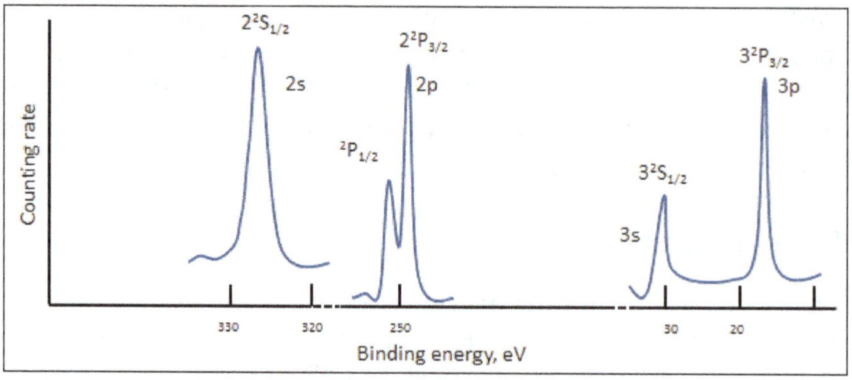

Photoelectron spectrum of Ar.

Ionization from the 2s and 3s orbitals results in $2^2S_{1/2}$ and $3^2S_{1/2}$ states of $Ar^{+\bullet}$, respectively. Thus, the photoelectron spectrum of argon is completely explained.

Relative Heights of the Peaks

The first is that the area under a peak is proportional to the number of electrons in the orbital from which the electron is ejected. Thus, the ratio of areas under the peak arising from ionization from 3s and 3p orbitals is 1:3. Moreover, since the degeneracy of every J value is 2J + 1, the areas under the $^2P_{1/2}$ and $^2P_{3/2}$ peaks should be in the ratio 1:2, and this can be clearly seen for the $2^2P_{1/2}$ and $2^2P_{3/2}$ peals near 250 eV. Assuming similar shapes, instead of the area, we may take the ratios of heights of peaks to determine the relative degeneracies.

The relative positions of the two peaks also confirm Hund's third rule, according to which the $2^2P_{3/2}$ level should be higher in energy (which translates to lower binding energy) than $2^2P_{1/2}$, since the 2p orbital is more than half-filled.

Koopmans' Theorem

The third aspect of photoelectron spectroscopy is Koopmans' theorem which bridges the gap between experiment and theory. The energy of an electron and the ionized atom is taken as zero when they are at infinite distance from each other and there is no interaction between the two. As they are brought together, the mutual attraction lowers the potential energy. Thus, the energy of a bound electron in an orbital is always negative. In order to remove this electron against the attraction of the positively charged ion, energy equal to the orbital energy has to be supplied so that the two moves apart and the total energy becomes zero.

The orbital energies can be obtained from quantum mechanical calculations, and we may write that the ionization (or binding) energy of an electron is the negative of the orbital energy:

$$IE_i = -\varepsilon_i$$

where ε_i is the orbital energy of the ith atomic orbital.

Strictly speaking, Koopmans' theorem is only an approximation. As we have seen, there are two energies associated with a 2p or 3p orbital because of spin-orbit coupling, but there is only one orbital energy. Furthermore, removal of an electron from an atom changes the entire electron distribution and Koopmans' theorem does not account for this relaxation of the electronic structure upon ionization. We can thus assign the entire photoelectron spectrum of an atom. Example: Identify the element in the photoelectron spectrum shown below. Briefly explain your reasoning.

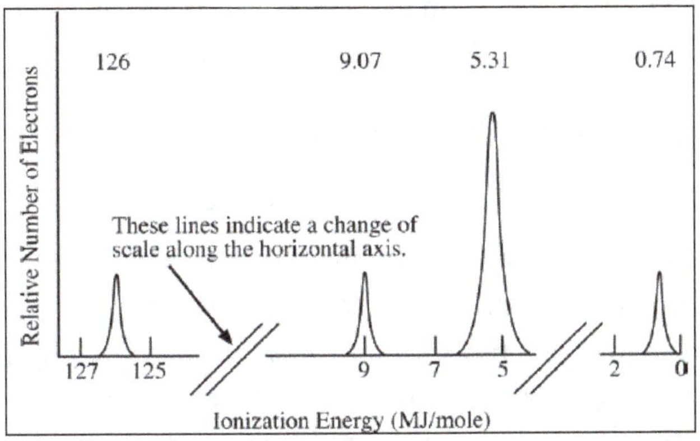

The highest binding energy is for $1s^2$ electrons and the peak at 126 MJ mol^{-1} may be identified with these electrons. The next two peaks at 9.07 MJ mol^{-1} and 5.31 MJ mol^{-1} respectively are associated with $2s^2$ and $2p^6$ electrons. The 1:3 ratio of the two peaks confirms this, as also the equal heights of the peaks at 126 MJ mol^{-1} and 9.07 MJ mol^{-1}. The same height of the peak at 0.74 MJ mol^{-1} indicates that there are two electrons in the 3s orbital too. The electron configuration is therefore $1s^2$ $2s^2$ $2p^6$ $3s^2$ and the atomic number is hence 12 and the element is magnesium.

Photoelectron Spectroscopy of Diatomic Molecules

In the case of molecules, there are other ways such as by changing its vibrational energy that the molecule ion may change its energy. Again ignoring the kinetic energy of the ion, we may rewrite equation $IE = h\nu_o - E_k$ as,

$$h\nu_o = IE + E_{M^+} + E_k(e^-)$$

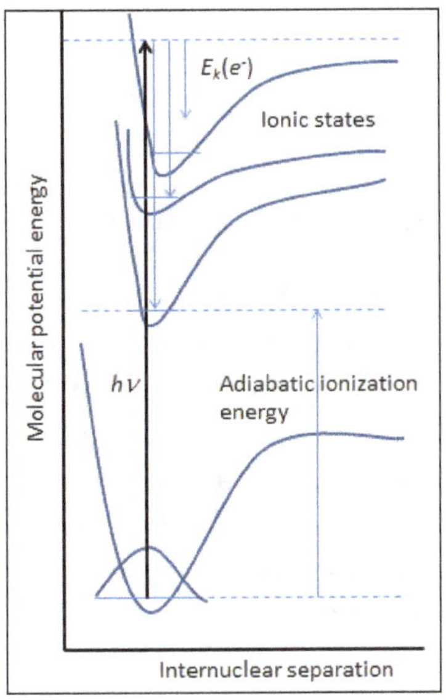

Photoelectron spectroscopy.

Where E_{M^+} is the change in internal (vibrational) energy of the ion. Therefore, photoelectron spectroscopy provides information about vibrational energy changes of the molecule ion. The (0, 0) transition from the ground state molecule to the ion does not involve any vibrational energy change and is called the adiabatic ionization energy. It refers to the least energy required to ionize the molecule. This is obviously the lowest energy transition.

Photoelectron Spectrum of Hydrogen Molecule

Let us first consider the photoelectron spectrum of hydrogen molecule. As shown in Figure, the electron configuration is $(1\sigma_g)^2$ and, since this is a closed shell system, the ground state is $^1\Sigma_g^+$. When an electron gets ionized, the resulting configuration is $(1\sigma_g)^1$ and the state is $^2\Sigma_g^+$. The photoelectron spectrum of hydrogen molecule is shown schematically in figure. The first line is observed at 15.41 eV, followed by some equally spaced lines as shown in the right side diagram of figure. These lines correspond to the vibrational spacings of the hydrogen molecule ion (H_2^+).

The bond order in hydrogen molecule is 1.0. Removal of a bonding electron, as in the ion, decreases the bond order to 0.5 and consequently the bond length increases. The photoelectron transitions are also governed by the Franck-Condon principle and are vertical transitions. Figure shows that the most intense transition should be the (2, 0) transition, as observed experimentally. The corresponding ionization energy (15.83 eV) is called the vertical ionization energy and is higher than the adiabatic ionization energy, as it also contains a vibrational contribution.

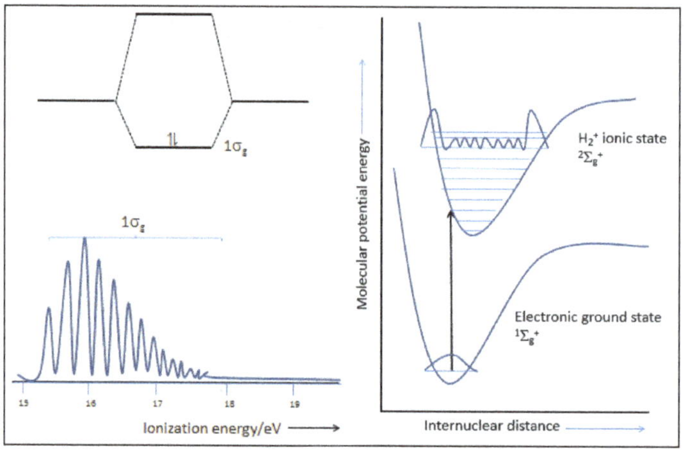

Photoelectron spectrum of H_2.

The spacing between the individual lines is 2322 cm^{-1}, much less than the vibrational spacing in hydrogen (3115 cm^{-1}). This implies that the electron is less tightly bound in the ion, as expected from the fact that the electron has been ionized from a bonding orbital, which weakens the bonding in the resultant ion. The binding energy of the hydrogen molecule may also be estimated from the photoelectron spectrum. It may be noted that Koopmans' theorem refers to the vertical ionization energy. The vertical ionization energy of hydrogen molecule is the ionization energy of a hydrogen atom (13.61 eV) plus half the binding energy of a hydrogen molecule – half because two hydrogen atoms participate to form a bond. Thus,

$$IE_v = 13.61 + ½ E_b$$

Or,

$$E_b = 2\times(15.83-13.61) = 4.44\text{ eV}$$

which compares well with the thermochemical value of 4.48 eV.

From Figure, it is clear that if the inter-nuclear distance in the ion is not appreciably different from the molecule, the (0, 0) transition would be the most intense, and very few other transitions could occur with diminishing intensity. Also, the vibrational spacing would be largely unchanged from that in the molecule. This could happen when the electron is ionized from a nonbonding orbital because that would leave the bond order unchanged.

If, on the other hand, the electron is ionized from a bonding orbital, the bonding in the ion would be reduced, increasing the inter-nuclear distance. According to the Franck-Condon principle, the vertical transition would end at a high quantum number, with quite a few other transitions observed. Due to the weakening in the bonding, the vibrational spacing in the ion would be smaller than that in the neutral molecule. Finally, if ionization is out of an anti-bonding orbital, again there is considerable vibrational fine structure because the bonding is stronger in the ionized state and hence the bond length is shorter. This situation can be distinguished from that of ionization from a bonding orbital by the fact that the vibrational spacing is larger in this case.

Photoelectron Spectrum of Nitrogen Molecule

Nitrogen is an illustrative example, because it verifies the concept of s-p mixing. Recall that in this molecule and all other molecules that have fewer than 15 electrons, the σ and π energy level order is reversed compared to O_2 and the bulk of diatomic molecules, due to s-p mixing. Photoelectron spectroscopy is experimental verification of this.

If we refer to Figure, the electron configuration of molecular nitrogen is $(1\sigma g)^2$ $(1\sigma u)^2$ $(2\sigma g)^2$ $(2\sigma u)^2$ $(1\pi u)^4$ $(3\sigma g)^2$. The ground state is a closed shell $^1\Sigma_g^+$ state. The 3σg electrons are the easiest to ionize. The next higher ionization potential is for the 1πu electrons, followed by 2σu. Ionization leaves the molecule in its cationic state (N_2^+).

Three bands are observed in the PES of dinitrogen. The first, with an adiabatic ionization energy of 15.58 eV, is due to the $^2\Sigma_g^+$ state resulting from ionization of a 3σg electron. There is very little vibrational structure, indicating that the 3σg orbital is not deeply involved in bonding. The vibrational spacing of 2150 cm^{-1} compared to 2379 cm^{-1} for molecular nitrogen clearly indicates that it is a weak bonding orbital. This is also seen from the equilibrium bond length of the ion, which increases slightly to 0.1116 nm from 0.1098 nm in the molecule. Contrast this with the band at 16.69 eV, which shows considerable vibrational structure and a significantly decreased vibrational spacing of 1810 cm^{-1}. All these facts indicate that the π_u orbital is a strongly bonding orbital. The bond length also increases considerably by ~0.08 nm to 0.1175 nm. Finally, the band at 18.76 eV again has sparse vibrational structure, and a vibrational spacing of 2390 cm^{-1}, which indicates that $2\sigma_{u,}$ is a weakly bonding orbital, with only a slight decrease in bond length to 0.1074 nm.

Another point to be noted is that the area under the peak for the central band is twice that of the other two bands. This is because the π bands are doubly degenerate, whereas the σ orbitals are nondegenerate. The vibrational data obtained from the vibrational fine structure for the three bands and their interpretation. Here $\tilde{v}_o$ is the vibrational wavenumber for molecular nitrogen (= 2379 cm^{-1}).

Band	$\tilde{v}$	$\tilde{v}/\tilde{v}_o$	Interpretation
$3\sigma_g$	2150	0.904	Weakly bonding
$1\pi_u$	1810	0.761	Strongly bonding
$2\sigma_u$	2390	1.005	Weakly antibonding

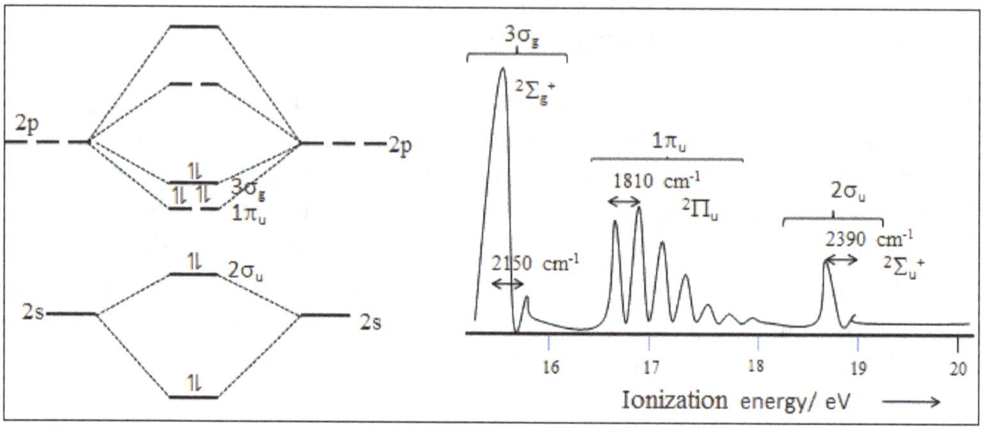

Photoelectron spectrum of N_2 showing the vibrational spacings in various N_2^+ states.

The above discussion validates our MO diagram for N_2 and proves that photoelectron spectroscopy provides valuable insights into the bonding in molecules and the nature of orbitals involved.

Vibrational Spectroscopy

Vibrational spectroscopy or infrared spectroscopy uses the fact that vibrational energy levels are quantized and molecules vibrate at discrete frequencies determined by their reduced mass and strength of bonds, which are determined by the shape of their potential energy diagram.

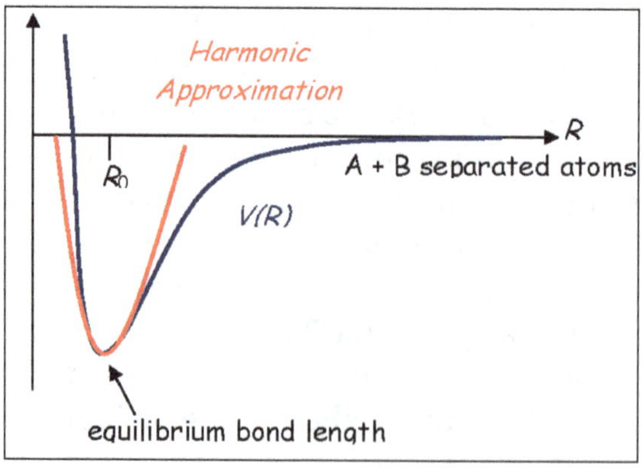

Within the Born-Oppenheimer approximation, the nuclei move on a potential energy surface (PES) determined by the electrons. For example, the potential felt by the nuclei in a diatomic molecule is shown in cartoon form at right. At low energies, the molecule will sit near the bottom of this potential energy surface. In this case, no matter what the detailed structure of the potential is, locally the nuclei will "feel" a nearly harmonic potential. Generally, the motion of the nuclei along the PES is called vibrational motion, and clearly at low energies a good model for the nuclear motion is a Harmonic oscillator.

Vibrational Spectroscopy of a Diatomic

If we just have a diatomic molecule, there is only one degree of freedom (the bond length), and so it is reasonable to model diatomic vibrations using a 1D harmonic oscillator:

$$\hat{H} = \frac{\hat{P}^2}{2\mu} + \frac{1}{2}k_o\hat{R}^2 = \frac{\hat{P}^2}{2\mu} + \frac{1}{2}m\omega_o^2\hat{R}^2$$

where k_o is a force constant that measures how stiff the bond is and can be approximately related to the second derivative to the true (anharmonic) PES near equilibrium:

$$k_o \approx \left.\frac{\partial^2 V}{\partial R^2}\right|_{R_o}$$

Applying Fermi's Golden Rule, we find that when we irradiate the molecule, the probability of a transition between the ith and f[th] Harmonic oscialltor states is:

$$W_{fi} \propto \frac{|V_{fi}|^2}{4\hbar^2}\left[\delta(E_i - E_f - \hbar\omega) + \delta(E_i - E_f - \hbar\omega)\right]$$

where ω is the frequency of the light (not to be confused with the frequency of the oscillator, ω_o). Because the vibrational Eigen states involve spatial degrees of freedom and not spin, we immediately recognize that it is the electric field (and not magnetic) that is important here. Thus, we can write the transition matrix element as:

$$|V_{fi}|^2 = \left|\int \phi_f^* \hat{\mu}\cdot\mathbf{E}_o\phi_i d\tau\right|^2 = \left|\mathbf{E}_o\cdot\int \phi_f^*\hat{\mu}\phi d\tau\right|^2 = \left|\mathbf{E}_o\cdot\int \phi_f^* e\hat{R}\phi_i d\tau\right|^2$$

Now, we define the component of the electric field, E_R, that is along the bond axis which gives,

$$|V_{fi}|^2 = \left|E_R\int \phi_f^* e\hat{R}\phi_i d\tau\right|^2 = e^2|E_R|^2\left|\int \phi_f^*\hat{R}\phi_i d\tau\right|^2$$

So the rate of transitions is proportional to the square of the strength of the electric field (first two terms) as well as the square of the transition dipole matrix element (third term). Now, because of what we know about the Harmonic oscillator eigen functions, we can simplify this. First, we rewrite the position operator, **R**, in terms of raising and lowering operators:

$$|V_{fi}|^2 = e^2|E_R|^2\left|\int \phi_f^*\sqrt{\frac{\hbar}{2\mu\omega}}(\hat{a}_+ + \hat{a}_-)\phi_i d\tau\right|^2 = \frac{\hbar e^2}{2\mu\omega}|E_R|^2\left|\int \phi_f^*(\hat{a}_+ + \hat{a}_-)\phi_i d\tau\right|^2$$

$$\Rightarrow |V_{fi}|^2 = \frac{\hbar e^2}{2\mu\omega}|E_R|^2\left((i+1)\delta_{f,i+1} + i\,\delta_{f,i-1}\right) \qquad \sqrt{i+1}\,\phi_{i+1} \qquad \sqrt{i}\phi_{i-1}$$

where above it should be clarified that in this expression "i" never refers to √-1- it always refers to the initial quantum number of the system. Thus, we immediately see that a transition will be forbidden unless the initial and final states differ by one quantum of excitation. Further, we see that the transitions become more probable for more highly excited states. That is, V_{fi} gets bigger as i gets bigger. Combining the explicit expression for the transition matrix element with Fermi's Golden Rule again gives:

$$W_{fi} \propto \frac{e^2}{8\lambda\!\!\!^{-}\mu\omega}|E_R|^2\left((i+1)\delta_{f,i+1} + i\delta_{f,i-1}\right)\left[\delta(E_i - E_f - \hbar\omega) + \delta(E_i - E_f + \hbar\omega)\right]$$

$$\Rightarrow W_{fi} \propto |E_R|^2 \begin{Bmatrix} (i+1)\delta_{f,i+1}\left[\delta(E_i - E_{i+1} - \hbar\omega) + \delta(E_i - E_{i+1} + \hbar\omega)\right] \\ +i\,\delta_{f,i+1}\left[\delta(E_i - E_{i-1} - \hbar\omega) + \delta(E_i - E_{i-1} + \hbar\omega)\right] \end{Bmatrix}$$

$$\Rightarrow W_{fi} \propto |E_R|^2 \{(i+1)\delta_{f,i+1}\left[\cancelto{0}{\delta(-\hbar\omega_o - \hbar\omega)} + \delta(-\hbar\omega_o + \hbar\omega)\right] + i\,\delta_{f,i-1}\left[\delta(\hbar\omega_o - \hbar\omega) + \cancelto{0}{\delta(\hbar\omega_o + \hbar\omega)}\right]\}$$

$$\Rightarrow W_{fi} \propto |E_R|^2 \{(i+1)\delta_{f,i+1} + i\,\delta_{f,i-1}\}\delta(\hbar\omega_o - \hbar\omega)$$

Thus, we see that a harmonic oscillator will only absorb or emit photons of frequency $\hbar\omega_o$, where ω_o is the frequency of the oscillator. Thus, if we look at the absorption spectrum, for example, we will see absorption at only one frequency:

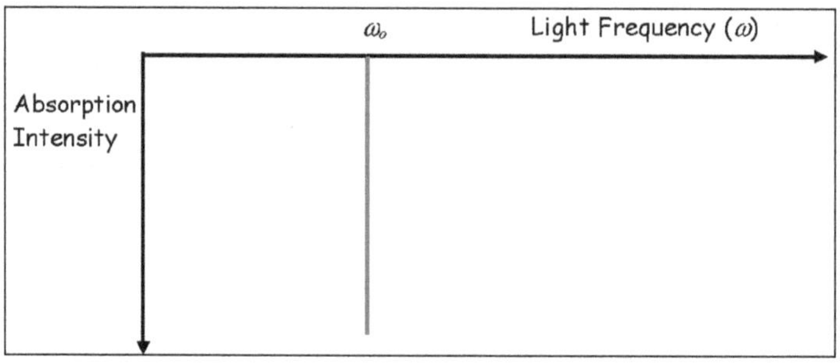

Molecular force constants are typically on the order of an eV per Å, which leads to vibrational frequencies that are typically between 5003500 cm1 and places these absorption features in the infrared. As a result, this form of spectroscopy is traditionally called IR spectroscopy. We associate the spectrum above as arising from all the $n \rightarrow n+1$ transitions in the Harmonic oscillator. Of course, most of the time the molecule will start in its ground state, so that the major contribution comes from the $0 \rightarrow 1$ transition. However, the other transitions occur at the same frequency and also contribute to the absorption.

This is the classic paradigm for IR vibrational spectroscopy: each diatomic molecule absorbs radiation only at one frequency that is characteristic of the curvature of the PES near its minimum. Thus, in a collection of different molecules one expects to be able differentiate one from the other by looking for the frequency appropriate to each one. In particular there is a nice correlation between the "strength" of the bond and the frequency at which it will absorb.

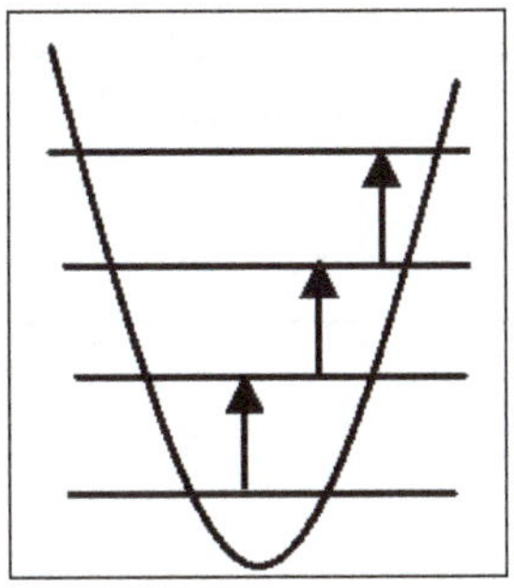

Of course, we are not always or even usually interested in diatomics, and even diatomics are not perfect Harmonic oscillators. Thus, there are a number of reasons why IR absorption spectra do not really look like the classic Harmonic oscillator spectrum shown above, but more like:

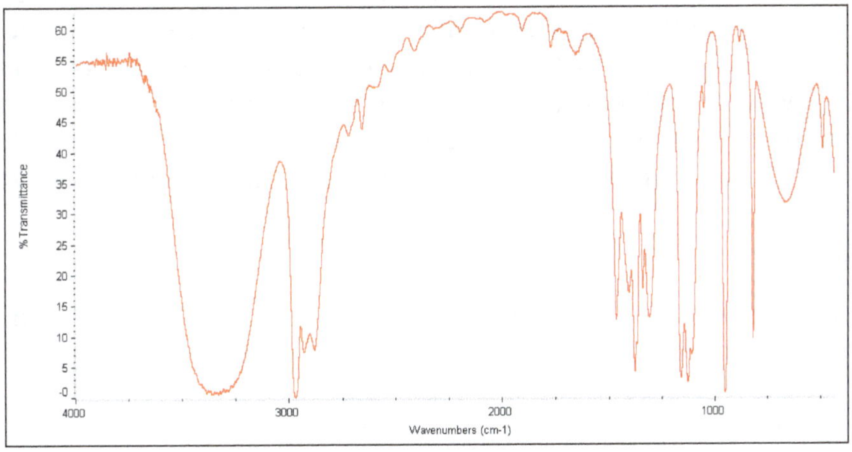

Heterogeneity

The primary reason the real spectrum above looks different than the model is because the real spectrum was taken in solution. The model is correct for a single diatomic, or for many, many copies of identical diatomic molecules. However, in solution, ever molecule is just slightly different, because every molecule has a slightly different arrangement of solvent molecules around it. These solvent molecules subtly change the PES, slightly shifting the vibrational frequency of each molecule and also modifying the transition dipole a bit. Thus, while a single hydrogen fluoride molecule might have a spectrum like the model above, a solution with many HF molecules would look something like:

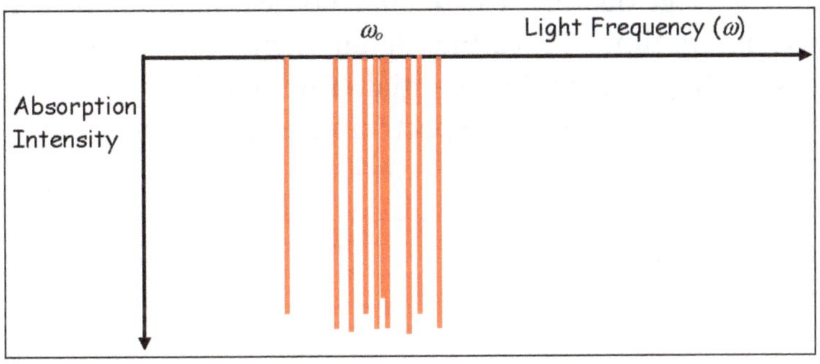

Going over to the situation where there are 10^{23} HF molecules and recognizing that our spectra will tend to add the intensity of lines that are closer together than our spectrometer can differentiate, we anticipate that for a diatomic molecule in solution, the vibrational spectrum should look something like:

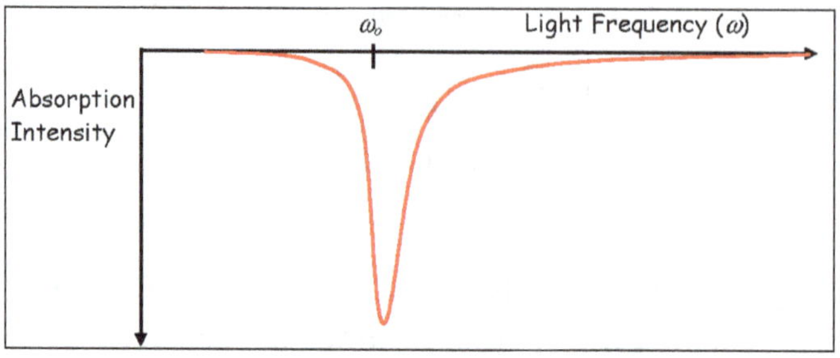

The resulting feature in the spectrum is usually called a line shape. It primarily reflects the distribution of different environments surrounding your oscillators. Thus, by analyzing the line shape of a well-known type of vibration (such as a C = O stretch) one can get an idea about the environments those CO groups live in: How polar are the surroundings? Are they near electron withdrawing groups? What conformations give rise to the spectrum? Finally, we should note that vibrational spectra recorded in the gas phase have very narrow line widths, qualitatively resembling our model above.

Anharmonicity

Another reason real spectra differ from our model is that assuming the PES is harmonic is only a model. If we want high accuracy, we need to account for anharmonic terms in the potential:

$$V(R)=\frac{1}{1}m\omega_o^2(R-R_o)^2+\frac{1}{6}\alpha(R-R_o)^3+\frac{1}{24}\beta(R-R_o)^4+....$$

One can investigate the quantitative effects of the anharmonic terms on the spectrum by performing variational calculations. However, at a basic level there are two ways that anharmonic terms impact vibrational spectra:

1. The energy differences between adjacent states are no longer constant. Clearly, the eigenvalues of an anharmonic Hamiltonian will not be equally spaced – this was a special feature of the Harmonic system. Thus, for a real system we should expect the $0 \rightarrow 1$ transition to have a slightly different frequency than 1→2, which in turn will be different, that 2→3 .... Generally, the higher transitions have lower (i.e. redshifted) energies because of the shape of the molecular PES – rather than tend toward infinity at large distances as the harmonic potential does, a molecular PES tends toward a constant dissociation limit. Thus, the higher eigen states are lower in energy than they would be for the corresponding harmonic potential. Taking this information, we would then expect a single anharmonic oscillator to have a spectrum something like:

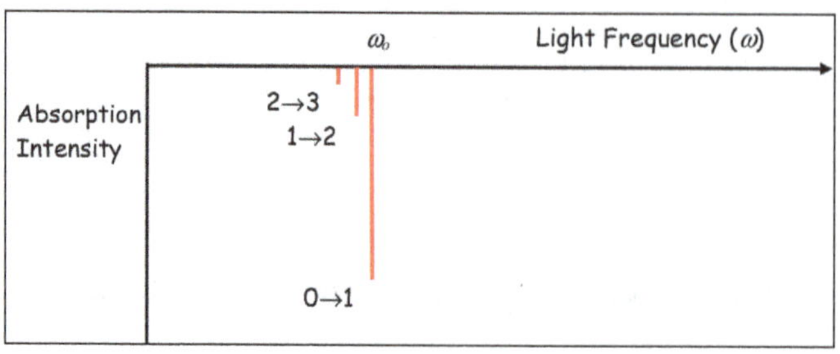

where we note that while the rate of, say, 1→2 is about twice that of 0→1 (because the transition dipole is twice as big) the intensity of 0→1 is greater because the intensity is (Probability of i being occupied)x(Rate of i→f) and at room temperature the system spends most of its time in the ground vibrational state. The 1→2, 2→3 ... lines in the spectrum are called hot bands.

2. Anharmonicity relaxes the $\Delta n = \pm 1$ selection rule. Note that the rules we arrived at were based on the fact that $\hat{a}^{+}\phi_i \propto \phi_{i+1}$. This is only true for the Harmonic oscillator states. For anharmonic eigen states $\hat{a}^{+}\phi_i \propto \phi_{i+1} + \varepsilon_1\phi_{i+2} + \ldots\ldots$ Thus transitions with $\Delta n = \pm 2, \pm 3\ldots$ will no longer be forbidden for anharmonic oscillators. Rather, in the presence of a bit of anharmonicity, they will be weakly allowed. Combining this observation with point 1) above results in a more complete picture for what the IR spectrum of an anharmonic oscillator should look like:

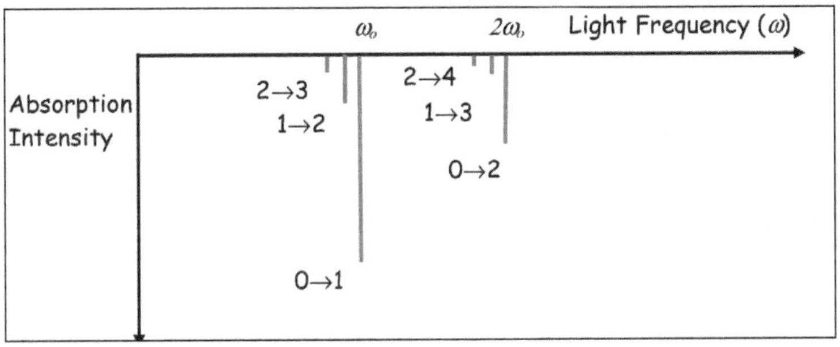

The peaks at around $2\omega_o$ are called overtones. Meanwhile, those at around ω_o are called fundamentals.

Polyatomic Molecules

The final difference between the model above and a general IR spectrum is that in chemistry, we are not always dealing with diatomic molecules. For a polyatomic molecule, we can still think of the potential as a Harmonic potential, but it has to be many dimensional – it has to depend on several variables R_1, R_2, R_3. The most general Harmonic potential we can come up with is then of the form:

$$V(R_1,R_2,R_3,\ldots) = \frac{1}{2}k_{11}R_1^2 + \frac{1}{2}k_{12}R_1R_2 + \frac{1}{2}k_{13}R_1R_3 + \ldots + \frac{1}{2}k_{21}R_2R_1 + \frac{1}{2}k_{22}R_2^2 + \frac{1}{2}k_{23}R_2R_3 + \ldots$$
$$+\frac{1}{2}k_{31}R_3R_1 + \frac{1}{2}k_{32}R_3R_2 + \frac{1}{2}k_{33}R_3^2 + \ldots$$

Where it is important to notice the cross terms involving, say R_1 and R_2, which couple the different vibrations. At first sight, it seems like we can't solve this Hamiltonian; the only many dimensional Harmonic potential we would know how to solve would be one that is separable:

$$\tilde{V}(R_1,R_2,R_3,\ldots) = \frac{1}{2}k_{11}R_1^2 + \frac{1}{2}k_{22}R_2^2 + \frac{1}{2}k_{33}R_3^2 + \ldots$$

If the Harmonic potential were of this form, we would be able to write down the eigen states as products of the 1D eigen states and get the energies as sums of the 1D eigen energies. As it turns out, by changing coordinates we can turn a quadratic system with off diagonal cross terms (like the first potential) into one with no cross terms (like the second). These new coordinates, in terms of which the Hamiltonian separates, are called normal modes and they allow us to reduce a polyatomic molecule to a collection of independent 1D oscillator.

First, we note that V can be rewritten concisely in matrix notation [It may be useful to consult McQuarrie's supplement on Matrix Eigenvalue problems if the following seems unfamiliar]:

$$V(R_1, R_2, R_3, \ldots) = \frac{1}{2}\mathbf{R}^T \bullet \mathbf{K} \bullet \mathbf{R}$$

$$\mathbf{R} \equiv \begin{pmatrix} R_1 \\ R_2 \\ R_3 \\ \vdots \end{pmatrix} \qquad \mathbf{K} \equiv \begin{pmatrix} k_{11} & k_{12} & k_{13} & \cdots \\ k_{21} & k_{22} & k_{23} & \cdots \\ k_{31} & k_{23} & k_{33} & \cdots \\ \vdots & \vdots & \vdots & \ddots \end{pmatrix}$$

Now, the Hamiltonian is of the form:

$$\hat{H} = \sum_i \frac{\hat{p}_i^2}{2\mu_i} + \frac{1}{2}\hat{\mathbf{R}}^T \bullet \mathbf{K} \bullet \hat{\mathbf{R}}$$

It is convenient to first transform to mass-weighted coordinates:

$$\hat{\tilde{p}}_i \equiv \frac{\hat{p}_i}{\sqrt{\mu_i}} \qquad \hat{\tilde{x}}_i \equiv \sqrt{\mu_i}\hat{R}_\rightarrow \qquad \hat{\tilde{k}}_{ij} \equiv \frac{k_{ij}}{\sqrt{\mu_i \mu_j}}$$

in terms of which we can write the Hamiltonian:

$$\hat{H} = \sum_i \frac{\hat{\tilde{p}}_i^2}{2} + \frac{1}{2}\hat{\tilde{x}}^T \bullet \tilde{\mathbf{K}} \bullet \hat{\mathbf{x}}$$

As is clear from the kinetic energy above, in these coordinates, every degree of freedom has the same reduced mass. Now we perform the normal mode transformation. We want to write:

$$\hat{x}^T \bullet \tilde{K} \bullet \tilde{x} = \hat{y}^T \bullet K' \bullet y \qquad K' = \begin{pmatrix} k_{11}' & 0 & 0 & \cdots \\ 0 & k_{22}' & 0 & \cdots \\ 0 & 0 & k_{33}' & \cdots \\ \cdots & \cdots & \cdots & \cdots \end{pmatrix}$$

Further, we will assume there is a matrix U that transforms from x to y:

$$y = U \bullet x \quad \Rightarrow y^T = x^T \bullet U^T$$

where in the second equality, we recall the general rule that the transpose of a product is the product of the transposes, but in the opposite order. Combining these two equations:

$$\hat{x}^T \bullet \tilde{K} \bullet \hat{x} = \hat{y}^T \bullet K' \bullet y = \hat{x}^T \bullet U^T \bullet K' \bullet U \bullet \hat{x}$$

$$\Rightarrow \tilde{K} = U^T \bullet K' \bullet U$$

The last equation is a common problem encountered in linear algebra: the quest to take a given matrix $(\tilde{K})$ and place it in diagonal form (the right hand side). For a symmetric matrix like $\tilde{K}$ the solution to this problem is well known: the diagonal entries of K' are the eigenvalues of $\tilde{K}$ and the columns of the transformation matrix U are the eigenvectors of $\tilde{K}$. The transformed variables y are called the normal modes. These modes are linear combinations of the local degrees of freedom R_1, R_2, R_3 ... that we started out with. Thus, while the initial motions might correspond clearly to local stretching of one bond or bending of an angle, the normal modes will generally be complicated mixtures of different molecular motions. We can visualize this in the simple case of two degrees of freedom. The local modes R_1, R_2 can be thought of as the two orthogonal axes in a plane. Meanwhile, the normal modes y_1, y_2 are also orthogonal axes, but rotated from the original set:

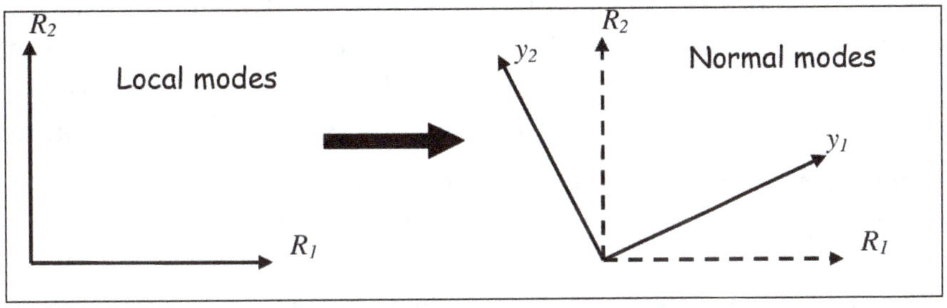

The local modes have interactions between each other: local stretches are coupled to local bends, etc. As a result, the Hamiltonian is not separable in terms of the local modes R_1, R_2. However, by design the Hamiltonian is separable when written in terms of the normal modes:

$$\hat{H} = \sum_i \frac{\hat{p}_i^2}{2} + \frac{1}{2} k'_{ii} \hat{y}_i^2$$

The eigenvalues, k'_{ii}, tell us the "stiffness" of the PES in the particular direction y_i. Note that the Hamiltonian above is exactly the same Hamiltonian as the one we started with. The coupling terms have simply been rotated away by changing the coordinates. We can immediately interpret the spectra of this Hamiltonian in terms of a sum of many independent oscillators. Thus, for a polyatomic molecule within the harmonic approximation we expect to see lines at each of the normal mode frequencies:

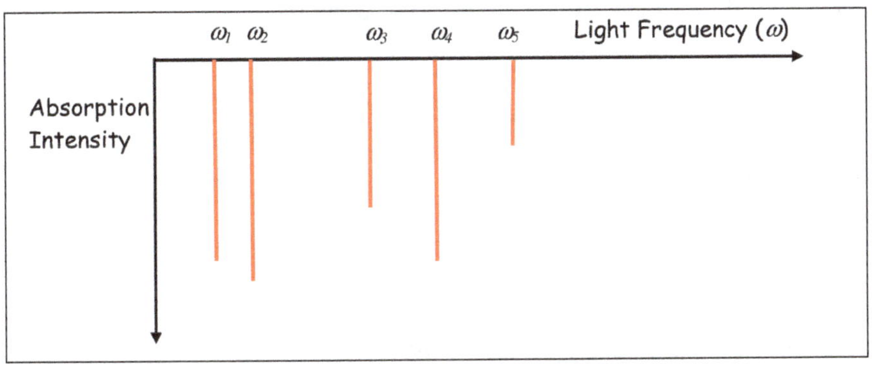

Where, we have noted that the different oscillators will typically also have different transition dipoles. (For obvious reasons, in vibrational spectroscopy the square of the transition dipole is often called the oscillator strength) We can, of course, combine this polyatomic picture with the anharmonicity effects above to get a more general picture that looks like:

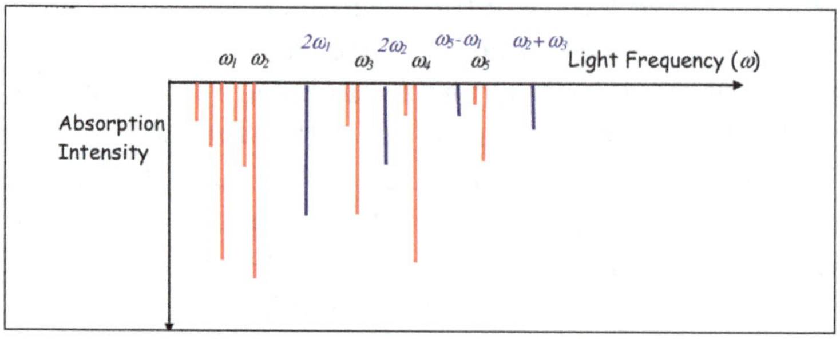

Where, we predict the existence of various hot bands and overtones for each of the normal mode oscillators in the molecule. Note that while the overtones always involve multiple quanta, the quanta need not come from the same normal mode – hence we expect not only overtones at $2\omega_1$, but also a combination bands at $\omega_2+\omega_3$ and $\omega_5-\omega_1$. the picture above is qualitatively correct for the IR spectrum of a single molecule. In solution, heterogeneity leads to a smearing out and broadening of the peaks, leading to the complex IR fingerprints we are used to.

Normal Coordinates and Linear Transformations

Let us first attempt to visualize the vibrations of a heteronuclear diatomic molecule. We look at the problem from the angle of diagonalization of the kinetic and potential energy matrices. The kinetic energy, T, for the two-mass system is given by,

$$T=\frac{1}{2}m_1v_1^2+\frac{1}{2}m_2v_2^2=\frac{1}{2}m_1\dot{x}_1^2+\frac{1}{2}m_2\dot{x}_2^2$$

Where, the dot on x represents the first order derivative of x (and hence the velocity). In matrix form, this equation can be represented as:

$$2T=\begin{pmatrix}\dot{x}_1 & \dot{x}_2\end{pmatrix}\begin{pmatrix}m_1 & o\\ o & m_2\end{pmatrix}\begin{pmatrix}\dot{x}_1\\ \dot{x}_2\end{pmatrix}$$

Since the force is the negative of the potential energy derivative, i.e.

$$f=-\frac{dV}{dx}$$

and the force is –kx, integration gives:

$$V=\frac{1}{2}kx^2$$

Where, $x=x_2-x_1$. The potential energy is given by,

$$V=\frac{1}{2}k(x_2-x_1)^2$$

This can be written as:

$$2V = \begin{pmatrix} x_1 & x_2 \end{pmatrix} \begin{pmatrix} k & -k \\ -k & k \end{pmatrix} \begin{pmatrix} x_1 \\ x_2 \end{pmatrix}$$

where, we have symmetrized the matrix.

We find that the kinetic energy matrix is in diagonal form, but the potential energy is not. We require both of these to be in diagonal form so that we can separate the two coupled equations to two independent, easily solvable equations. If we could somehow convert the kinetic energy matrix to a unit matrix, then it would also be possible to put the potential energy in a diagonal form by the usual procedure without disturbing the kinetic energy matrix.

Such a transformation is possible by converting to mass-weighted coordinates $\xi_i = \sqrt{m_i}x_i$. We, then have:

$$T = \frac{1}{2}m_1\dot{x}_1^2 + \frac{1}{2}m_2\dot{x}_2^2 = \frac{1}{2}m_1\left(\frac{\dot{\xi}_1}{\sqrt{m_1}}\right)^2 + \frac{1}{2}m_{21}\left(\frac{\dot{\xi}_2}{\sqrt{m_2}}\right)^2 = \frac{1}{2}\left(\dot{\xi}_1^2 + \dot{\xi}_2^2\right)$$

and

$$2T = \begin{pmatrix} \dot{\xi}_1 & \dot{\xi}_2 \end{pmatrix} \begin{pmatrix} 1 & 0 \\ 0 & 1 \end{pmatrix} \begin{pmatrix} \dot{\xi}_1 \\ \dot{\xi}_2 \end{pmatrix}$$

In a similar fashion, it can be shown that:

$$2V = \begin{pmatrix} \xi_1 & \xi_1 \end{pmatrix} \begin{pmatrix} \frac{k}{m_1} & -\frac{k}{\sqrt{m_1 m_2}} \\ -\frac{k}{\sqrt{m_1 m_2}} & \frac{k}{m_2} \end{pmatrix} \begin{pmatrix} \xi_1 \\ \xi_2 \end{pmatrix}$$

Diagonalization of the matrix in equation above in the usual way $|V - \lambda I| = 0$ gives:

$$\begin{pmatrix} \frac{k}{m_1} - \lambda & -\frac{k}{\sqrt{m_1 m_2}} \\ -\frac{k}{\sqrt{m_1 m_2}} & \frac{k}{m_2} - \lambda \end{pmatrix} = 0$$

that leads to,

$$\left(\frac{k}{m_1} - \lambda\right)\left(\frac{k}{m_2} - \lambda\right) - \frac{k^2}{m_1 m_2} = 0$$

$$\Rightarrow \left(\frac{k^2}{m_1 m_2} - \lambda\left(\frac{k}{m_1} + \frac{k}{m_2}\right) + \lambda^2\right) - \frac{k^2}{m_1 m_2} = 0$$

$$\Rightarrow \lambda = 0, \frac{k}{m_1} + \frac{k}{m_2}$$

Eigenvectors

We now find the eigenvectors corresponding to each eigenvalue. This is done by inserting the obtained eigenvalue one by one into the following equation:

$$\begin{pmatrix} \frac{k}{m_1}-\lambda_i & -\frac{k}{\sqrt{m_1 m_2}} \\ -\frac{k}{\sqrt{m_1 m_2}} & \frac{k}{m_2}-\lambda_i \end{pmatrix}\begin{pmatrix} A_{i1} \\ A_{i2} \end{pmatrix}=0$$

1. $\lambda_1 = 0$

Inserting $\lambda = 0$ in equation above, we get,

$$\begin{pmatrix} \frac{k}{m_1} & -\frac{k}{\sqrt{m_1 m_2}} \\ -\frac{k}{\sqrt{m_1 m_2}} & \frac{k}{m_2} \end{pmatrix}\begin{pmatrix} A_{11} \\ A_{12} \end{pmatrix}=\begin{pmatrix} 0 \\ 0 \end{pmatrix}$$

Or,

$$\frac{k}{m_1}A_{11}-\frac{k}{\sqrt{m_1 m_2}}A_{12}=0$$

$$\Rightarrow A_{12}=\sqrt{\frac{m_2}{m_1}}A_{11}$$

Thus, the normalized eigenvector can be written as:

$$Q_1=\frac{1}{\sqrt{m_1+m_2}}\left(\sqrt{m_1}\xi_1+\sqrt{m_2}\xi_2\right)$$

This is the normal coordinate in terms of mass-weighted coordinates. To obtain the eigenvector in terms of the displacement coordinates, divide each coefficient by the square root of the respective mass, i.e. the first term by $\sqrt{m_1}$ and the second by $\sqrt{m_2}$. After normalization, we obtain:

$$Q_1=\frac{1}{\sqrt{2}}(x_1+x_2)$$

Similarly, for the other eigenvalue,

2. $\lambda_2=\frac{k}{m_1}+\frac{k}{m_2}$

$$\begin{pmatrix} \frac{k}{m_1}-\frac{k}{m_1}-\frac{k}{m_2} & -\frac{k}{\sqrt{m_1 m_2}} \\ -\frac{k}{\sqrt{m_1 m_2}} & \frac{k}{m_2}-\frac{k}{m_1}-\frac{k}{m_2} \end{pmatrix}\begin{pmatrix} A_{21} \\ A_{22} \end{pmatrix}=\begin{pmatrix} 0 \\ 0 \end{pmatrix}$$

Or,

$$A_{22} = -\sqrt{\frac{m_1}{m_2}}A_{21}$$

Hence,

$$Q_2 = \sqrt{\frac{m_2}{m_1 + m_2}}\xi_1 - \sqrt{\frac{m_1}{m_1 + m_2}}\xi_2$$

Again, dividing by the square root of the respective masses and normalization leads to the equation,

$$Q_2 = \frac{1}{\sqrt{m_1 + m_2}}\left(\sqrt{\frac{m_2}{m_1}}x_1 - \sqrt{\frac{m_1}{m_2}}x_2\right)$$

Visualization of the Normal Modes

Let us now visualize the two normal modes:

1. Q_1 equation $Q_1 = \frac{1}{\sqrt{2}}(x_1 + x_2)$

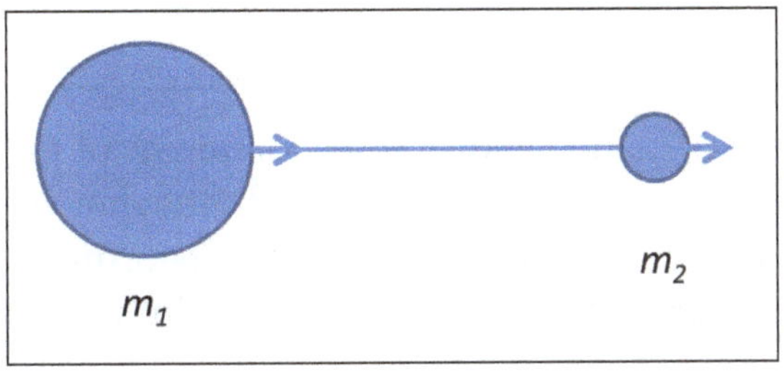

Notice that both atoms are moving in the same direction by the same amount, and so this is translational motion.

2. Q_2 equation $Q_2 = \frac{1}{\sqrt{m_1 + m_2}}\left(\sqrt{\frac{m_2}{m_1}}x_1 - \sqrt{\frac{m_1}{m_2}}x_2\right)$.

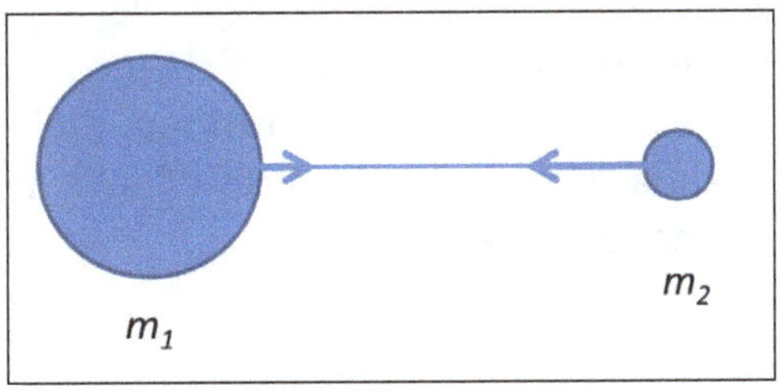

In this mode, both atoms move in opposite directions. Also, the lengths of the arrows are proportional to the square roots of the masses of the opposite atoms. For example, in HCl, the hydrogen atom moves almost six times as much as the chlorine atoms since the ratio of the mass of a chlorine atom to that of a hydrogen atom is 35.5.

This is the essence of normal modes. For each normal mode, all the atoms move with the same frequency and phase, but the amplitudes are different, depending on the relative masses. They move in such a way that they all reach their extremes at the same time and pass the equilibrium position at the same instant. This makes the lighter atom move further than the heavier atom.

Assignment of Normal Modes to Irreducible Representations

Another property of normal modes needs mention. Heteronuclear diatomic molecules belong to the $C_{\infty v}$ point group, whose character table is given in Table below.

Table: Character table of the $C_{\infty v}$ point group.

$C_{\infty v}$	E	C_{∞}^{Φ}	$\cdots$	$\infty\sigma_{\upsilon}$		
$A_1 \equiv \Sigma^+$	1	1	$\cdots$	1	z	x^2+y^2, z^2
$A_2 \equiv \Sigma^-$	1	1	$\cdots$	−1	R_z	
$E_1 \equiv \Pi$	2	$2\cos\Phi$	$\cdots$	0	$(x,y);(R_x,R_y)$	(xz,yz)
$E_2 \equiv \Delta$	2	$2\cos 2\Phi$	$\cdots$	0		(x^2-y^2,xy)
$E_3 \equiv \Phi$	2	$2\cos 3\Phi$	$\cdots$	0		
$\vdots$	$\vdots$	$\vdots$	$\ddots$	$\vdots$		

Look at the first mode. The arrows do not change direction in any of the symmetry operations, E, C_{∞} and σ_v, and hence the character is +1 for each of these, and the symmetry species is A_1 or Σ^+. The situation is similar for Q_2 and that too is a basis for Σ^+. In fact, all normal modes of vibration form a basis for one or the other of the irreducible representations of the point group of the molecule.

Born-Oppenheimer Approximation

When considering pure vibrational or pure rotational transitions, we assume that the two are independent. This is essentially the Born-Oppenheimer approximation, which states that since the energies involved are very different, they do not interact. Consider a molecule that is simultaneously undergoing vibration and rotation. The rotational energy level separations are of the order 1-10 cm^{-1} while the vibrational energy level separations are of the order of 3000 cm^{-1}. Thus, to a good approximation, a molecule executes rotational and vibrational motions independently. According to the Born-Oppenheimer approximation, the total energy of such a molecule can be written as the sum of the rotational and vibrational energies:

$$E_{Total} = E_{rot} + E_{vib}$$

In wavenumber terms, the total energy would be as:

$$S(\upsilon,J) = \tilde{B}J(J+1) - \tilde{D}J^2(J+1)^2 + \tilde{H}J^3(J+1)^3 + \ldots + \left(\upsilon+\frac{1}{2}\right)\omega_e - \left(\upsilon+\frac{1}{2}\right)^2 \omega_e\chi_e$$

The rotational constants are much smaller than the vibrational frequencies and the centrifugal distortion constants are still smaller and can be ignored in comparison with the vibrational energies. Thus we have:

$$S(\upsilon,J)=\tilde{B}J(J+1)+\left(\upsilon+\frac{1}{2}\right)\omega_e-\left(\upsilon+\frac{1}{2}\right)^2\omega_e\chi_e$$

The selection rules are the same as those for the individual transitions, i.e.

$$\Delta\upsilon=\pm1,\pm2,....$$

$$\Delta J=\pm1$$

An analytical expression for the spectrum may be obtained by applying the selection rules to equation $S(\upsilon,J)=\tilde{B}J(J+1)+\left(\upsilon+\frac{1}{2}\right)\omega_e-\left(\upsilon+\frac{1}{2}\right)^2\omega_e\chi_e$. Considering only the $\upsilon=0,\ J=J'' \rightarrow \upsilon=1,\ J=J'$ transition, we can have:

$$\Delta S(J,\upsilon)=S(J',1)-S(J'',0)$$

$$=\tilde{B}J'(J'+1)+(3/2)\omega_e-(9/4)\omega_e\chi_e-\left(\tilde{B}J''(J''+1)+(1/2)\omega_e-(1/4)\omega_e\chi_e\right)$$

$$=\omega_e(1-2\chi_e)+\tilde{B}(J'-J'')(J'+J''+1)$$

$$=\omega_o+\tilde{B}(J'-J'')(J'+J''+1)$$

Where, $\omega_o=\omega_e(1-2\chi_e)$.

In the case where $\Delta J=+1$, i.e. $J'=J''+1$, we obtain:

$$\Delta S(J'',\upsilon)=\omega_o+2\tilde{B}(J''+1),J''=0,1,2,.....$$

In the case where $\Delta J=-1$, i.e. $J''=J'+1$, we get:

$$\Delta S(J',\upsilon)=\omega_o-2\tilde{B}(J'+1),J'=0,1,2,....$$

These may be combined to yield:

$$\Delta S(J',\upsilon)=\omega_o+2\tilde{B}m,\ m=\pm1,\pm2,....$$

Where, ω_o is known as the band origin or band centre. Equation above represents the combined vibration-rotation spectrum. It will consist of evenly spaced lines (spacing = $2\tilde{B}$) on each side of the band origin ω_o but since $m \neq 0$, the line at ωo will not appear. Lines to the low side of ω_o (when m is negative, i.e. ΔJ = -1) are referred to as the P branch, while those to the high frequency side (m positive, ΔJ = +1) are referred to as the R branch.

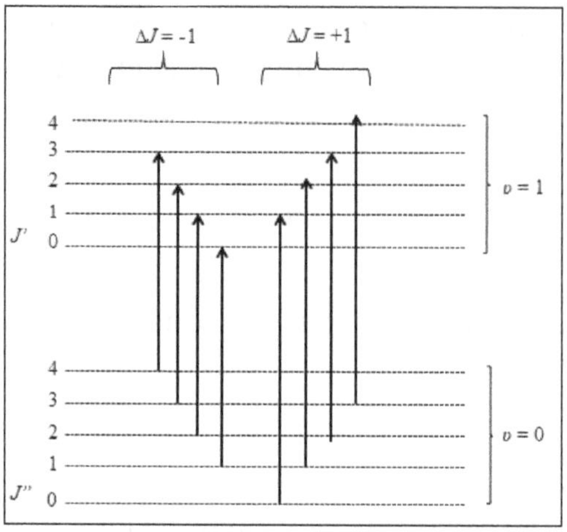

Schematic diagram showing the P and R branches of a diatomic molecule. Lines on the left side are the P branches and those on the right side are R branches. The band centre is missing.

From Figure, it is clear that the band centre is the average of $P_{(1)}$ and $R_{(0)}$. Table gives part of the IR spectrum of CO. From the table we see that the band centre is at 2143.255 cm^{-1}, while the average line separation near the centre is 3.83 cm^{-1}. This immediately gives:

$$2\tilde{B} = 3.83 \text{ cm}^{-1} \text{ and } \tilde{B} = 1.915 \text{ cm}^{-1}.$$

Table: Part of the IR spectrum of CO.

Line	cm^{-1}	separation, cm^{-1}	Line	cm^{-1}	separation, cm^{-1}
$P_{(1)}$	2139.43	3.88	$R_{(0)}$	2147.08	3.78
$P_{(2)}$	2135.55	3.92	$R_{(1)}$	2150.86	3.73
$P_{(3)}$	2131.63		$R_{(2)}$	2154.59	

This is in satisfactory agreement with the more precise value of $\tilde{B}$ = 1.92118 cm^{-1} derived from microwave spectroscopy. From $\tilde{B}$, the moment of inertia, I, and the bond length can be determined.

Example: The four central lines in the high resolution $\upsilon = 1 \leftarrow \upsilon = 0$ infrared spectrum of $H^{37}Cl$ occur at 2837.6, 2858.8, 2899.2 and 2918.6 cm^{-1}. Deduce as much as possible about the molecule. Would the corresponding lines in $H^{35}Cl$ lie at the same spectral positions?

The band centre is at the average of the two central lines, i.e. 2879.0 cm^{-1}. This is equal to the fundamental frequency, ω_0. The 4 $\tilde{B}$ separation at the centre is 40.4 cm^{-1}, giving a value of 10.1 cm^{-1} for $\tilde{B}$. The moment of inertia is,

$$I = \frac{h}{8\pi^2 \tilde{B}_c} = 2.771 \times 10^{-47} \text{ kg m}^2.$$

The reduced mass is given by,

$$\mu = \frac{m_H m_{Cl}}{(m_H + m_{Cl}) N_A} = \frac{1.0078 \times 36.9659 \times 10^{-3}}{37.9737 \times 6.022 \times 10^{23}} = 1.629 \times 10^{-27} \text{ kg}$$

Therefore,

$$r = \sqrt{\frac{I}{\mu}} = \sqrt{\frac{2.771\times10^{-47}}{1.629\times10^{-27}}} = 1.304\times10^{-10}\ m = 0.1304\ nm.$$

Since the positions of the lines depend on the reduced mass, the lines for $H^{35}Cl$ will be at a different position. The reduced mass of $H^{35}Cl$ is smaller and hence its moment of inertia is smaller and rotational constant larger, so the lines will have larger separation. The fundamental vibration frequency will also be higher since it is inversely proportional to the square root of the reduced mass.

The Interaction of Rotations and Vibrations

According to the previous discussion, the rotational spacing should be the same. However, this is not the case. A crowding of lines is seen in the R branch with increasing J, whereas the spacing increases in the P branch. Apparently, the assumption of independence of vibration and rotation is in error and there is an interaction between the two.

Since the frequency of vibration is higher than that of rotation, a molecule vibrates 10^3 times during the course of a single rotation and so the bond length (and hence the moment of inertia I and $\tilde{B}$) also changes continually during the rotation. The rotational constant $\tilde{B}$ depends on the average value of $\frac{1}{r^2}$. Further, an increase in vibrational energy is accompanied by an increase in vibrational amplitude because of the widening of the potential well, and hence the value of $\tilde{B}$ will depend on the vibrational quantum number.

This increase is even more pronounced for anharmonic vibrations. An increase in vibrational energy will lead to an increase in the average bond length. The vibrational constant then varies even more with vibrational energy. In general, since r_{av} increases with the vibrational energy, $\tilde{B}$ is smaller in the upper vibrations than in the lower ones. We may express this dependence in terms of an empirical equation,

$$B\upsilon = B_e - \alpha\left(\upsilon + \frac{1}{2}\right)$$

Where, B_υ is the rotational constant in the vibrational level υ, B_e is the equilibrium rotational constant at the potential minimum, and α is a small positive constant.

Consider the fundamental vibrational change $\upsilon = 0 \rightarrow \upsilon = 1$. The respective $\tilde{B}$ values are $\tilde{B}_0$ and $\tilde{B}_1$. For this transition, we will have:

$$\begin{aligned}\tilde{v} &= G(J', \upsilon = 1) - G(J'', \upsilon = 0) \\ &= \omega_0 + \tilde{B}_1 J'(J'+1) - \tilde{B}_0 J''(J''+1)\end{aligned}$$

Where:

$$\omega_0 = \omega_e(1 - 2\chi_e)$$

We then have the two cases:

Case 1:

$$\Delta J = +1,\ J' = J'' + 1$$

$$\tilde{v} = \omega_0 + (\tilde{B}_1 + \tilde{B}_0)(J''+1) + (\tilde{B}_1 - \tilde{B}_0)(J''+1)^2$$

Where, J" = 0, 1, 2 ...

Case 2:

$$\Delta J = -1,\ J'' = J' + 1$$

$$\tilde{v} = \omega_0 - (\tilde{B}_1 + \tilde{B}_0)(J'+1) + (\tilde{B}_1 - \tilde{B}_0)(J'+1)^2$$

Where, J' = 0, 1, 2 ...

These two equations may be combined to yield:

$$\tilde{v}_{P,R} = \omega_0 + (\tilde{B}_1 + \tilde{B}_0)m + (\tilde{B}_1 - \tilde{B}_0)m^2$$

With m = ±1, ±2, ±3 ...

The positive m values refer to R branches and the negative ones to P branches. Equation $\tilde{v}_{P,R} = \omega_0 + (\tilde{B}_1 + \tilde{B}_0)m + (\tilde{B}_1 - \tilde{B}_0)m^2$ allows us to fit the observed transitions to a quadratic equation, from which the various terms can be determined.

Since $\tilde{B}_1 < \tilde{B}_0$ because the internuclear separation is higher in higher vibrational states, the spacing in the R branches decreases with increasing J and the opposite is observed for the P branches. This gives a distinctive shape to the spectrum.

Microwave Spectroscopy

The Microwave Spectroscopy or Rotational Spectroscopy is a high frequency spectroscopic technique using microwaves study molecules with dipole moments.

Zeeman Effect

Absorption plots of σ- Zeeman components of the 5_{23}-6_{16} rotational transition of H_2O has been obtained which demonstrate clearly the τ dependence of the molecular magnetic moment. Figure shows the lower frequency σ components as observed on a slow sweep, Stark-modulation spectrograph. Figure is a plot of expected absorption component structure for an arbitrary difference in magnetic moment for the two states involved in the transition, assuming that the lower state has the larger g value. The individual components have not been resolved, since with a half-width at half-intensity of approximately 220 kc/sec the line width is too great. The asymmetry is clearly in agreement with the calculated intensity pattern, indicating that $g(5_{23}) > g(6_{16})$. In the present case, the maximum is split 2. 39 Mc/sec from the undisplaced line in a field of 4900 g, yielding an effective g value of 0. 531. Since the components are unresolved, this g value has no precise interpretation. Work is continuing to determine the true g values of both rotational levels by a process of curve fitting and other means.

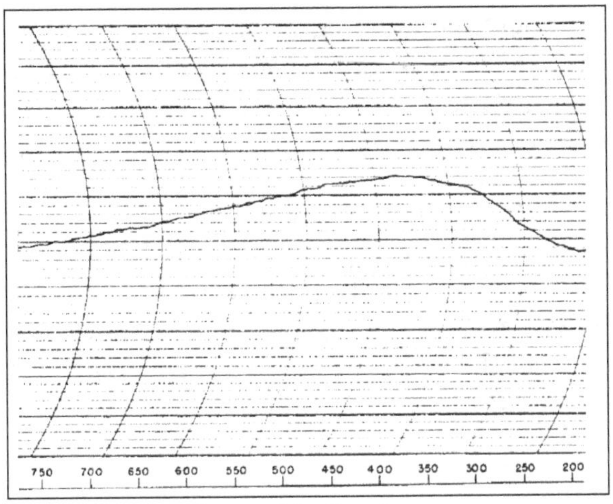

Absorption pattern observed for low frequency σ components of 5_{23}–6_{16} transition in H_2O.

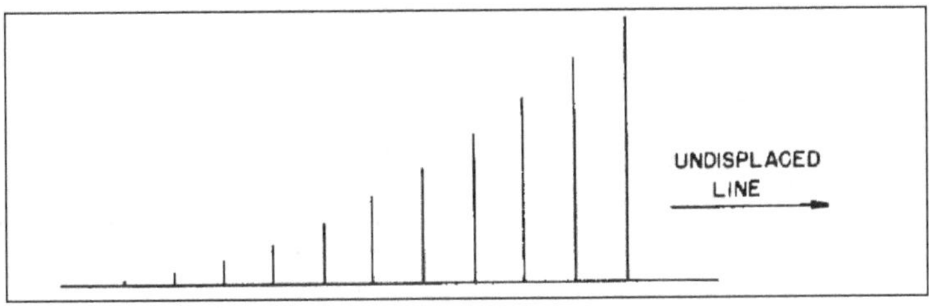

Intensities of individual σ Zeeman components for 5_{23}–6_{16} rotational transition.

Paramagnetic Resonance

One of the principal changes was the introduction of a magnetic field control servo. In this, the voltage induced in a synchronous flip coil is subtracted from a fraction of that produced by a 30-cycle generator (on the same shaft) to produce an error signal. After amplification and phase detection in a Brown converter, this signal controls the magnetic field by means of a dc amplifier and electronic field current supply for the motor generator supplying the magnet current. The response of these highly inductive components is speeded by the use of an inner dc feedback loop from the magnet current back to the input of the dc amplifier. This allows a stable loop gain of about 80 for the complete ac feedback loop, with a corresponding decrease in drift. This stabilized field can now be swept at a very uniform rate by sweeping the reference-signal amplitude with a helipot across the reference generator. At present field marker pips are provided manually with the aid of an auxiliary null-comparator circuit using 4-decade resistance boxes to allow setting precisely to one gauss.

To increase the sensitivity of the apparatus, we have adopted a modification of a method used by Beringer in which a 47-cycle oscillating field of amplitude small compared to the absorption line width is superimposed on the slowly sweeping field. This gives a 47-cycle component of crystal current proportional to the derivative $^{d}\chi''/dB$ of the imaginary part of the susceptibility of the sample. This eliminates the crystal modulator of the previous apparatus which chopped the entire carrier rather than tagging only the desired signal with sa certain frequency. This 47-cycle component of crystal current is amplified and phase detected to give a deflection proportional to $^{d}\chi''/dB$. The dX''/dB signal may also be integrated electronically to give X" itself, if this quantity is more easily interpreted.

At present we are engaged in perfecting the integration arrangement to remove baseline shifts and in testing the performance of the apparatus on various samples which have been studied previously with the other apparatus. The accompanying specimen charts demonstrate the versatility of the sweep scheme in handling patterns from 5 gauss to 4000 gauss in width.

The irregularity in the pattern from the free radical indicates that the field servo does leave a small amount of rapid, irregular fluctuations in the field, although it does take out the slower variations. The sharpness of this absorption line and the stability of the field have allowed the homogeneity of the field to be studied. It appears that, with the pole piece separation required for low temperature work, the field changes by about 15 gauss over the width of the cavity. This is greatly reduced with a smaller gap.

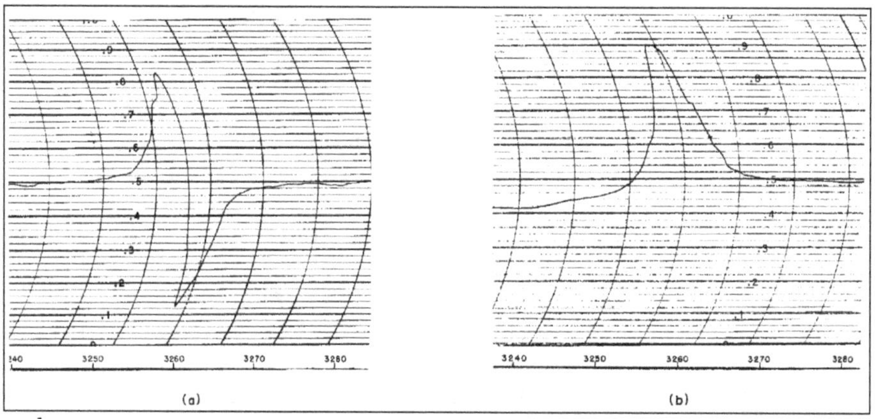

$^{d}\chi''/dB$ and χ'' for the organic free radical Diphenyl-trinitrophenyl-hydrazyl.

The increased sensitivity is shown by the trace of the 0.01 molar $MnCl_2$ solution. This concentration is only one tenth that of the most dilute sample which could be studied with the previous apparatus. Since a rather rapid sweep and broad band pass was used in this trace, it is clear that another factor of 10 can be expected. The chrome alum trace demonstrates the convenience of the very long linear sweep which this field control arrangement makes possible.

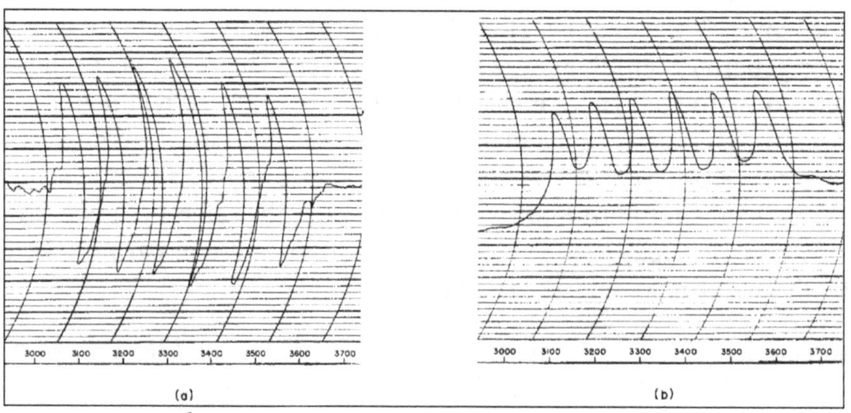

$^{d}\chi''/dB$ and χ"for 0. 01 molar $MnCl_2$ solution.

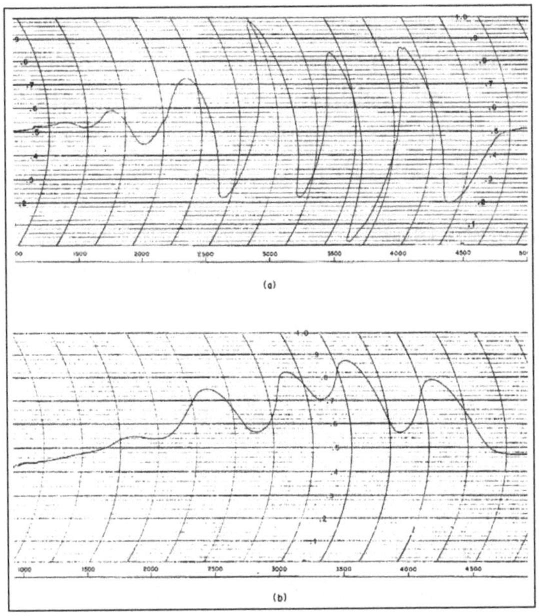

$^{d}\chi''/dB$ and χ'' for rubidium chrome alum crystal.

The Microwave Spectrum of Methyl Alcohol

The Stark effect of CH_3oH is being studied in the region of 40, 000 Mc/sec to 50, 000 Mc/sec. A series of lines with second-order Stark effect has been found and is assumed to be due to the $J_\tau = 0_0 \rightarrow 1_{-1}$ transition. Accurate Stark measurements are being made in order to calculate and compare the Stark coefficients of this series. One of the fundamental vibrations of methyl alcohol is that of torsional motion of the hydroxyl group with respect to the methyl group. Depending on the potential due to the methyl group, the hydroxyl group will either oscillate in one of three identical minima or perform a hindered rotation relative to the methyl group. Many of the absorption lines in the microwave spectrum are due to changes in the rotational state of the molecule and/or to changes in this internal rotational state.

Microwave absorption lines of this molecule were first discovered by Hershberger and Turkevich and were divided by Dailey according to their Stark effect patterns. Coles re-measured these lines and their Stark components and was able to establish J values for a series of nine lines. Dennison and Burkhard have identified this series as the $J \rightarrow J$, $K = 1 \rightarrow 2$, $\tau = 2 \rightarrow 1$ transition. A much more extensive study has been carried out by Hughes, Good and Coles, who have listed about 85 lines between 16, 000 Mc/sec and 39, 000 Mc/sec, as well as lines of $C^{13}H_3OH$ and $CH_3o\ ^{18}H$. Approximate J values and Stark patterns are given for most of the lines in their report. Edwards, Gilliam and Gordy have measured 27 lines in the region 35, 000 Mc/sec to 50, 000 Mc/sec, but these are in general unidentified.

Calculations of the energy of an asymmetric rotator show that the transition $J_\tau = 0_0 \rightarrow 1_{-1}$ (corresponding to the $J = 0 \rightarrow 1$, $\Delta K = 0$ transition of the limiting symmetric rotator) should give an energy difference in the range 40, 000 Mc/sec to 50, 000 Mc/sec. This calculation depends only on the rotational constants determined from infrared observations and so can be accepted with little doubt. However, many other strong lines occur in this region due to near degeneracies of higher order J transitions. Identification in the absence of additional information would seem impossible.

By inspection of the Stark effect of the lines in this region, they may be separated into a group with first-order Stark effect (symmetrical components), and one with secondorder Stark effect (a single component). Of this latter group, a series of lines has been found which are assumed to be due to the $J_\tau = 0_0 \rightarrow 1_{-1}$ transition. The line at 49, 867. 2 Mc/sec, assigned by Edwards, Gilliam and Gordy to this transition, has a first-order Stark effect and thus would appear to be improperly identified.

Accurate line and Stark component measurements are being made at the present time in order to compare the spacing of the lines in the series and the Stark coefficients of each of the lines.

Rotational Spectra

Solution of the rotational Schrödinger equation gives the allowed energy levels as:

$$E_J = BJ\left(J+1\right)$$

Where, B is a quantity known as the rotational constant and J is the rotational quantum number, which can take on positive integer values, including zero. The rotational constant B has the dimension of energy (Joule) and is expressed as:

$$B = \frac{\hbar^2}{2I}$$

Where, I is the moment of inertia of the molecule, which is equal to μR^2, μ being the reduced mass of the molecule and R the inter-nuclear distance.

Like other transitions, rotational transitions are also governed by selection rules. For a transition from a state with rotational quantum number J" to another state J', the transition dipole moment integral is given by:

$$\left|M_{J'J''}\right| = \int \psi_{J'}^{*}\,\hat{\mu}\,\psi_{J''}d\tau = \mu_0 \int \psi_{J'}^{*}\psi_{J''}d\tau$$

Where, μ_0 is the permanent dipole moment of the molecule. Since the intensity of a transition is proportional to the square of the transition dipole moment, it is obvious that molecules that do not possess a dipole moment do not exhibit rotational spectra. In other words, they cannot interact with microwave radiation and are said to be microwave inactive. Examples are all homo-nuclear diatomic molecules like H_2, N_2, O_2, and F_2. All tetrahedral and octahedral molecules like CH_4 and SF_6 are also microwave inactive, as are other symmetrical molecules like benzene. However, molecules like H_2O, which possesses a dipole moment are microwave active. The domestic microwave oven operates at the microwave rotational frequency of water. The water in the food gets rotationally excited and heats up, cooking the food.

Example: Which of the molecules H_2, NO, N_2O can have a pure rotational spectrum?

Solution: All except H_2.

Example: Which of these molecules will show a pure rotational (microwave) spectrum: CH_4, CH_3Cl, CH_2Cl_2, $CH_2 = CH_2$, benzene, SF_6?

Solution: CH_3Cl and CH_2Cl_2. For rotational spectra the gross selection rule is: All molecules that do not possess a permanent dipole moment are microwave inactive.

Even for molecules that possess a permanent dipole moment, all transitions are not allowed. Detailed calculations show that the only transitions that are allowed are those for which:

$$\Delta J = \pm 1$$

This selection rule can be rationalized on the basis of the conservation of angular momentum. The photon has an intrinsic angular momentum of one unit. Absorption or emission of one photon can therefore increase or decrease the rotational quantum number by one unit.

Let us first consider absorption spectra. The energy of the photon that is absorbed is the difference in energies between these two states,

$$h\nu_{J\rightarrow J+1} = E_{rot}(J+1) - E_{rot}(J) = B(J+1)(J+2) - BJ(J+1) = 2B(J+1)$$

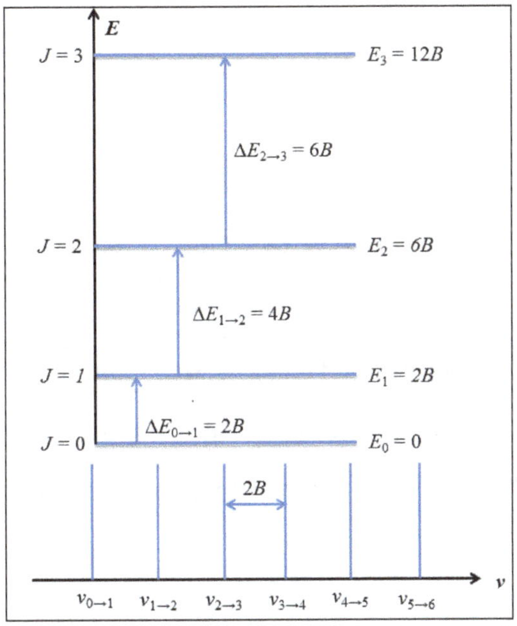

Rotational energy levels and transitions between them.

The lowest value of the rotational quantum number is zero and absorption of a photon can take the molecule to the J = 1 level, resulting in absorption of a photon of energy 2B. The J = 1 → 2 transition occurs at 4B. Subsequent lines are observed at 6B, 8B, 10B, etc. Rotational spectra should therefore consist of equally spaced lines with spacing of 2B between them, from which the rotational constant can be obtained.

Appearance of Rotational Spectra

After the selection rules, the next important consideration is the relative populations of the various levels. We have already seen that rotational energy spacings are small, and so a number of rotational energy levels are occupied at room temperature, since $\Delta E_{rot} << k_B T$, where kB is the Boltmann constant and T is the temperature in Kelvin. At room temperature, the value of kBT is

~ 4 × 10^{-21} J, while rotational energy spacings are much smaller (~10^{-23} s). According to the Boltzmann distribution, the ratio of populations of the excited and ground state is given by:

$$\frac{N_J}{N_0} = g_J e^{-\Delta E_J/k_B T}$$

Where, g_J is the degeneracy of the Jth level (degeneracy of the J = 0 level is 1) and ΔE_J is the energy difference between the Jth level and the J = 0 level, which is equal to E_J, since the J = 0 level has zero energy.

There are two opposite factors operating. One is the rapid linear increase in the degeneracy, 2J + 1; with increasing J and the other is the exponential decrease with increasing J. The net result is that the population initially increases because of the initial dominance of the degeneracy term, goes through a maximum and then decreases.

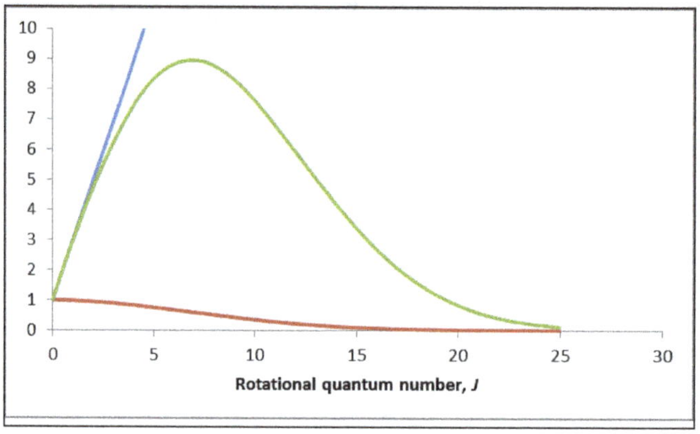

Variation of the population of rotational states with quantum number.

We can analytically derive an expression for J_{max}, the value of J for the most populated state, by differentiating expression $\frac{N_J}{N_0} = g_J e^{-\Delta E_J/k_B T}$ with respect to J and setting it equal to zero. Rewriting equation $\frac{N_J}{N_0} = g_J e^{-\Delta E_J/k_B T}$,

$$N_J = N_0 (2J+1) \exp\left(\frac{-E_J}{k_B T}\right)$$

$$\frac{dN_J}{dJ} = e^{-E_J/k_B T}\left[2-(2J+1)\frac{B}{k_B T}(2J+1)\right] = 0$$

$$\Rightarrow (2J+1)^2 \frac{B}{k_B T} = 2$$

$$J_{max} = \frac{1}{2}\sqrt{\frac{2k_B T}{B}} - \frac{1}{2} = \sqrt{\frac{k_B T}{2B}} - \frac{1}{2}$$

Example: Calculate the ratio of the number N_J to N_0 of molecules in a sample of HI at (a) 300 K and (b) 1000 K having rotational quantum numbers J = 5 and J = 0 (I = 4.31 × 10^{-47} kg m^2) (c) Calculate the value of J at the maximum N_J at these two temperatures.

Solution: (a) T = 300 K,

$$\frac{N_J}{N_o} = (2J+1)\exp(-E_J / k_B T) = 11\exp\left(-\frac{30h^2}{8\pi^2 I k_B 300}\right) = 0.584$$

(b) At 1000 K,

$$\frac{N_J}{N_o} = (2J+1)\exp(-E_J / k_B T) = 11\exp\left(-\frac{30h^2}{8\pi^2 I k_B 1000}\right) = 0.756$$

(c) $B = \dfrac{\hbar^2}{2I} = 1.29\times 10^{-22}\,J$,

$$J_{max} = \sqrt{\frac{k_B T}{2B}} - \frac{1}{2}$$

$$= \sqrt{\frac{1.38\times 10^{-23}T}{2\times 1.29\times 10^{-22}}} - \frac{1}{2}$$

$$= \sqrt{5.35\times 10^{-2}T} - 0.5$$

At 300 K, J_{max} = 4.00-9.5 = 3.5 ≈ 4

At 1000 K, J_{max} = 6.8 ≈ 7

We note that J_{max} moves to a higher value when the temperature is raised. The intensities of the spectral lines therefore depend on the temperature.

Example: The moment of inertia of the $H^{35}Cl$ molecule is 2.644 × 10^{-47} kg m². Prepare relative population curves, N_J/N_o versus J for HCl at 300 K and 1000 K.

Solution: $\dfrac{N_J}{N_o} = (2J+1)\exp\left(-\dfrac{J(J+1)\hbar^2}{2Ik_B T}\right) = (2J+1)\exp(-15.23J(J+1)/T)$

The relative population versus J graph is shown below:

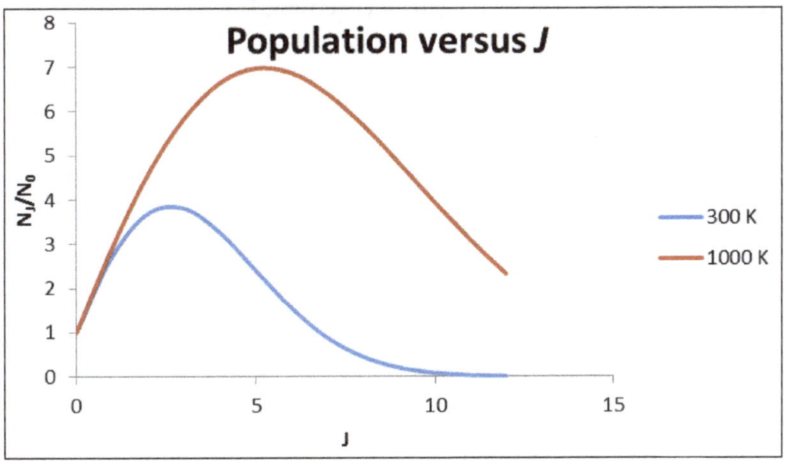

The rotational energy of the level which has the highest population is, therefore,

$$E_{rot}(max) = J_{max}(J_{max}+1)B = \left(\sqrt{\frac{k_BT}{2B}} - \frac{1}{2}\right)\left(\sqrt{\frac{k_BT}{2B}} + \frac{1}{2}\right)B = \left(\frac{k_BT}{2B} - \frac{1}{4}\right)B = \frac{k_BT}{2} - \frac{B}{4}$$

Except at very low temperatures, $\frac{k_BT}{2} >> \frac{\tilde{B}}{4}$; the energy of the most populated level is, therefore, $\approx \frac{k_BT}{2}$.

Besides population, the other factor that governs the intensity maximum is the magnitude of B. Since it appears in the exponential term, larger the value of B, faster is the exponential decay.

Example: What is the value of J in the most highly populated rotational level of ICl at room temperature? $\left(\tilde{B} = 0.114\ cm^{-1}\right)$?

Solution: Since $\tilde{B} = B/hc$, the expression for the most populated level becomes,

$$J_{max} = \sqrt{\frac{k_BT}{2\tilde{B}hc}} - \frac{1}{2}$$

Substitution of the various values gives:

$$J_{max} = 29.65.$$

The nearest whole number is 30.

We notice that the J_{max} value is extremely large because the rotational constant is extremely small in this case.

References

- Photoelectron-spectroscopy, accelerators-beamlines, techniques: maxiv.lu.se, Retrieved 20, June 2020
- Microwave-Spectroscopy-41186977: researchgate.net, Retrieved 11, January 2020

3

Organic Spectroscopy

Organic spectroscopy derives structural information by using various techniques. A few of these techniques are UV-Visible spectroscopy, infrared spectroscopy and nuclear magnetic resonance spectroscopy. The aim of this chapter is to explore these techniques of organic spectroscopy which are closely related to its progress.

Organic Spectroscopy presents the derivation of structural information from UV, IR, Raman, 1H NMR, 13C NMR, Mass and ESR spectral data in such a way that stimulates interest of students and researchers alike. There are several spectroscopic techniques which can be used to identify organic molecules: infrared (IR), mass spectroscopy (MS) UV/visible spectroscopy (UV/Vis) and nuclear magnetic resonance (NMR). IR, NMR and UV/vis spectroscopy are based on observing the frequencies of electromagnetic radiation absorbed and emitted by molecules. MS is based on measuring the mass of the molecule and any fragments of the molecule which may be produced in the MS instrument.

UV-Visible Spectroscopy

UV-Visible spectroscopy deals with the study of the electronic transitions of molecules as they absorb light in the UV (190-400 nm) and visible regions (400-800 nm) of the electromagnetic spectrum. The absorption of ultraviolet or visible radiation lead to transition among electronic energy levels, hence it is also often called electronic spectroscopy.

As a rule, the energetically favored electron promotion will be from the highest occupied molecular orbital (HOMO) to the lowest unoccupied molecular orbital (LUMO). The resulting species is said to be in an excited state. The wavelengths at which absorption occurs, together with the degree of absorption at each wavelength is recorded by optical spectrometer. A spectrum is obtained as a result. It commonly provides the knowledge about π-electron systems, conjugated systems, aromatic compounds and conjugated non-bonding electron systems etc.

Types of Electronic Transitions

The ground state of an organic molecule contains valence electrons in three principal types of molecular orbitals, namely -sigma (σ) orbitals, pi (π) orbitals and filled but nonbonding orbitals (n).

Both σ and π orbitals are formed from the overlap of two atomic or hybrid orbitals. Each of these molecular orbitals therefore has an antibonding σ* or π* orbital associated with it. An orbital containing n electron does not have an antibonding orbital because it is not formed from two orbitals. Sigma bonding orbitals have lower energy than π bonding orbitals, which in turn have lower energy than non-bonding orbitals. In the Electronic transitions, promotion of an electron from one of the three ground states (σ, π, or n) to one of the two excited states (σ*, or π*) takes place. As a result, there are six possible transitions- σ to σ*, σ to π*, π to π*, π to σ*, n to π*, n to σ*. The most commonly observed transitions and their relative energies are summarized in Figure below. The exact energy differences between the orbitals depend on the nature of atoms present and the nature of the bonding system.

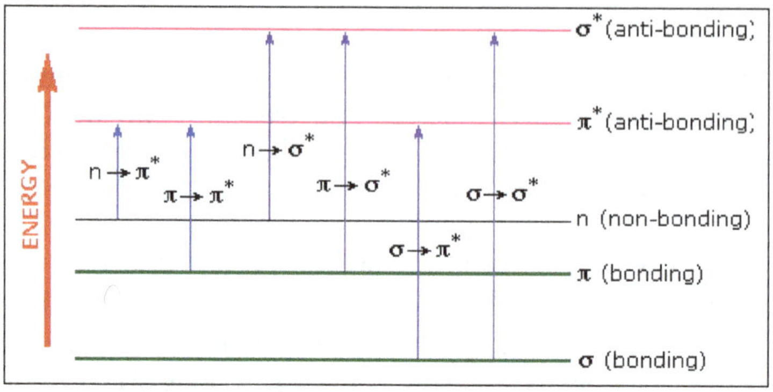

Electronic Transitions in UV/VIS spectroscopy.

You can see, in each possible case, an electron is excited from a low energy, ground state orbital into a higher energy, excited state anti-bonding orbital. Each transition requires a defined amount of energy. The larger the gap between the energy levels, the greater the energy required to promote the electron to the higher energy level, resulting in electromagnetic radiation of higher frequency, and therefore shorter wavelength, being absorbed.

σ to σ*

A transition of electrons from a bonding sigma orbital to the antibonding sigma orbital is designated as σ to σ* transition. These are high energy transitions as σ bonds are generally very strong. Thus these transitions involve very short wavelength ultraviolet light (< 150 nm) and usually fall outside the range of UV-visible spectrophotometers (200-800 nm). Alkanes can only undergo σ to σ* transitions. Methane and ethane undergo σ → σ * transitions with an absorbance maximum at 122 and 135 nm, respectively. Study of such transitions is usually done in an evacuated spectrophotometer (< 200 nm) since oxygen present in air absorbs strongly at 200 nm and below. Similarly nitrogen absorbs at ~150 nm and below.

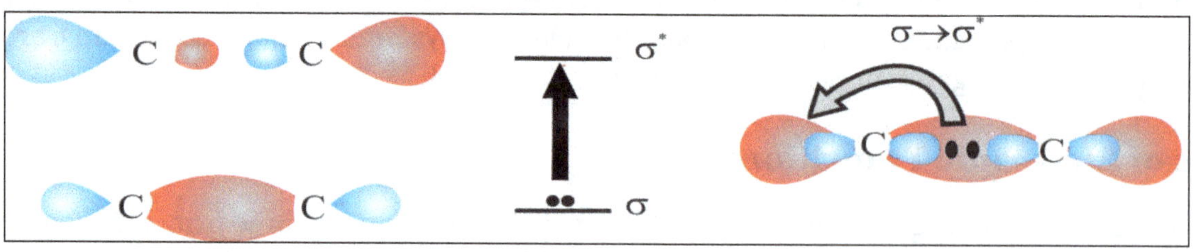

π to π*

The transition of an electron from a π bonding orbital to a π* antibonding orbital is designated as π to π* transition. These types of transitions take place in compounds containing one or more unsaturated groups like simple alkenes, carbonyl, aromatics, nitriles, nitro etc. These transitions require lesser energy than n to σ* transitions. In non-conjugated alkenes, this type of transition occurs in the range 170-190 nm e.g. ethane shows absorption maxima at 171 nm. Similarly, π to π* transition in the range of 180-190 nm occurs in non-conjugated carbonyl compounds e.g. acetone shows absorption maxima at 188 nm.

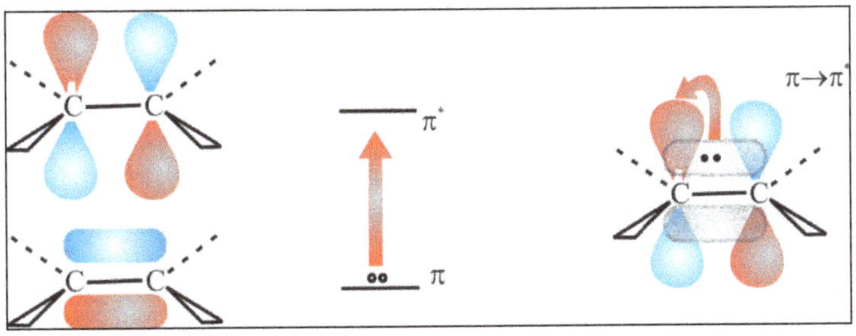

n to σ*

The transition of an electron from a non-bonding orbital to the antibonding sigma orbital is designated as n to σ* transition. Saturated compounds containing atoms with lone pairs (nonbonding electrons) like saturated alcohols, amines, halides, ethers etc are capable of showing n to σ* transitions. Energy required for these transitions is usually less than σ to σ* transitions. Such compounds absorb light having wavelength in the range 150-250 nm. For example the absorption maxima for water, methyl alcohol, methyl chloride and methyl iodide are 167 nm, 174 nm, 169 nm and 258 nm, respectively.

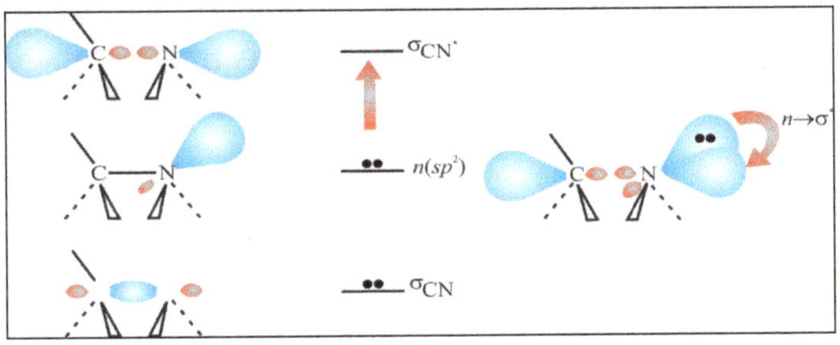

n to π*

The transition of an electron from a non-bonding orbital to a π* antibonding orbital is designated as n to π* transition. This transition involves the least amount of energy in comparison to all other transitions and therefore gives rise to an absorption band at longer wavelength.

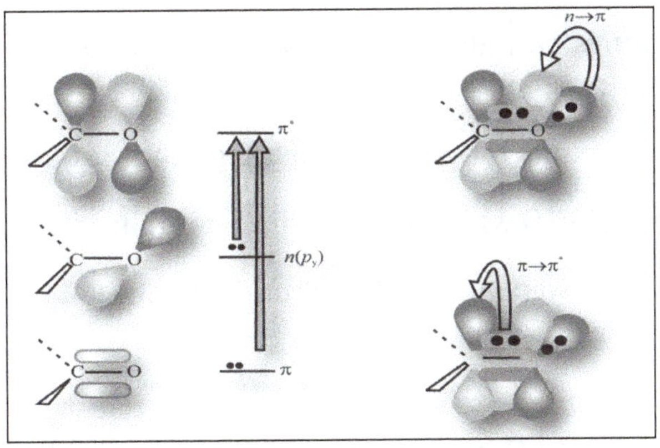

Saturated carbonyl compounds show two types of transitions, low energy n to π* (270-300 nm) and high energy π to π* (180-190 nm). The transition n to π* is of lowest energy but is of low intensity as it is symmetry forbidden. Thus the most intense band for these compounds is always due to π to π* transition.

Selection Rule

Some electronic transitions, which are otherwise theoretically possible, are generally not observed in the UV/VIS spectroscopy. Therefore there are some restrictions which govern the observable transitions:

- Transitions, which involve change in the spin quantum number of an electron during the transitions do not occur, i.e. singlet-triplet transitions are not allowed.
- Transitions between orbitals of different symmetry do not occur. For example, transition n to π* is forbidden because the symmetry of n and π* do not match.

UV-spectrophotometer

Spectrophotometer is a kind of spectrometer, which measures the transmittance or absorbance of a sample as a function of wavelength, when light of certain intensity and frequency range is passed through the sample. Unlike a spectrometer (which is any instrument that can measure the properties of light over a range of wavelengths), a spectrophotometer measures only the intensity of light as a function of its wavelength. The key components of a spectrophotometer are:

- Light source,
- Monochromator,
- Sample area,
- Detector and Recorder.

Light Source

The most suitable sources of light are:

- Deuterium lamp which emits the light in the UV-region (160-375 nm).
- Tungsten lamp or tungsten-halogen lamp which emits radiation in the Visible and near infrared regions (350-2500 nm).
- Xenon arc lamp which emits radiation in the range 190-800 nm.
- Light emitting diodes (LED) which emits radiation in the visible range 400-800 nm.

The instruments automatically swap lamps when scanning between the UV and UV-VIS regions.

Monochromator

The main function of the monochromator is to disperse the beam of light obtained from the primary source, into its component. The principle components of monochromator are:

- An entrance slit.
- A collimating lens.
- A dispersing device.
- A focusing lens.
- An exit slit.

The radiation emitted from the primary source (polychromatic radiation) enters the monochromator through the entrance slit. The beam is collimated and then strikes the dispersing element (Prism or grit) at a particular angle. Two types of dispersion devices viz. prisms and holographic gratings are commonly used in UV-visible spectrophotometers.

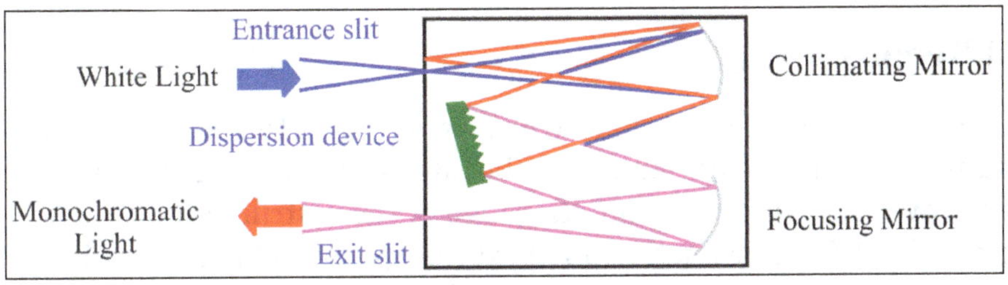

Functioning of Monochromator.

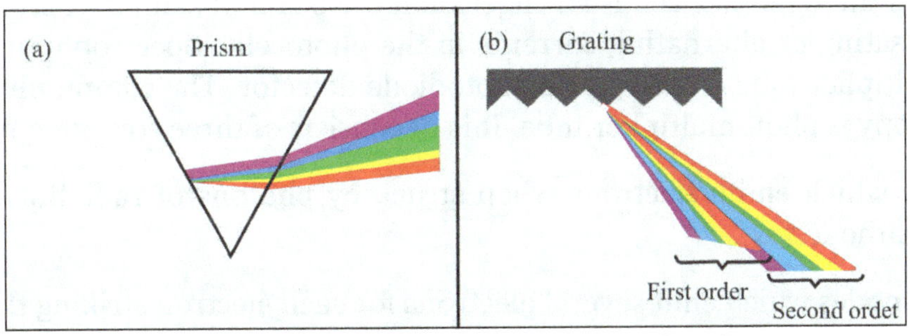

Dispersion of white light through prism and grating.

Light falling on the prism is reflected at different angles, depending on the wavelength. The beam is split into its component colors. By moving the dispersing element or the exit slit, radiation of only a particular wavelength can be obtained, which leaves from the exit slit and can be used for the recording purpose. The beam selected by the slit is monochromatic and further divided into two beams with the help of another prism, which then passes through the sample and reference solutions.

Sample Area

One of the two divided beams is passed through the sample solution and the other beam is passed through the reference solution. Although, the samples for recording spectra are most commonly liquids, but the absorbance of gases and even of solids can be measured. Both sample and reference solutions are placed in a transparent cell, known as cuvette. The cuvettes are rectangular in shape, and usually have an internal width of 1 cm.

A quartz cuvette.

It is important that the material of cells must be transparent to the radiation throughout the region under study. The cells are usually made of glass, plastic as well as silica or quartz. Of these, glass cells cannot be used for the UV region as they absorb light in the UV region but can be used satisfactorily in the visible region. Quartz is transparent in all (200-700 nm) ranges and is the best choice and hence can be easily used in UV as well as visible region.

Detector and Recorder

A detector converts a light signal into an electrical signal. After the beams are passed through the sample under study as well as the reference cell, the intensities of the respective transmitted beams are then compared over the whole wavelength range of the instrument. Generally two photocells are used as detector in UV spectrometer to record the spectra. One of the photocell receives the beam from sample cell and second detector receives the beam from the reference. The intensity of the radiation from the reference cell is stronger than the beam of sample cell. This results in the generation of pulsating or alternating currents in the photocells. Spectrophotometers consist of either a photomultiplier tube detector or a photodiode detector. The commonly used detector in UV-Vis spectroscopy is photomultiplier tube. It is m consists of three components:

- A cathode which emits electrons when struck by photons of radiation known as photo emissive cathode.
- Several dynodes which emit several electrons for each electron striking them.
- An anode.

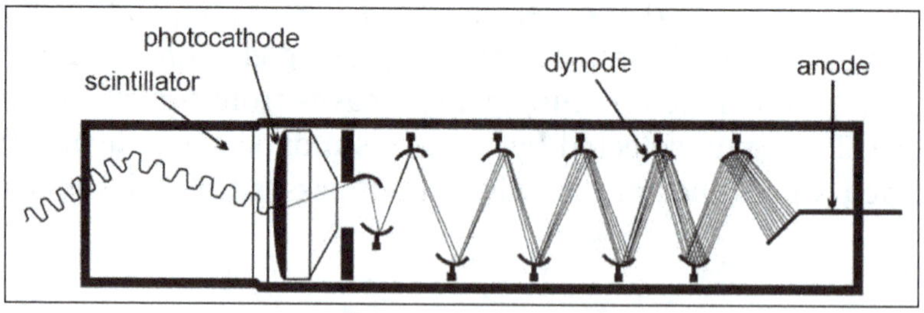

A photomultiplier tube.

In its functioning, when a photon of radiation strikes the cathode, emission of several electrons take place. These emitted electrons are then accelerated towards the many dynodes. The first dynode is 90V more positive than the cathode. The electrons strike the first dynode, causing the emission of several electrons for each incident electron. These electrons are then accelerated towards the second dynode, to produce more electrons which are accelerated towards dynode three and so on. Finally all the electrons are collected at the anode. Several dynodes are arranged between the anode and the cathode to produce an amplification effect. Each photon usually produces $10^6 - 10^7$ electrons, resulting in the amplified current that can be measured.

Photomultipliers are very sensitive to both UV and visible radiations and have fast response times. It is significant to note that Intense light may damage photomultipliers; hence they are limited to measuring low power radiation.

Photodiodes are increasingly being used as detectors in modern spectrophotometers. Photodiode detectors have a wider dynamic range and are more robust than photomultiplier tube detectors. In a photodiode, light falling on the semiconductor material allows electrons to flow through it, thereby depleting the charge in a capacitor connected across the material. The amount of charge that is required to recharge the capacitor at regular intervals is proportional to the intensity of the light. The limits of detection are approximately 170–1100 nm for silicon-based detectors.

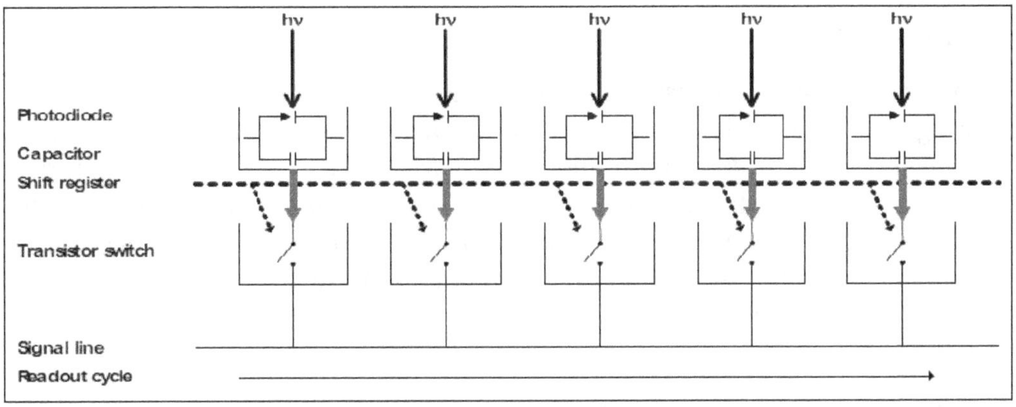

An array of photodiodes.

Some modern spectrophotometers contain an array of photodiode detectors instead of a single detector. A diode array consists of a series of photodiode detectors positioned side by side on a silicon chip. Each photodiode is connected to a transistor switch via a charged capacitor When photons strike the diode, free electrical charge carriers are generated that discharge the capacitors. The capacitors are recharged at regular intervals. The amount of charge needed to recharge the capacitors is proportional to the number of photons detected by each diode, which in turn is proportional to the light intensity. The absorption spectrum is obtained by measuring the variation in light intensity over the entire wavelength range.

Types of UV/Vis Spectrophotometer

There are two types of spectrophotometers, namely, single beam spectrophotometer or double beam spectrophotometer. Each has its advantages and disadvantages.

A single beam spectrophotometer has only one beam of light, which passes through the sample, therefore it requires taking reading for the reference and sampling separately. On the other hand, in a double-beam instrument, the light is split into two beams before it reaches the sample. The two beams move simultaneously, one passing through a reference solution and the other through the sample. The reference beam intensity is taken as 100% Transmission (or 0% Absorbance), and the measurement displayed is the ratio of the two beam intensities.

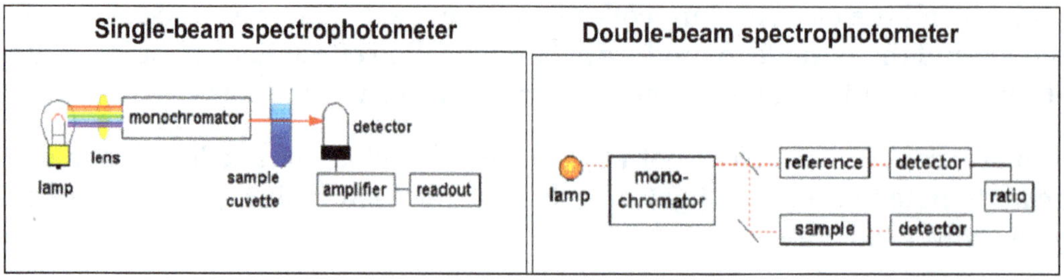

Types of Spectrophotometer.

Of the two types of spectrophotometer, double beam spectrophotometers are faster to operate and in their performance. They provide more reproducible results because they perform an automatic correction for the loss of light intensity as the beam passes through the sample and reference solution.

Sample Preparation

The UV-Vis spectra are usually measured in very dilute solutions. Usually 1 mg of the compound of molecular weight of 100-200 is dissolved in 100 ml of suitable solvent and only a portion is used for recording the spectra. The solvent should not absorb radiation and must be transparent over the desired range of wavelength. The solvents, which do not contain conjugated system, are most suitable for recording the UV-spectra. The solvent should also be inert to the sample. Some commonly used solvents are water, 95% ethanol, methanol and hexane.

Absorption Laws

When the radiation passes through a solution, some amount of light is absorbed. Suppose I_o is the intensity of the incident radiation and I, the intensity of the transmitted radiation. The amount of radiation absorbed can be measured by:

$$\text{Transmittance } T = I/I_o$$

$$\%\text{ Transmittance} = T \times 100$$

$$\text{Absorbance } A = \text{Log}_{10}\, I_o/I = \text{Log}_{10}\, 1/T$$

If the entire light passes through a solution without any absorption, then absorbance is zero. In this case, the percent transmittance is 100%. If all the light is absorbed, then absorption is infinite and the percent transmittance is zero %.

Beer-lambert Law

It gives a linear relationship between absorbance and concentration of an absorbing species. This law states that the absorption is directly proportional to the concentration of absorbing substance and the length of the path of radiation through the sample. It is independent of the intensity of the incident light and each successive unit layer absorbs an equal fraction of light incident on it.

$$A = \varepsilon cb$$

$$\text{Log}_{10}\, I_o / I = \varepsilon cb$$

Where,

- A is the sample's Absorbance value at specific wavelength or frequency.
- ε is the molar absorptivity or the molar extinction coefficient of the substance.
- b is the path length of the sample in cm.
- c is the concentration of the solute in mol L^{-1}.

The value of ε absorptivity coefficient is constant for a particular material at a particular wavelength.

Limitations of the Beer-lambert Law

Under certain conditions Beer-lambert law fails to maintain a linear relationship between absorbance and concentration of the solution.

- Deviations in absorptivity coefficients at high concentrations (>0.01M) due to electrostatic interactions between molecules in close proximity.
- Scattering of light due to particles present in the sample.
- Chemical deviations due to the specific chemical species of the sample under investigation.
- Non-monochromatic radiation.
- Fluorescence or phosphorescence of the sample.

Applications of UV Spectroscopy

As such UV spectroscopy is used extensively in teaching, research and analytical laboratories for the quantitative analysis of all molecules that absorb ultraviolet and visible electromagnetic radiation.

Qualitative Analysis

UV-Vis spectroscopy has been exploited for the structure elucidation of organic compounds, detection of different organic compounds present in a mixture and their separation by several analytical techniques such as thin layer chromatography.

Structure Elucidation of Organic Compounds

UV spectroscopy is useful in the structure elucidation of organic compounds. The presence or absence of a particular absorption band at a particular wavelength may be regarded as an evidence for the presence or absence of a particular chromophores in the compound.

Determination of Impurities

TLC is a very simple technique and is commonly used for the qualitative analysis of the reaction mixture. Many organic compounds absorb UV light of various wavelengths. This can be used for the determination of impurities in organic compounds. A small spot is kept on the TLC plate which is then allowed to run in appropriate solvent media. Different compounds travel different distance based on their polarity which can be visualized on the TLC plate under a UV lamp. Extra spots on TLC plate can be observed due to impurities in the sample and it can be compared with the spot of standard.

A concept of UV/Vis spectroscopy may be used as a detector for HPLC (High-performance liquid chromatography formerly referred to as high-pressure liquid chromatography). UV detectors are usually of variable wavelength and can be used to detect the molecules with absorption maxima above 210 nm by measuring the absorbance of the eluent. It provides a very fast detection and separation of a reaction mixture. When a compound is eluted from the HPLC column, it absorbs UV radiation at the appropriate wavelength. The amount of UV radiation absorbed is directly proportional to the amount of a particular compound that is passing through the column at that time.

Quantitative Analysis

Determination of Concentration of the Compound

UV absorption spectroscopy can be used for the quantitative determination of compounds which absorb UV radiation. This determination is based on Beer's law which relates the absorbance A of a substance at a particular wavelength to the concentration, c as follows:

$$A = \log_{10} I_o / I = \log 1/T = -\log T = \varepsilon bc$$

Where,

- ε is extinction coefficient.
- c is concentration.
- b is the length of the cell that is used in UV spectrophotometer.
- I_o is the intensity of the incident radiation.
- I is the intensity of transmitted radiation.

The use of UV-spectroscopy in quantitative analysis can be understood by a simple example of the estimation of a mixture of anthracene and naphthalene. The UV spectrum of ethanolic solution of anthracene shows λmax at 375 nm, whereas naphthalene does not absorb in this region. Thus from a mixture of anthracene-naphthalene, we can calculate the amount of anthracene by using Lambert Beer's law. Measure the absorption of the mixture at 375 nm wavelength and thus by using the equation $A = \varepsilon bc$, we can calculate the concentration of pure anthracene and hence the proportion of anthracene in the mixture.

UV-Vis spectroscopy has been used extensively in industrial analytical organic chemistry. The percentage of Vit A_1 and Vit A_2 in natural fats or oils can be estimated by the measurement of the intensities of peaks at 325 nm and at 351 nm, respectively. Thus by comparing the intensities of the solution of these two vitamins with the intensities of the known solutions of these two enables the concentration to be calculated.

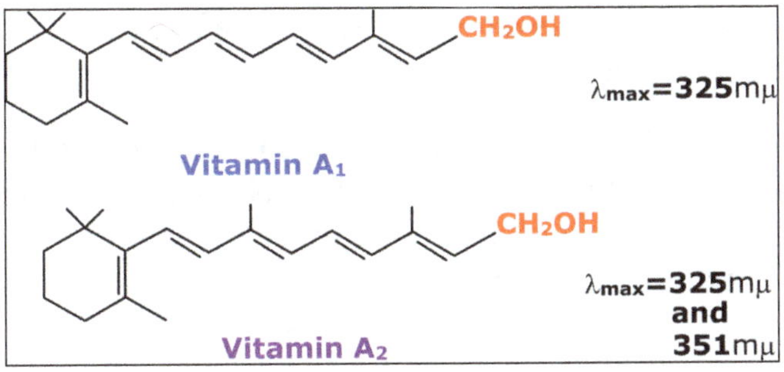

A similar procedure may be used for the estimation of ergosterol in fats, anthracene in benzene, carbon disulphide in carbon tetrachloride, chlorophyll in plant material etc.

Chemical kinetics: Study of Chemical Reactions

Ultraviolet/visible spectroscopy can also be used to study reaction rates. Spectrometric rate determination involves the measurement of absorption of either the reactants or products at a fixed wavelength, when a reaction occurs. If a reactant or reagent or product of the reaction absorbs radiation at a particular wavelength, the spectrophotometer can be set to measure the absorption at that wavelength as a function of time and hence the rate of a reaction can be measured.

To explain this let us a take an example which shows the removal of proton from nitroethane with the help of hydroxide ion. Here only the anion of nitroethane absorbs in the UV region at 240 nm, while other reactants or products do not show any significant absorbance at this wavelength. In order to measure the rate at which hydroxide ion removes a proton from nitroethane (i.e., the rate at which the nitroethane anion is formed), the UV spectrophotometer is adjusted to measure absorbance at 240 nm as a function of time. Nitroethane is taken in a cuvette containing a basic solution, and the rate of the reaction is determined by monitoring the increase in absorbance at 240 nm.

$$\underset{\text{nitroethane}}{CH_3CH_2NO_2} + HO^- \rightleftharpoons \underset{\substack{\text{nitroethane anion} \\ \lambda_{max}=240\text{nm}}}{CH_3\bar{C}HNO_2} + H_2O$$

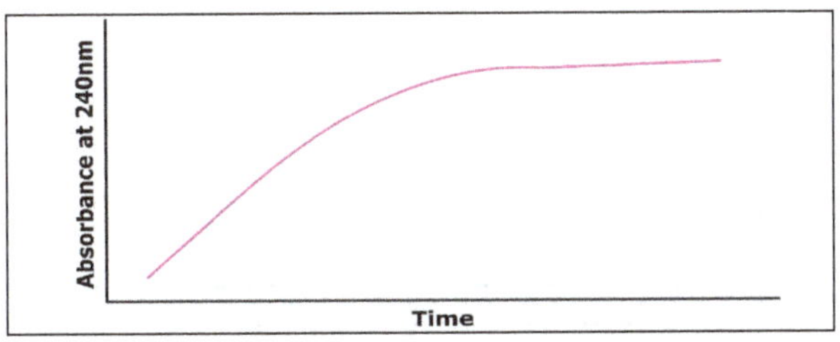

The rate at which a proton is removed from nitroethane is determined by monitoring the increase in absorbance at 240 nm.

The most widely used application of spectrometric rate measurements is for study of enzymes (proteins that are present in all living tissue). Enzymes cannot be measured directly but their catalytic properties allow their estimation from the speed of the reactions which they catalyse. Enzymes have many uses as reagents or as labels that can be attached to other molecules to permit their indirect detection and measurement. The widest use in the field of clinical diagnostics is as an indicator of tissue damage. When cells are damaged by disease, enzymes leak into the bloodstream and the amount present indicates the severity of the tissue damage. The relative proportions of different enzymes can be used to diagnose disease, say of the liver, pancreas or other organs which otherwise exhibit similar symptoms.

The most common reaction used in these clinical assays is the reduction of nicotinamide adenine dinucleotide (NAD) to $NADH_2$. The spectra of NAD and $NADH_2$ are shown in Figure given below. It is clear that if the absorbance is measured at 340 nm, the readings will increase as the reaction progresses towards the formation of $NADH_2$.

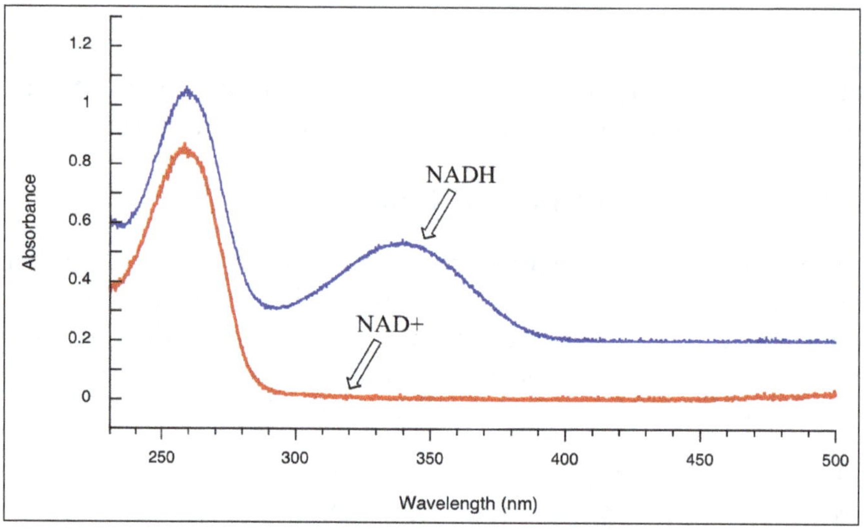

Absorption spectra of NAD and NADH.

Similarly, the enzyme lactate dehydrogenase catalyses the reduction of pyruvate by NADH to form lactate. Here, NADH is the only species in the reaction mixture which absorbs light at 340 nm hence by measuring the decrease in absorbance at 340 nm, the rate of reaction can be determined.

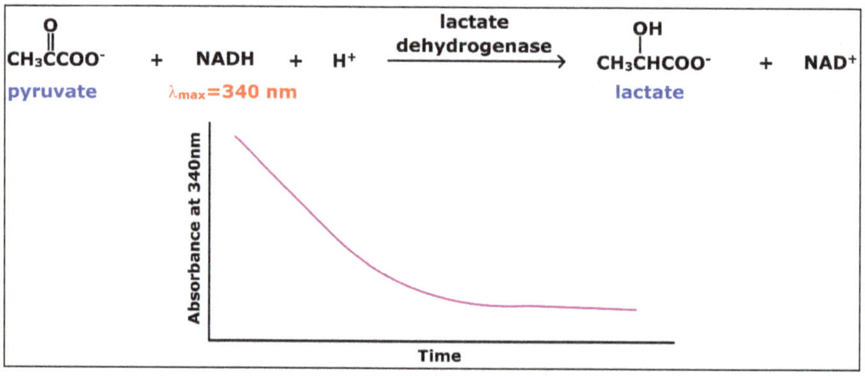

The rate of reduction of pyruvate by NADH is measured by monitoring the decrease in absorbance at 340 nm.

Dissociation Constants of Acids and Bases

The pK_a of a compound can be determined by UV/Vis spectroscopy if either the acidic form or the basic form of the compound absorbs UV or visible light. For example, the phenoxide ion has a λ_{max} at 287 nm. If the absorbance at 287 nm is determined as a function of pH, the pK_a of phenol can be calculated by using Henderson–Hasselbalch equation:

$$pH = pK_a + \log [A^-] / [HA]$$

By plotting a graph between absorbance versus wavelength at different pH values, the ratio of $[A^-] / [HA]$ can be determined and hence the pK_a value of the compounds can be calculated.

In the Quantification and Thermal Denaturation of DNA

UV spectroscopy can also be used to estimate the nucleotide composition of DNA. The two strands of DNA are held together by both A–T base pairs and G–C base pairs. When DNA is heated, the double stranded DNA breaks down. Single-stranded DNA has a greater molar absorptivity at 260 nm than double-stranded DNA. The melting temperature (Tm) of DNA is the midpoint of an absorbance versus temperature curve. For double-stranded DNA, Tm increases with increasing numbers of G–C base pairs because they are held together by three hydrogen bonds, whereas A–T base pairs are held together by only two hydrogen bonds. And hence can be used to estimate the number of G–C base pairs.

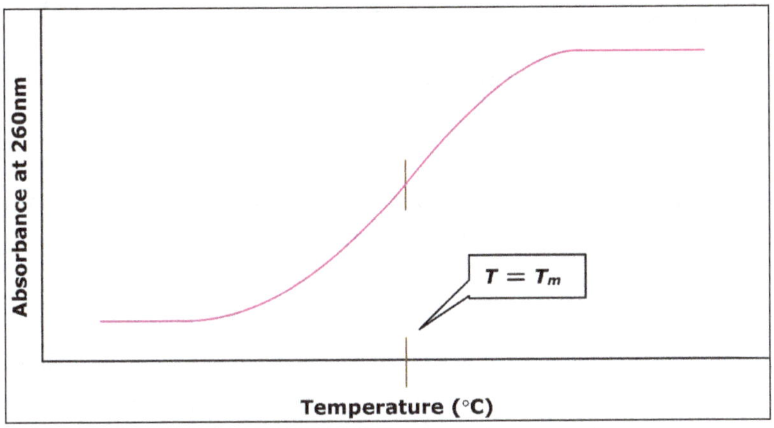

The absorbance of a solution of DNA as a function of temperature.

Study of Charge Transfer Complexes

The formation of charge-transfer complex occurs between molecules which, when mixed, allow the transfer of electronic charge through space from an electron rich molecule to an electron deficient molecule with molecular orbitals of suitable energy and symmetry.

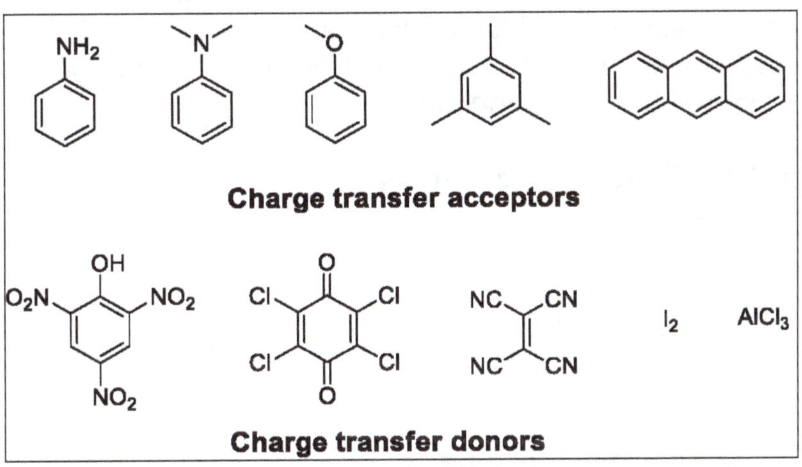

The filled π-orbitals in the donor molecule overlap with the depleted orbital in the acceptor molecule and generate two new molecular orbitals. Thus transition between these newly formed orbitals are responsible for the new absorption bands observed in the charge transfer complexes. The two new molecular orbitals formed and the transitions between them are shown in the figure given below.

The brown colour of iodine in benzene or the appearance of deep blue colour when tetracyanoethylene is added to a chloroform solution of aniline may be explained due to the formation of charge transfer complex. The structure of most charge transfer complexes can be visualized as a face to face association on a 1:1, donor: acceptor basis to provide maximum overlapping of π-orbitals of benzene ring.

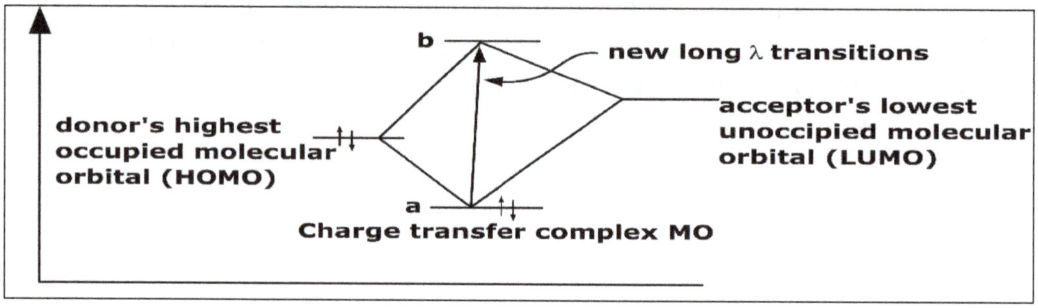

Electronic transitions for charge-transfer complexes. Donor and acceptor orbitals combine to form two new orbits (a and b) for the complex. New electronic transitions of long are then possible between a and b.

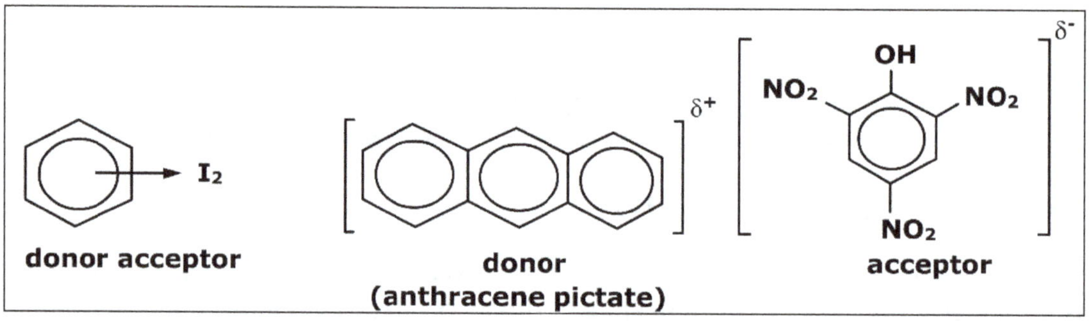

The λ_{max} of benzene is 255 nm while for iodine in hexane is 500 nm. The charge transfer complex (benzene-iodine) displays an intense additional band at 290 nm. Similarly in the anilinetetracyanoethylene complex, λ_{max} for aniline and tetracyanoethylene are 280 nm and 300 nm, respectively, while the deep blue complex has λ_{max} at 600 nm.

Preference over Two Tautomeric Forms

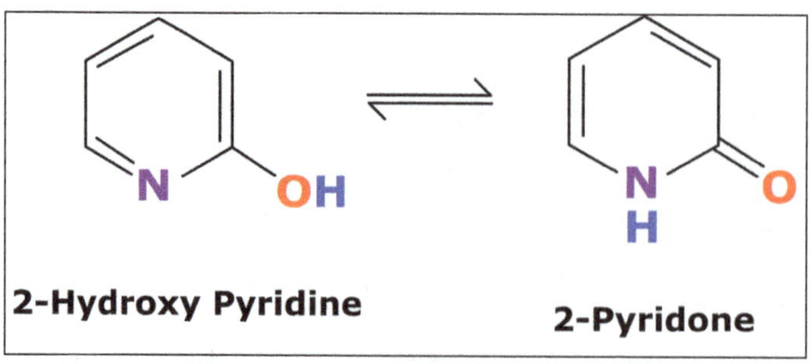

If a molecule has two tautomeric forms then the preference of one structure over the other can be detected by UV-Vis spectroscopy. For example, in case of tautomeric forms of 2-hydroxy pyridine and pyridone-2, the latter is preferred because it is a α, β unsaturated ketone and clearly the equilibrium is shifted towards the right hand side.

Determination of Configurations of Geometrical Isomers

The cis geometrical isomer suffers distortion and is forced to be non-coplanar and thus absorbs at lower wavelength. In comparison the Trans geometrical isomer suffers no such distortion and thus absorbs at higher wavelength.

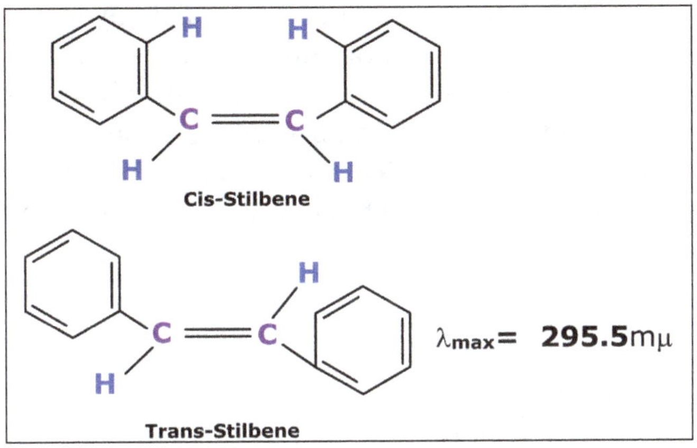

Distinction in Conjugated and Non-conjugated Compounds

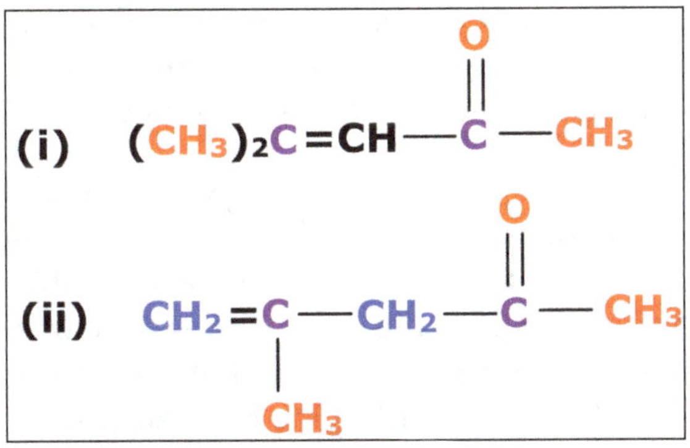

One can easily distinguish between conjugated and non-conjugated compounds by using UV-Vis spectroscopy very effectively. The n to π* band for the carbonyl group (i) will appear at a longer wavelength as it is α,β unsaturated ketone. In comparison to its absence in compound (ii), the same band appears at a shorter wavelength.

Differentiating between Equatorial and Axial Conformations

The n to π* (R band) which appears at longer wavelength in α,β unsaturated ketones is influenced by the presence of polar group in the γ-position. It has been noted that the effect of an axial substituent to displace the R band to longer wavelength is greater compared to that observed in its equatorial isomer.

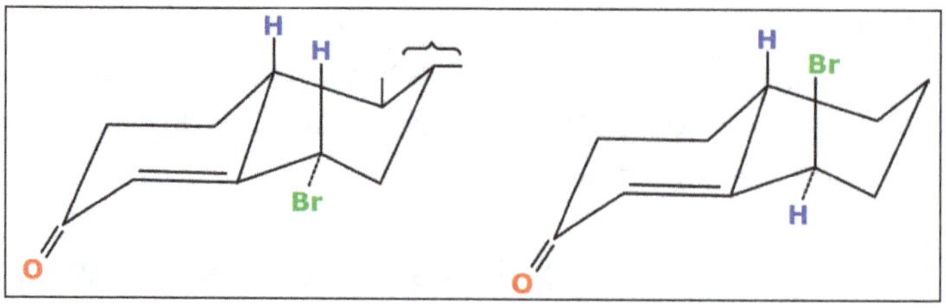

Other Uses

UV-visible spectroscopy is also used in the quality control in the development and production of dyeing reagents, inks and paints and the analysis of intermediate dyeing reagents. In environmental and agricultural fields the quantification of organic materials and heavy metals in fresh water can be carried out using UV-visible spectroscopy. A special type of UV-watermark is kept on many sensitive documents such as credit cards, driving licenses, passports to prevent forgery. The watermark can only be seen in UV light. The optical whiteners absorb ultraviolet light and re-emit it in the visible range. This feature is used in washing powders. Optical whiteners are also added to many toothpastes and detergent powders.

Infrared Spectroscopy

Infrared spectroscopy (IR spectroscopy) is the spectroscopy that deals with the infrared region of the electromagnetic spectrum that is light with a longer wavelength and lower frequency than visible light. It covers a range of techniques, mostly based on absorption spectroscopy. As with all spectroscopic techniques, it can be used to identify and study chemicals. A common laboratory instrument that uses this technique is a Fourier transform infrared (FTIR) spectrometer.

The infrared portion of the electromagnetic spectrum is usually divided into three regions; the near-, mid- and far- infrared, named for their relation to the visible spectrum. The higher energy near-IR, approximately 14000-4000 cm^{-1} (0.8-2.5 μm wavelength) can excite overtone or harmonic vibrations. The mid-infrared, approximately 4000-400 cm^{-1} (2.5-25 μm) may be used to study the fundamental vibrations and associated rotational-vibrational structure. The far-infrared, approximately 400-10 cm^{-1} (25-1000 μm), lying adjacent to the microwave region, has low energy and may be used for rotational spectroscopy. The names and classifications of these sub-regions are conventions, and are only loosely based on the relative molecular or electromagnetic properties.

Infrared spectroscopy exploits the fact that molecules absorb specific frequencies that are characteristic of their structure. These absorptions are resonant frequencies, i.e. the frequency of the absorbed radiation matches the frequency of the bond or group that vibrates. The energies are determined by the shape of the molecular potential energy surfaces, the masses of the atoms, and the associated vibronic coupling.

In particular, in the Born-Oppenheimer and harmonic approximations, i.e. when the molecular Hamiltonian corresponding to the electronic ground state can be approximated by a harmonic oscillator in the neighborhood of the equilibrium molecular geometry, the resonant frequencies are determined by the normal modes corresponding to the molecular electronic ground state potential energy surface. Nevertheless, the resonant frequencies can be in a first approach related to the strength of the bond, and the mass of the atoms at either end of it. Thus, the frequency of the vibrations can be associated with a particular bond type.

In order for a vibrational mode in a molecule to be "IR active," it must be associated with changes in the permanent dipole. A molecule can vibrate in many ways, and each way is called a vibrational mode. Linear molecules have 3N - 5 degrees of vibrational modes whereas nonlinear molecules have 3N - 6 degrees of vibrational modes (also called vibrational degrees of freedom). As an example H_2O, a non-linear molecule, will have 3 × 3 - 6 = 3 degrees of vibrational freedom, or modes.

Simple diatomic molecules have only one bond and only one vibrational band. If the molecule is symmetrical, e.g. N_2, the band is not observed in the IR spectrum, but only in the Raman spectrum. Unsymmetrical diatomic molecules, e.g. CO, absorb in the IR spectrum. More complex molecules have many bonds, and their vibrational spectra are correspondingly more complex, i.e. big molecules have many peaks in their IR spectra.

Zones of the IR region

Three main zones can be identified in the IR region:

- The far-IR (FIR, 400–10 cm−1, 25–300 μm).
- The mid-IR (MIR, 4000–400 cm−1, 2.5–25 μm).
- The near-IR (NIR, 14,000–4000 cm−1, 0.7–2.5 μm).

Ranges given between parentheses are identified on the wavenumber (ῦ) and wavelength (λ) scales, respectively, Figure below. It is noteworthy to mention that different schemes for the IR division exist depending on the application. For example, a sensor response division scheme classifies the IR region into five zones instead of the three shown above depending on the detector's sensitivity. Some classifications add a region of long-wave IR (LWIR) or thermal IR (TIR: 8–15 μm).

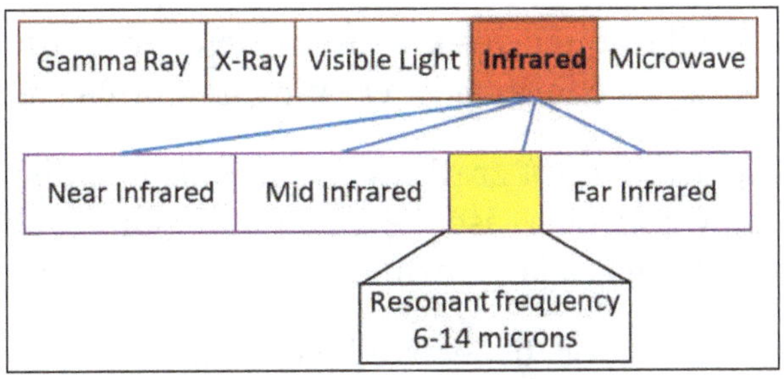

Section from the electromagnetic spectrum showing regions of IR. The diagram is replicated.

The MIR region can be further divided into the fingerprint region (400–1400 cm^{-1}) and the functional groups' region (1400–4000 cm^{-1}). For that reason, the MIR region is the most commonly used where most compounds would have a signature absorption/emission in this region.

Absorption in the NIR region is an outcome of two major processes: molecular overtones and combinations. For simplification, an overtone is a frequency that is higher than the fundamental frequency (lowest frequency). Both frequencies (overtone + fundamental) are known as harmonic partials. In simpler words, each fundamental frequency will generate a succession of absorptions at integer multiples of this frequency, and a molecule experiencing an overtone is the one getting excited from the ground state to the second excited state. Consequently, NIR absorptions need higher energy than a fundamental, and few overtones can be observed in this region. On the contrary to overtones arising from a group of fundamental absorptions, combinations, appearing because of the allocation of energy between two or more fundamental MIR bands that get excited simultaneously, are largely seen. Therefore, the NIR region is basically a region of broad peaks with limited use (except for bulk materials keeping in mind that NIR can penetrate much further compared to MIR). Thus, application of multivariate calibration techniques would resolve the complexity of the spectra and help extracting the needed information.

Instrumentations and Sampling Techniques

In general, this field has seen a major progress especially after grating was first introduced in 1823, and after the first commercial IR spectrometer came to the scene. The conventional IR spectrophotometer, first introduced in the 1940s, was a dispersive instrument. Rudimentary parts of this instrument were a radiation source, a monochromator, and a detector, most frequently in a double-beam setup. Monochromator, the dispersive device, serves to separate the broad spectrum of IR radiations into a continuous sequence of IR bands with resolved frequencies. These instruments work by tracing only a sole frequency at a time.

Consequently, a whole spectrum needs a long time to be recorded. It was in the 1960s when Fourier transform (FT) instruments came to the scene. The key difference between FT and dispersive instruments was the presence of interferometers. The output of the FT-IR instrument is then called an interferogram. FT-IR instruments, though were intended to extend the use of IR, had limited applications and were used only for advanced research. This was mainly because of the expensive component electronics and the need for supercomputers to record the generated data. However, after the microelectronic revolution, the capabilities and availability of these instruments have been greatly improved. In addition to the noticeable speed in acquiring spectra, improved signal to noise ratio, high resolution, accuracy, and reproducibility, FT-IR offers two major pluses over the classical dispersive instruments:

- Multiplex plus: Where all frequencies fall on the detector in unison. Therefore, each resolution component is grasped continually, creating a multifaceted spectrum.
- Throughput plus: The presence of gratings and prisms in a dispersive instrument and the need for an ingress slit would decrease the amount of light reaching the detector. Quite the reverse, an interferometer has a large orifice that in turn increases the output.

Samples to be measured with FT-IR or dispersive instruments (transmission techniques) need a prior treatment. A solid sample is mixed with potassium bromide, packed together into a compressed disc, and inserted into the pathway of radiation. Liquid samples are placed in special holders or mounted on the surface of a KBR disc. Other sampling techniques, e.g., reflectance techniques, however, permit IR radiations to be applied on a larger assortment of sample forms without additional treatment. As their name implies, radiations in these techniques are reflected from the sample surface rather than being transmitted through it. Reflections might be total internal, specular, or diffuse. Details on speculations of each technique can be summarized as follows and as shown in figure below.

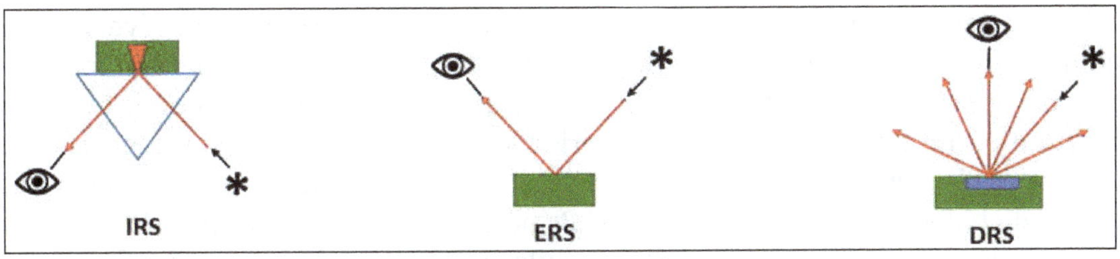

FT-IR reflection modes.

- Total reflection IR or attenuated total reflectance IR (ATR-IR): Where light undergoes several internal reflections when passed through an ATR crystal of high refractive index (RI), which in turn is in contact with the sample. The resultant evanescent wave spreads to the sample and infiltrates to a depth that is dependent on a variety of controls such as light wavelength, RI of the crystal and the medium being examined, and incidence angle. An ATR crystal might be made of Diamond, Zinc Selenide (ZnSe), Germanium (Ge), or Silicon (Si). The resulting spectrum is analogous to that produced by dispersive instruments especially for thin films. ATR-IR is suitable for a variety of sample forms with no or minimal need for sample preparation. Yet, ATR-IR is less sensitive compared to transmission-based techniques.

- Specular reflection IR spectroscopy (SRS): Also known as external reflection IR occurs when light is reflected from a specular surface (mirror-like) at a well-defined angle that is equal to the angle of incidence of IR radiation. Analogous to ATR-IR, thin films' reflectance spectra are like the transmission spectra and are identified as a reflection-absorption mode. For ultra-thin films (monomolecular layers), however, using a grazing incidence mode is needed where the path of radiation becomes parallel to the metal surface augmenting the absorption intensity. In general, coupling of an IR spectrophotometer with an optical microscope is known as FT-IR microscopy. Collection of the specular reflection spectrum is performed using such a setup and is commonly known as micro FT-IR.

- Diffuse reflectance spectroscopy (DRS): This is a combination between internal and external reflections and originates from rough surfaces (powders) reflecting light in different directions. Having diffuse reflectance exploited in IR FT spectroscopy is known as DRIFTs.

Data Treatment in IR Spectroscopy

Chemical data encompassed in the MIR spectra exist in peak positions, peak intensities, and peak shape. Therefore, while spectra from MIR can be easily interpreted and information can be easily extracted, the spectra obtained from NIR, have the absorption bands from the overtones and combinations of fundamental MIR bands and therefore need special treatment to resolve this overlapping. Analysis of data from NIR implementing multivariate data analysis or chemometrics would be a suitable approach. With the advances in data sciences, coupling of IR spectroscopy to chemometrics serves to combine the advantages of both in terms of time, effort, and multicomponent analysis. Application of a certain chemometric method in NIR data analysis would depend greatly on whether the performed analysis is qualitative or quantitative. In general, three approaches are used to examine data from NIR:

- Arithmetic data pre-processing: This approach helps diminution of the impact of side data and keeps the focus on the main data in a spectrum. Normalization, derivatization, and smoothing are common techniques.

- Classification techniques: Mainly used for data obtained from qualitative analysis where samples are congregated based on their spectra. Classification might be unsupervised where only the spectra are known for the researcher and no other information is available, or supervised pattern recognition where a prior knowledge of samples' categories exists, and spectroscopists would create a classification model then evaluate this model by comparing the predicted values to the actual ones. Principle component analysis (PCA) is a common unsupervised classification approach, while partial least squares discriminant analysis (PLSDA) is a classical supervised classification technique.
- Regression methods: Used mainly in quantitative analysis. Common approaches include principle component regression (PCR), multi-linear regression (MLR), partial-least squares (PLS), artificial neural network (ANN), and support vector machines (SVMs).

The Position of an IR-absorption Band

The amount of energy required to stretch a bond depends on the strength of the bond and the masses of the bonded atoms. The stronger the bond, the greater the energy required to stretch it. The frequency of the vibration is inversely proportional to the mass of the atoms, so heavier atoms vibrate at lower frequencies. The approximate vibration frequency of a bond can be calculated from the following equation derived from Hooke's law.

$$\bar{v} = \frac{1}{2\pi c}\sqrt{\frac{K}{\mu}}$$

Where,

- μ is the reduced mass.
- $\mu = m_1 \times m_2 / m_1 + m_2$
- m_1 and m_2 are the masses of the two atoms.
- C is the velocity of light (2.998×10^{10} cms^{-1}).
- K = force constant of the bond.

[For a single C-C bond, it is approximately 5×10^5 gs^{-2} and it doubles and triples for double and triple bond respectively.]

The force constant is a measure of the strength of a bond. Thus the value of stretching frequency of a bond increases with increasing the bond strength and decreasing the reduced mass of the bonded atoms. The triple bonds are stronger than the double bonds followed by single bonds between the same two atoms and hence vibrational frequency is in the order of triple>double>single.

$$\begin{array}{ccc} C\equiv C & C=C & C-C \\ 2150\ cm^{-1} & 1650\ cm^{-1} & 1200\ cm^{-1} \end{array}$$

$$\xleftarrow{\text{Increasing K}}$$

As the atom attached to carbon a particular atom say carbon increases in mass, the reduced mass increases and hence the frequency of vibration decreases.

C–H	C–C	C–O	C–Cl	C–Br	C–I
3000 cm^{-1}	1200 cm^{-1}	1100 cm^{-1}	750 cm^{-1}	600 cm^{-1}	500 cm^{-1}

$\xrightarrow{\text{Increasing } \mu}$

The Intensity of Absorption Bands

The intensity of an absorption band depends on the change in the dipole moment of the bond associated with the vibration. The greater the change in dipole moment, the more intense is the absorption. The dipole moment of a bond is equal to the magnitude of the charge on one of the bonded atoms, multiplied by the distance between the two charges. If we have two different atoms, there will be an electro-negativity difference and larger the electro-negativity difference, the more intense the absorption will be. The stretching vibration of an O-H bond will be associated with a greater change in dipole moment than that of an N-H bond because O-H bond is more polar. Consequently, the stretching vibration of the O-H bond will be more intense. Likewise, the stretching vibration of an N-H bond is more intense than that of a C-H bond because the N-H bond is more polar.

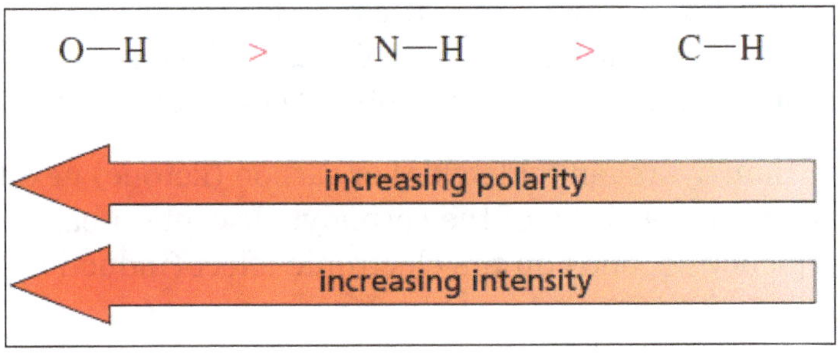

The intensity of an absorption band also depends on the number of bonds responsible for the absorption. For example, if you compare the IR spectra of methane to that of octane, the octane molecule will have a much more intense C-H peak because it has many more C-H bonds than methane. The concentration of the sample used to obtain an IR spectrum also affects the intensity of the absorption bands. Concentrated samples have greater numbers of absorbing molecules and, therefore, more intense absorption bands.

The Shape of Absorption Bands

The shape of an absorption band can be helpful in identifying the compound responsible for an IR spectrum. For example, both O-H and N-H bonds stretch at wave numbers above 3100 cm^{-1} but the shapes of their bands are distinct. An N-H absorption band (3300 cm^{-1}) is comparatively narrower and less intense than an O-H absorption band in an alcohol and the O-H absorption band of a carboxylic acid (3300-2500 cm^{-1}) is broader than the O-H absorption band of an alcohol.

Factors Influencing Vibrational Frequencies

The value of vibrational frequency of a bond calculated by Hooke's Law is not always equal to their observed value. The force constant is changed with the electronic and steric effects caused by other groups present in the surroundings. Following are some important factors affecting the vibrational frequency of a bond.

Effect of Bond Order

Bond order affects the position of absorption bands. Higher the bond order larger is the band frequency. A C-C triple bond is stronger than a C=C bond, so a C-C triple bond has higher stretching frequency than does a C = C bond. The C-C bonds show stretching vibrations in the region from 1200-800 cm^{-1} but these vibrations are weak and of little value in identifying compounds. Similarly, a C = O bond stretches at a higher frequency than does a C-O bond and a C-N triple bond stretches at a higher frequency than does a C = N bond which in turn stretches at a higher frequency than does a C-N bond.

Resonance and Inductive Electronic Effects

Whenever we discuss about the IR band, a range of frequency for each stretch is usually assigned as the exact position of the absorption band depends on other structural features of the molecule, such as electron delocalization, the electronic effect of neighbouring substituents, and hydrogen bonding. Important details about the structure of a compound can be revealed by the exact position of the absorption band. For example, the IR bands for the carbonyl group in 2-pentanone, 2-cyclohexenone appears at different frequencies. The 2-Cyclohexenone absorbs at a lower frequency because the carbonyl group has less double-bond character due to electron delocalization.

If you put an atom other than hydrogen (aldehyde) or carbon (ketone) next to the carbonyl group, it also causes the shifting of the position of the carbonyl absorption band. Whether it shifts to a lower or to a higher frequency depends on the electronic effect (inductive and mesomeric effect) of the atom.

Different substituents on carbonyl carbon changes the electronegativity of carbonyl group due to the inductive effect arises due to the different electro-negativities of the carbonyl carbon and of the substituent in compounds of the type RCOZ. It involves the electrons in the sigma bonds.

The mesomeric effect involves the electrons in the pie and nonbonding orbitals and it operates in general opposite to the inductive effect. These effects cannot be isolated from one another and at a time only one of them can be applied to determine approximate results.

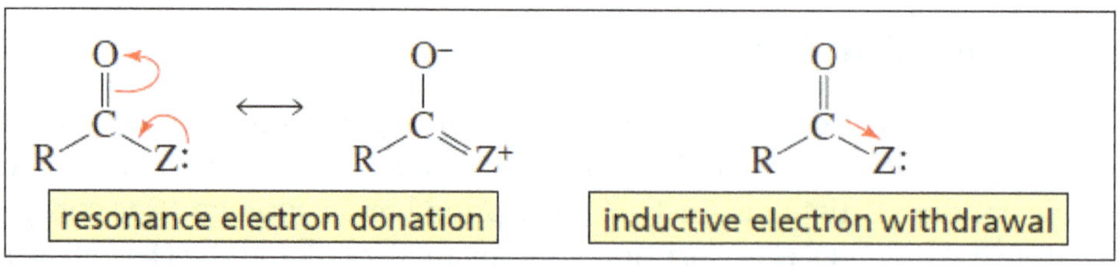

Electron releasing groups attached to the carbonyl group tend to favour the polar contribution by mesomeric effect and thus lower the bond order of the C=O bond (less double bond character) and hence resulting in a decrease of the carbonyl stretching frequency. Electron withdrawing groups suppress the polar contribution with an effective increase in the double bond character and hence resulting in the increase of the frequency of absorption.

The predominant effect of the nitrogen of an amide is electron donation by resonance. Therefore, the carbonyl group of an amide has less double bond character than the carbonyl group of a ketone and hence amide has low C=O bond IR frequency than the ketone. In contrast, the predominant effect of the oxygen of an ester is inductive electron withdrawal; therefore the carbonyl group of an ester has more double bond character than the carbonyl group of a ketone, so the former appears at larger frequency value.

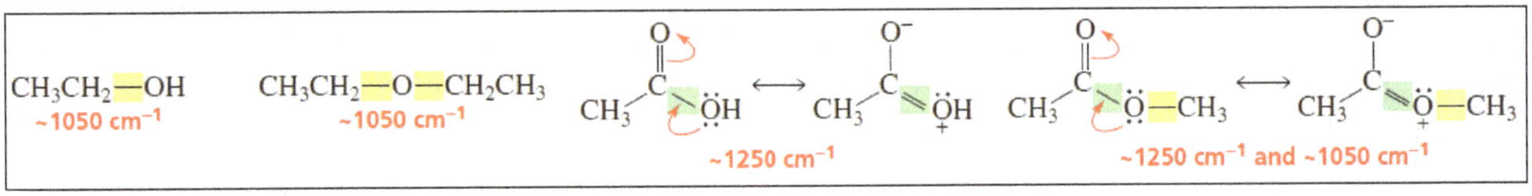

If we talk about the C-O stretching frequency it appears in the range of 1250-1050 cm^{-1}. If C-O bond is in an alcohol or in ether, the stretching frequency will be toward the lower end of the range. If this C-O bond is a part of a carboxylic group, the stretching frequency will occur at the higher end of the range. The position of the absorption varies because the bond in an alcohol is a pure single bond, whereas the bond in a carboxylic acid has partial double bond character that is due to resonance electron donation by OH group in acids. Esters show stretches at both ends of the range because esters have two single bonds, one that is a pure single bond and one that has partial double-bond character.

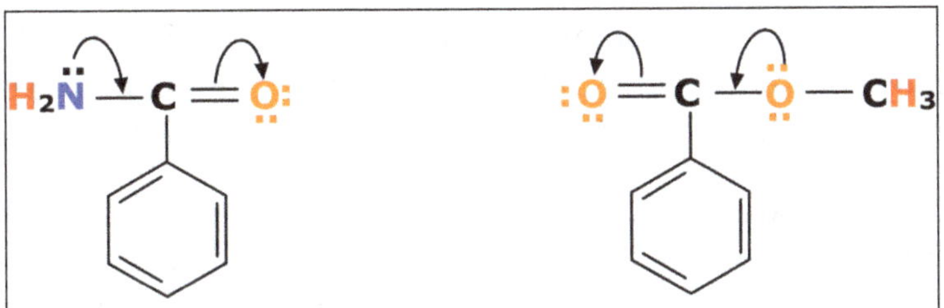

Benzamide ν(C = O) 1693 cm^{-1} | **Phenyl Acetate** ν(C = O) 1730 cm^{-1}

In a specific example of benzamide and phenyl acetate, as the nitrogen atom is less electronegative than the oxygen atom, in amide and is more labile and participates to a greater extent in conjugation. The presence of greater degree of conjugation, C=O absorption frequency is relatively lesser than in esters.

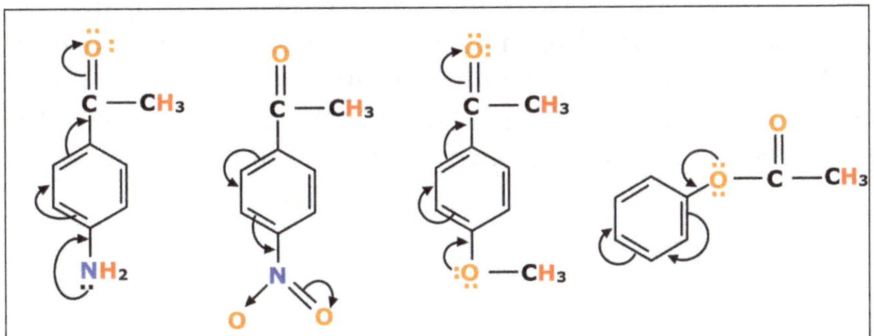

I p-Amino Acetophenone ν(C = O) 1677 cm^{-1} | **II** p-Nitro Acetophenone ν(C = O) 1700 cm^{-1} | **III** p-Methoxy Acetophenone ν(C = O) 1684 cm^{-1} | **IV** Phenyl Acetate ν(C = O) 1770 cm^{-1}

In most of the cases, mesomeric effect works along with inductive effect and cannot be ignored. It is important to note that only inductive effect is considered when the compound in meta substituted. In para-substitution, both inductive and mesomeric effects become important and the domination of one over the other will decide the wavenumber of the absorption.

In the above examples, due to the low electronegativity of nitrogen atom, the lone pair of electrons participates more in conjugation in compound I as compared to that compound III. Thus, in compound I, ν(C=O) absorption occurs at lower wave number compared to that in compound III. In compounds II and IV, inductive effect dominates over mesomeric effect and hence absorption takes place at comparatively higher frequencies.

In ortho-substitution, inductive effect, mesomeric effect along with steric effect is considered. In ortho substituted compounds, the lone pairs of electrons on two atoms influence each other through space interactions and change the vibrational frequencies of both the groups. This effect is called field effect.

Hydrogen Bonding

The presence of hydrogen bonding changes the position and shape of an infrared absorption band. Frequencies of both stretching as well as bending vibrations are changed because of hydrogen bonding. The X-H stretching bands move to lower frequency usually with increased intensity and band widening. The X-H bending vibration usually shifts to higher frequencies. Stronger the hydrogen bonding, greater is the absorption shift from the normal values. The two types of hydrogen bonding (intra-molecular and intermolecular) can be differentiated by the use of infrared spectroscopy.

Inter-molecular hydrogen bonding involves association of two or more molecules of the same or different compound, and it may result in dimer molecules as in carboxylic acids. Intra-molecular hydrogen bonds are formed when the proton donor and acceptor are present in a single molecule under special conditions that allow the required overlap of orbitals, e.g. formation of a 5 or 6 member rings.

The extent of inter-molecular hydrogen bonding depends upon the concentration of the solution and hence the position and the shape of an absorption band also depend on the concentration of the solution. The more concentrated the solution, the more likely it is for the OH-containing molecules to form intermolecular hydrogen bonds. It is easier to stretch an O-H bond if it is hydrogen bonded, because the hydrogen is attracted to the oxygen of neighbouring molecule. Therefore, the O-H stretching of a concentrated (hydrogen bonded) solution of an alcohol occurs at about 3550 cm^{-1}, whereas the O-H stretching band of a dilute solution (with little or no hydrogen bonding) appears at 3650 cm^{-1}. Additionally, hydrogen-bonded OH groups also have broader absorption bands whereas the absorption bands of non-hydrogen–bonded OH groups are sharper.

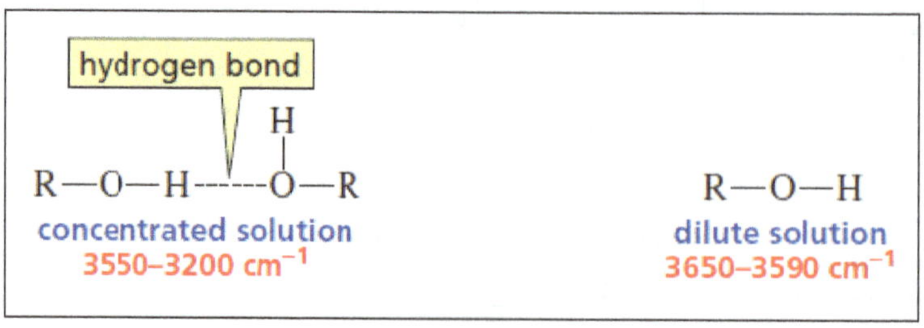

Hydrogen bonding interactions can also occur between functional groups of solvent and solute. Three classes of solvent exist which may lead to trouble when used for hydrogen-bonding studies is solution. First, compounds which contain hydrogen donor groups e.g. halogenated compounds which contain a sufficient number of halogens to activate the hydrogens present, such as chloroform. Secondly, compounds which contain non-bonded electron pairs, such as ethers, aldehydes and tertiary amines.

In addition, there are compounds which contain both types of group, e.g. water and alcohols. There are only a few solvents that do not have the above characteristics, such as carbon tetrachloride (CCl_4) and carbon disulfide (CS_2). These still contain lone electron pairs, but being on S and Cl are less available, and any interactions will be extremely weak. Apart from the concentration and solvent effects, temperature also affects the degree of hydrogen bonding in a compound. High temperature decreases the extent of hydrogen bonding.

Fermi Resonance

Fermi resonance is a phenomenon which was first explained by the Italian physicist Enrico Fermi to account for shifting of the energies and intensities of absorption bands in an infrared spectroscopy. Sometimes, it happens that two different vibrational levels have nearly the same energy. If such vibrations belong to the same species of molecules, then a mutual perturbation of energy may occur, resulting in the shift of one towards lower frequency and the other towards a higher frequency. It is also accompanied by a substantial increase in the intensity of the respective bands.

Fermi resonance was first discovered in carbon dioxide by Fermi and it is also found in the vibrational spectra of aldehydes, where the C-H bond in the CHO group interacts with the second harmonic level, derived from the fundamental frequency of the deformation vibration of the CHO group (2*1400 cm^{-1}). The result is a Fermi doublet with branches around 2830 cm-1 and 2730 cm^{1}. It is important for Fermi resonance that the vibrations connected with the two interacting levels be localized in the same part of the molecule.

Bond Angles

Smaller ring requires the use of more p-character to make the internal C-C bonds for the requisite small angles. This gives more s character to the C=O sigma bond which causes the strengthening and stiffening of the exocyclic double bond. The force constant K is then increased and the absorption frequency increases.

Characteristic Absorption Bands for Hydrocarbons

These compounds have carbon and hydrogens only and may be saturated, unsaturated, aliphatics, cyclics and aromatics. The characteristic vibrational modes of aliphatic and aromatic hydrocarbons and their absorption frequencies are described here:

Aliphatic Hydrocarbons

The stretching and bending vibrations frequencies of alkanes, alkenes and alkynes are described in table.

Table: Characteristic frequencies of absorption bands for aliphatic hydrocarbons.

Wavenumber (cm^{-1})	Assignment
	Alkanes
2960	Methyl symmetric C-H stretching
2930	Methylene asymmetric C-H stretching
2870	Methyl asymmetric C-H stretching
2850	Methylene symmetric C-H stretching
1470	Methyl asymmetrical C-H bending
1465	Methylene scissoring
1380	Methyl symmetrical C-H bending
1305	Methylene wagging
1300	Methylene twisting
720	Methylene rocking
	Alkenes
3100-3000	=C–H stretching
1680-1600	C=C stretching
1400	=C–H in-plane bending
1000-600	=C–H out-of-plane bending
	Alkynes
3300-3250	=C-H stretching
2260-2100	C=C stretching
700-600	=C–H bending

Alkanes

In alkanes only two types of bonds are present that is C–H and C–C bonds. The most useful bands in alkanes are those arising from C–H stretching and C–H bending. A methyl group usually shows two strong and distinct bands, one at 2960 cm-1 due to asymmetric stretching and the other at 2870 cm^{-1} due to symmetric stretching. While the methylene group show asymmetric stretching at 2930 cm^{-1} and symmetric stretching at 2850 cm^{-1}. The C-H bonds do not participate in Hbonding and hence the frequencies are not affected by chemical environment. The presence of several methyl groups in a molecule results in strong absorptions at these positions.

The C-H in-plane bending vibrations of a methyl group give two bands, i.e. a symmetrical band at 1380 cm^{-1} and an asymmetrical band at 1460 cm^{-1}. Methylene groups give rise to four bending vibrations: scissoring (1465 cm^{-1}), rocking (720 cm^{-1}), wagging (1305 cm^{-1}) and twisting (1300 cm^{-1}). The band at 1460 cm^{-1} indicates the C-H asymmetric bending of both methyl and methylene groups whereas the band at 1380 cm^{-1} represents symmetric bending of methyl group only. Thus the peak at 1380 cm^{-1} is extremely useful in detecting the presence of methyl group because this band is hardly overlapped to the absorptions due to the methylene and methane absorption bands.

The presence of two or more methyl groups, attached to the same carbon atom results in splitting of the C-H symmetric bending vibrations. For example the isopropyl group gives rise to two bands 1397 cm^{-1} and 1370 cm^{-1} and in t-butyl group the two bands appear at 1385 cm^{-1} and 1370 cm^{-1}. The methine C-H shows C-H stretching and bending vibrations at 2890 cm^{-1} and 1340 cm^{-1} respectively and are very weak with no practical significance.

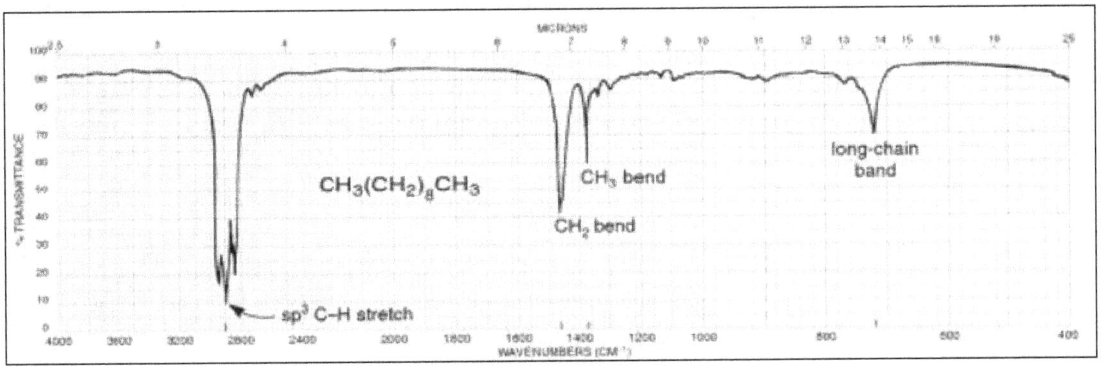

IR-spectrum of decane.

Cycloalkanes

In case of unstrained cycloalkanes, the C-H stretching frequencies are the same (2800 – 3000 cm^{-1}) as in the case of acyclic compounds. However, methylene scissoring bands shift slightly to smaller wavenumber (1470 cm^{-1} in hexane and 1448 cm^{-1} in cyclohexane). In sterically strained cyclic compounds, the C-H stretching normally occurs at slightly higher frequencies. For example the ring CH_2 and CH groups in a monoalkyl cylcopropane ring absorb in the range of 3100-2990 cm^{-1}.

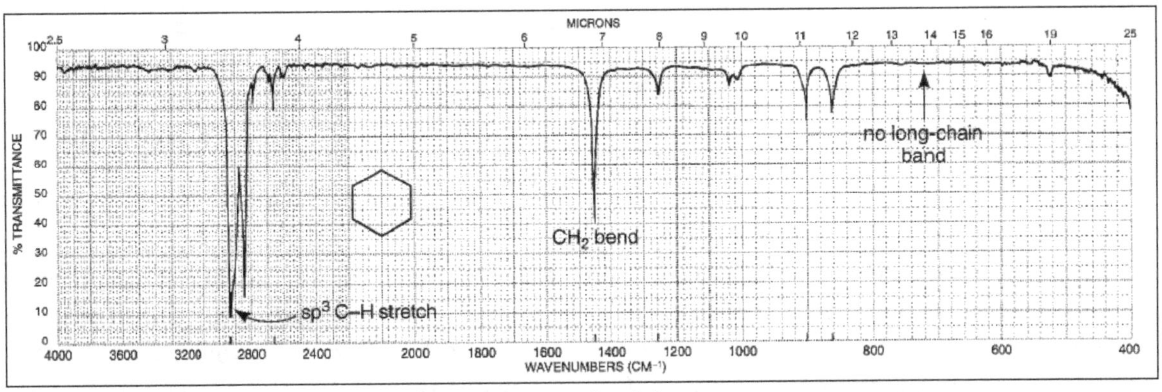

IR-spectrum of cyclohexane.

Methylene C-H in-plane bending vibrations in cyclic alkenes shift slightly to smaller frequency as compared to their acyclic counterparts. For example, cyclohexane absorb at 1452 cm^{-1} whereas nhexane absorb at 1468 cm^{-1}. The C-C bond vibrations appear as weak bands in 1200-800 cm^{-1} region and are seldom used for structural study. Whereas the C-C bending absorptions occurs at $<$ 500 cm^{-1} and are usually below the range of IR instrument.

Alkenes

There are four major bands associated with alkene molecular framework. These are the out-of-plane and in-plane =C–H bending, C=C stretching and =C–H stretching. In completely symmetrical alkenes, such as ethylene, tetrachloroethylene etc., C=C stretching band is absent, due to lack of change in dipole moment in completely symmetrical molecule. On the other hand, non-symmetrically substituted double bonds exhibit strong absorption bands.

Olefinic C-H stretching occurs near 3100-3000 cm^{-1}. C=C stretching band occurs near 1680-1600 cm^{-1}. The C = C absorption bands are more intense for cis isomers than for trans isomers. Also, terminal olefins show stronger C = C double bond stretching vibrations than internal double bonds. Conjugation moves C = C stretching bands towards lower frequency with increased intensity than in comparison to non-conjugated ones. In case of olefins, conjugated with an aromatic ring, the C = C stretching appears at 1625 cm^{-1} (s) and an additional band at ~1600 cm^{-1} is observed due to aromatic double bond.

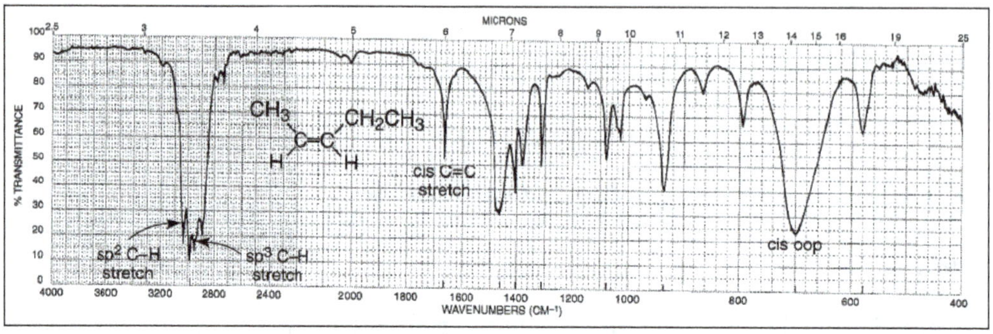

IR-spectrum of Cis-2-pentene.

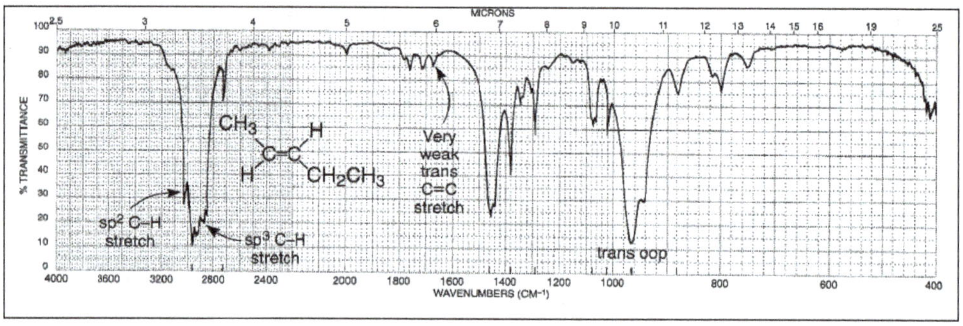

IR-spectrum of trans-2-pentene.

The =C-H in-plane bending vibrations occur near 1400 cm^{-1} whereas the =C-H out-of-plane bending band occur in the range from 1000-680 cm^{-1}. Note: In compounds containing both olefinic and alkyl C-H bonds, the bands above 3000 cm^{-1} are generally attributed to aromatic or aliphatic C-H stretching, whereas between 3000-2840 cm^{-1} are generally assigned to the alkyl C-H stretching.

Cycloalkenes

The absorption frequency of a double bond in a cyclic ring is very sensitive to ring size. The absorption frequency decreases as the internal bond angle decreases and is becomes lowest in cyclobutene (90° angle) but after that the frequency increases again for cyclopropene.

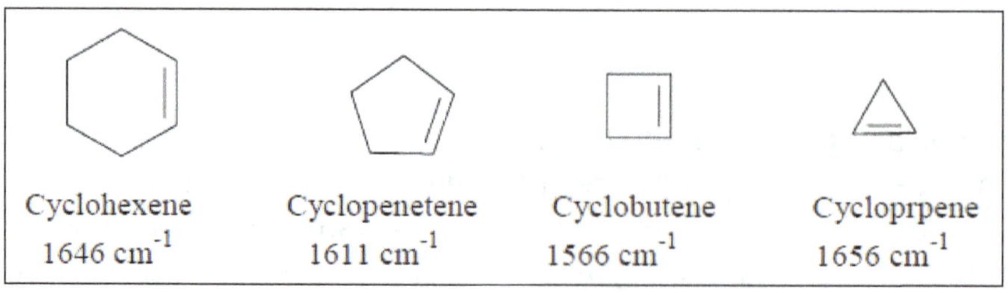

Vibrational frequencies of endocyclic C=C bonds.

The exocyclic double bonds exhibit an increase in frequency with decrease in ring size. The exocyclic double bond on six-membered ring absorbs at 1651 cm^{-1} and it is shifted to 1780 cm^{-1} in case of exocyclic double bond on cyclopropane. The allenes show the highest double bond absorptions at 1940 cm^{-1}

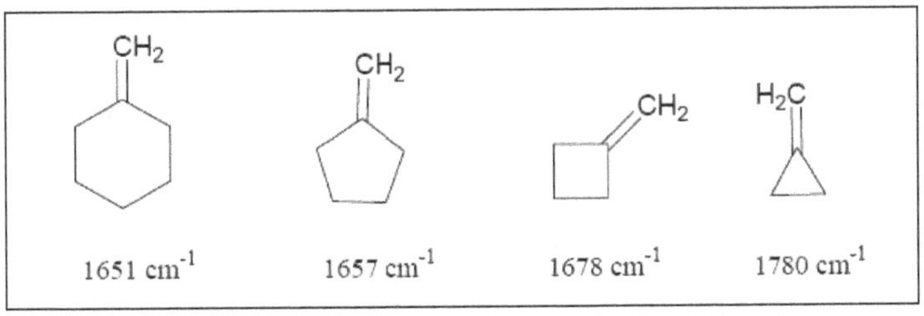

Vibrational frequencies of exocyclic C = C bonds.

Six-membered rings with carbonyl group e.g. cyclohexanone absorb at normal value i.e. 1715 cm^1. Decrease in ring size increases the C = O stretching frequency. Smaller rings require the use of more p- character to make C-C bonds for the requisite small angles.

Alkynes

Alkynes contain the C≡C group and three characteristic bands can be present, i.e. ≡C–H stretching, ≡C–H bending and C≡C stretching. As the force constant for a triple bond is greater than that for a double bond. Consequently, whereas a C-C stretching vibrations occur between 1300-800 cm^{-1} and the C=C stretching vibration occur in the region 1700-1500 cm^{-1}, the C triple C vibrations are observed at significantly higher frequencies in the region of 2300 to 2050 cm^{-1}. The terminal alkynes show C-C triple bond stretching vibration at 2140-2050 cm^{-1}, whereas the symmetrical or unsymmetrical internal alkynes show either no band or show a very weak C-C triple bond absorption at 2260-2190 cm^{-1}.

The terminal alkynes show a strong and sharp C-H stretching vibration bands in the region 3310-3200 cm^{-1}. The sharpness of this band prevents form the confusion with other absorptions in this region (NH, OH etc.).

The C-H bending vibrations in alkynes occur as a strong broad band in the region 700-600 cm^{-1}. The first overtone of this band appears as a weak broad band near 1375-1225 cm^{-1}. Therefore the frequency of the absorption of C-H bond is a function of the type of hybridization of the carbon to which hydrogen atom is attached. While moving from sp^3 to sp^2 and sp hybridized carbons, the s-character increases and so is the bond strength (force constant of C-H bond and the frequency of absorption.

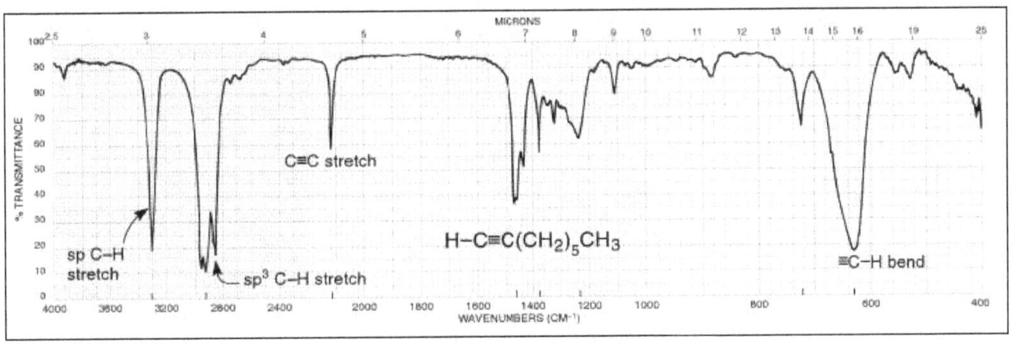

IR-spectrum of 1-octyne.

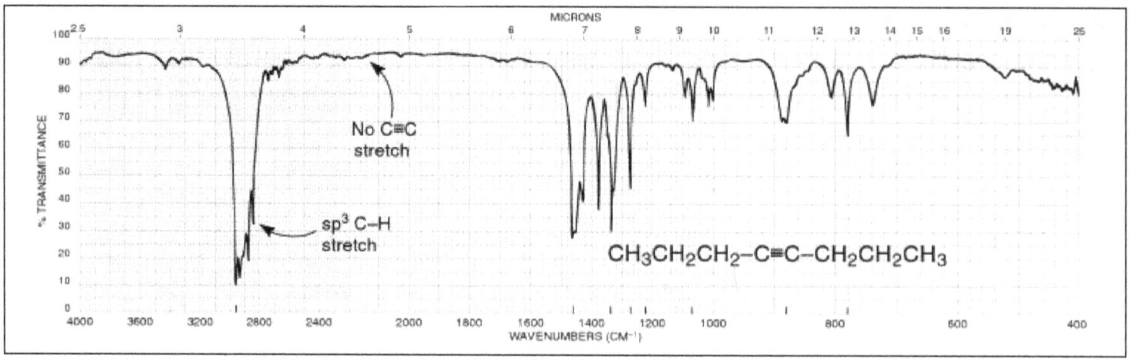

IR-spectrum of 4-octyne

Aromatic Hydrocarbons

The C–H stretching bands of aromatic compounds appear as a multiple weak bands in the 3100–3000 cm^{-1} range. It is important to note here that the aromatic C–H stretching bands occur at slightly higher frequency (as a shoulder on the stronger alkane C-H stretching bands) than is the C–H stretch in alkanes which appear below 3000 cm^{-1}. Besides the C–H stretch above 3000 cm^{-1}, two other regions of the infrared spectra of aromatic compounds distinguish aromatics from organic compounds that do not have an aromatic ring.

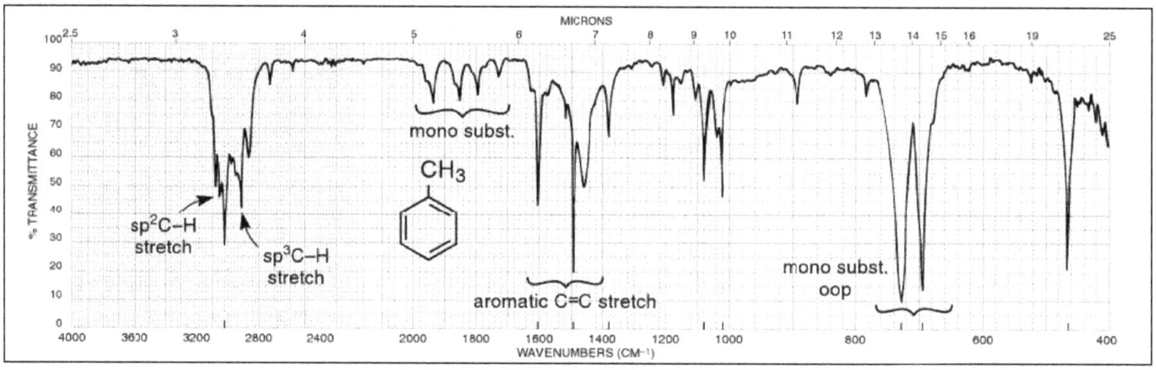

IR-spectrum of toluene.

In the region of 2000–1700 cm^{-1}, a series of weak combination and overtone bands appears. However, they are very weak and are observed only in the case of concentrated solutions of highly symmetric benzene derivatives.

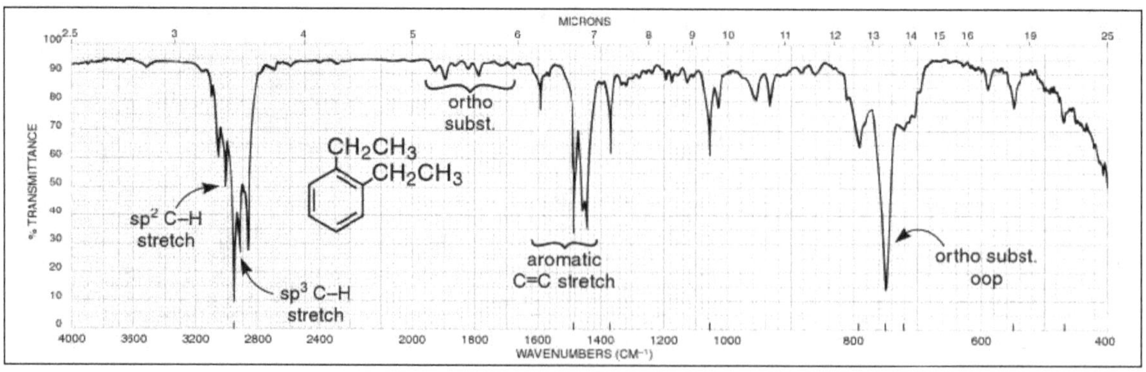

IR-spectrum of 1, 2-diethyl benzene.

The other most prominent bands are in the region of 900-650 cm^{-1} due to the out-of-plane bending of ring C-H bonds. The C–H in-plane bending vibrational bands appear in the regions 1275–1000 cm^{-1}. The bands of the out-of-plane bending vibrations of aromatic compounds are strong and characteristic of the number of hydrogens in the ring, and hence can be used to give the substitution pattern. Mono substituted benzene derivatives exhibit strong absorption bands in the regions 690-710 cm^{-1} and 730-770 cm^{-1}. The absence of band at 690 cm^{-1} indicates the absence of mono substituted phenyl rings in the sample. 1, 2-Disubstituted benzenes give one strong absorption band in the region 735-770 cm^{-1} but no band in the region 690-710 cm^{-1}. 1, 3- Di-substituted rings give two absorption bands in the region 680-725 cm-1 and 750-810 cm^{-1}. 1, 4- Di-substituted rings give only one strong absorption band in the region 800-850 cm^{-1}. The IR spectrums of all the possible di-substituted benzene derivatives are shown below:

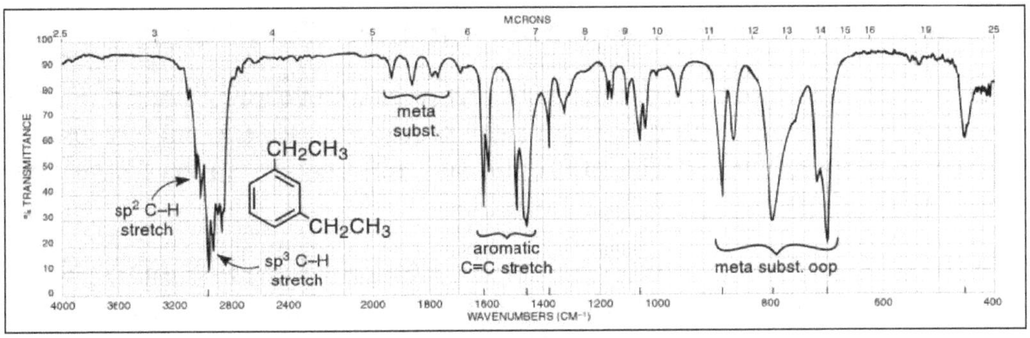

IR-spectrum of 1, 3-diethyl benzene.

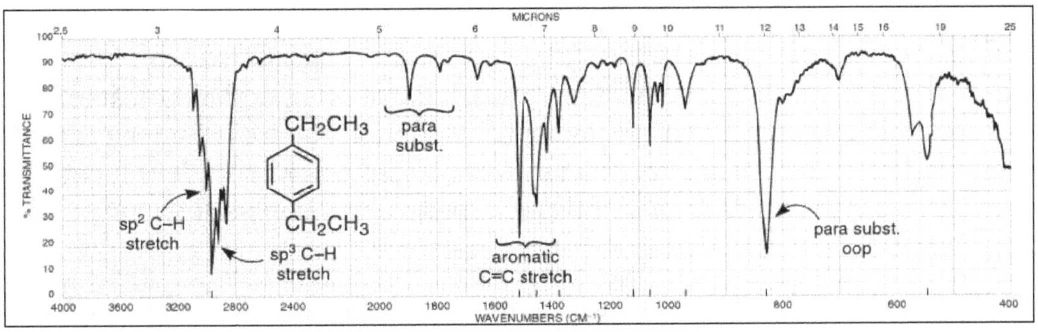

IR-spectrum of 1, 4-diethyl benzene.

The bands related to other substitution patterns are summarized in the figure given below:

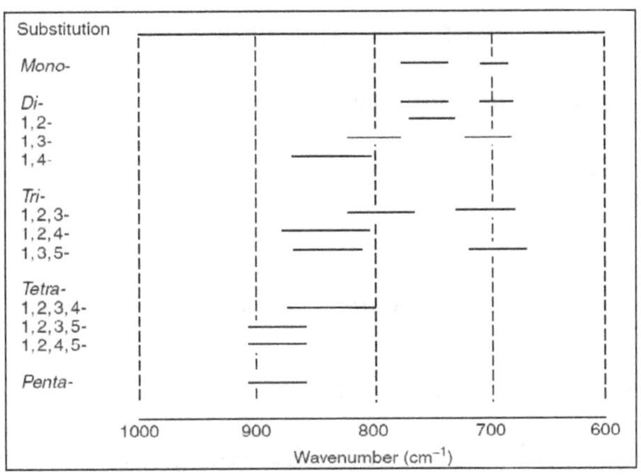

Out-of-plane bending vibrations of ring C-H bonds in substituted aromatic compounds.

Aromatic hydrocarbons show absorptions in the regions 1650-1400 cm^{-1} due to C=C stretching vibrations in the aromatic ring. There are normally four bands in this region at about 1600, 1585, 1500 and 1450 cm^{-1} and are due to C=C in-plane vibrations. The absence of these bands indicates that the compound is not aromatic.

Applications of IR Spectroscopy

Infrared spectroscopy is widely used in industry as well as in research. It is a simple and reliable technique for measurement, quality control and dynamic measurement. It is also employed in forensic analysis in civil and criminal analysis. Some of the major applications of IR spectroscopy are as follows:

Identification of Functional Group and Structure Elucidation

Entire IR region is divided into group frequency region and fingerprint region. Range of group frequency is 4000-1500 cm^{-1} while that of finger print region is 1500-400 cm^{-1}. In group frequency region, the peaks corresponding to different functional groups can be observed. According to corresponding peaks, functional group can be determined.

Each atom of the molecule is connected by bond and each bond requires different IR region so characteristic peaks are observed. This region of IR spectrum is called as finger print region of the molecule. It can be determined by characteristic peaks.

Identification of Substances

IR spectroscopy is used to establish whether a given sample of an organic substance is identical with another or not. This is because large number of absorption bands is observed in the IR spectra of organic molecules and the probability that any two compounds will produce identical spectra is almost zero. So if two compounds have identical IR spectra then both of them must be samples of the same substances.

IR spectra of two enatiomeric compound are identical. So IR spectroscopy fails to distinguish between enantiomers. For example, an IR spectrum of benzaldehyde is observed as follows.

- C-H stretching of aromatic ring- 3080 cm^{-1}.
- C-H stretching of aldehyde- 2860 cm-1 and 2775 cm^{-1}.
- C=O stretching of an aromatic aldehyde- 1700 cm^{-1}.
- C=C stretching of an aromatic ring- 1595 cm^{-1}.
- C-H bending- 745 cm-1 and 685 cm^{-1}.

No other compound then benzaldehyde produces same IR spectra as shown above.

Studying the Progress of the Reaction

Progress of chemical reaction can be determined by examining the small portion of the reaction mixure withdrawn from time to time. The rate of disappearance of a characteristic absorption band of the reactant group and/or the rate of appearance of the characteristic absorption band of the product group due to formation of product is observed.

Detection of Impurities

IR spectrum of the test sample to be determined is compared with the standard compound. If any additional peaks are observed in the IR spectrum, then it is due to impurities present in the compound.

Quantitative Analysis

The quantity of the substance can be determined either in pure form or as a mixure of two or more compounds. In this, characteristic peak corresponding to the drug substance is chosen and log Io/It of peaks for standard and test sample is compared. This is called base line technique to determine the quantity of the substance.

Nuclear Magnetic Resonance Spectroscopy

Nuclear Magnetic Resonance (NMR) spectroscopy is an analytical chemistry technique used in quality control and research for determining the content and purity of a sample as well as its molecular structure. For example, NMR can quantitatively analyze mixtures containing known compounds. For unknown compounds, NMR can either be used to match against spectral libraries or to infer the basic structure directly. Once the basic structure is known, NMR can be used to determine molecular conformation in solution as well as studying physical properties at the molecular level such as conformational exchange, phase changes, solubility, and diffusion. In order to achieve the desired results, a variety of NMR techniques are available.

Some types of atomic nuclei act as though they spin on their axis similar to the Earth. Since they are positively charged they generate an electromagnetic field just as the Earth does. So, in effect, they will act as tiny bar magnetics. Not all nuclei act this way, but fortunately both 1H and 13C do have nuclear spins and will respond to this technique.

NMR Spectrometer.

In the absence of an external magnetic field the direction of the spin of the nuclei will be randomly oriented. However, when a sample of these nuclei is place in an external magnetic field, the nuclear spins will adopt specific orientations much as a compass needle responses to the Earth's magnetic field and aligns with it. Two possible orientations are possible, with the external field (i.e. parallel to and in the same direction as the external field) or against the field (i.e. antiparallel to the external field).

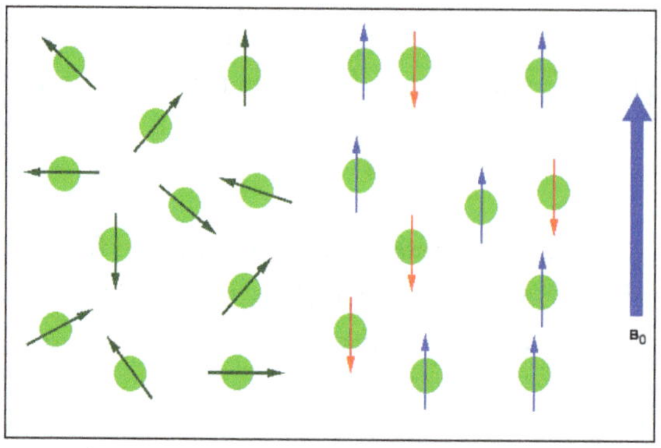

Random nuclear spin without an external magnetic field (Left).
Ordered nuclear spin in an external magnetic field (Right).

If the ordered nuclei are now subjected to EM radiation of the proper frequency the nuclei aligned with the field will absorb energy and "spin-flip" to align themselves against the field, a higher energy state. When this spin-flip occurs the nuclei are said to be in "resonance" with the field, hence the name for the technique, Nuclear Magnetic Resonance or NMR.

The amount of energy, and hence the exact frequency of EM radiation required for resonance to occur is dependent on both the strength of the magnetic field applied and the type of the nuclei being studied. As the strength of the magnetic field increases the energy difference between the two spin states increases and a higher frequency (more energy) EM radiation needs to be applied to achieve a spin-flip.

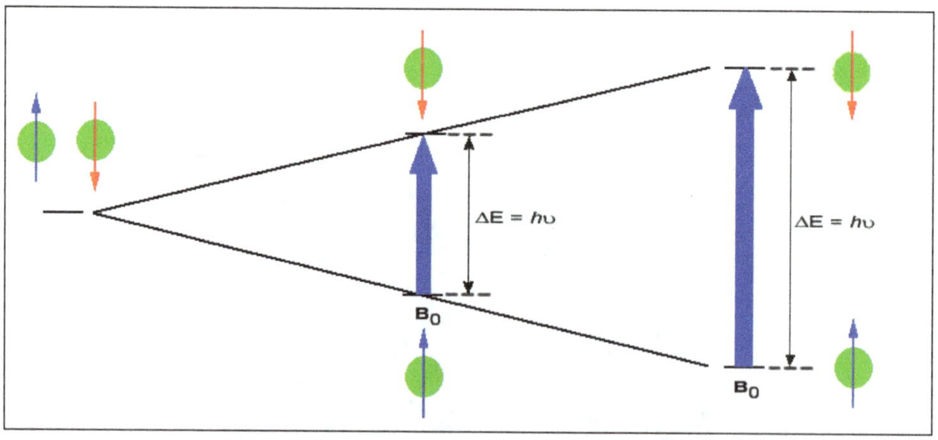

Superconducting magnets can be used to produce very strong magnetic field, on the order of 21 tesla (T). Lower field strengths can also be used, in the range of 4 - 7 T. At these levels the energy required to bring the nuclei into resonance is in the MHz range and corresponds to radio wavelength energies, i.e. at a field strength of 4.7 T 200 MHz bring ^{1}H nuclei into resonance and 50 MHz bring ^{13}C into resonance. This is considerably less energy than is required for IR spectroscopy, ~10^{-4} kJ/mol versus ~5 - ~50 kJ/mol.

^{1}H and ^{13}C are not unique in their ability to undergo NMR. All nuclei with an odd number of protons (^{1}H, ^{2}H, ^{14}N, ^{19}F, ^{31}P ...) or nuclei with an odd number of neutrons (i.e. ^{13}C) show the magnetic properties required for NMR. Only nuclei with even number of both protons and neutrons (^{12}C and ^{16}O) do not have the required magnetic properties.

The basic arrangement of an NMR spectrometer is displayed below. A sample (in a small glass tube) is placed between the poles of a strong magnetic. A radio frequency generator pulses the sample and excites the nuclei causing a spin-flip. The spin flip is detected by the detector and the signal sent to a computer where it is processed.

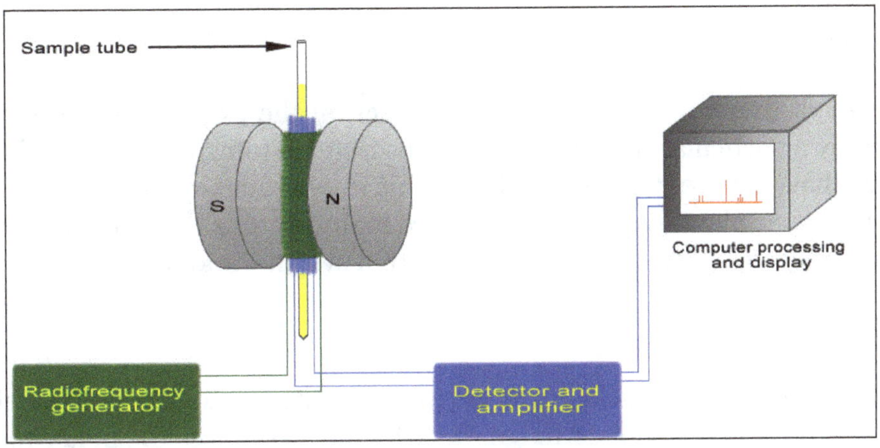

Chemical Shifts

The NMR spectrum is displayed as a plot of the applied radio frequency versus the absorption. The applied frequency increases from left to right, thus the left side of the plot is the low field, downfield or de-shielded side and the right side of the plot is the high field, upfield or shielded side. The concept of shielding will be explained shortly.

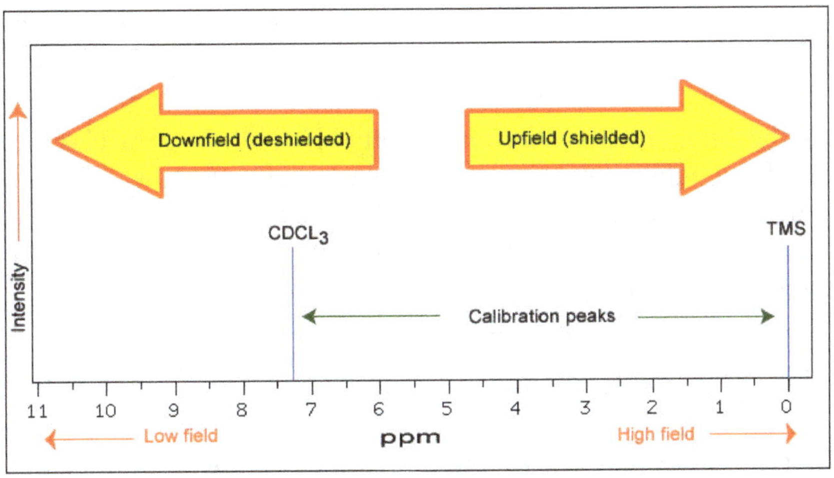

The position on the plot at which the nuclei absorbs is called the chemical shift. Since this has an arbitrary value a standard reference point must be used. The two most common standards are TMS (tetramethylsilane, ($Si(CH_3)_4$) which has been assigned a chemical shift of zero, and $CDCl_3$ (deuterochloroform) which has a chemical shift of 7.26 for ^{1}H NMR and 77 for ^{13}C NMR.

The scale is commonly expressed as parts per million (ppm) which is independent of the spectrometer frequency. The scale is the delta (δ) scale.

$$\delta = \frac{\text{frequency of signal} - \text{frequency of standard}}{\text{spectrometer frequency}} \times 10^6$$

The range at which most NMR absorptions occur is quite narrow. Almost all 1H absorptions occur downfield within 10 ppm of TMS. For ^{13}C NMR almost all absorptions occurs within 220 ppm downfield of the C atom in TMS.

Shielding in NMR

Structural features of the molecule will have an effect on the exact magnitude of the magnetic field experienced by a particular nucleus. This means that H atoms which have different chemical environments will have different chemical shifts. This is what makes NMR so useful for structure determination in organic chemistry. There are three main features that will affect the shielding of the nucleus, electronegativity, magnetic anisotropy of π systems and hydrogen bonding.

Electronegativity

The electrons that surround the nucleus are in motion so they created their own electromagnetic field. This field opposes the the applied magnetic field and so reduces the field experienced by the nucleus. Thus the electrons are said to shield the nucleus. Since the magnetic field experienced at the nucleus defines the energy difference between spin states it also defines what the chemical shift will be for that nucleus. Electron with-drawing groups can decrease the electron density at the nucleus, deshielding the nucleus and result in a larger chemical shift. Compare the data in the table below.

Compound, CH_3X	CH_3F	CH_3OH	CH_3Cl	CH_3Br	CH_3I	CH_4	$(CH_3)_4Si$
Electronegativity of X	4.0	3.5	3.1	2.8	2.5	2.1	1.8
Chemical shift δ (ppm)	4.26	3.4	3.05	2.68	2.16	0.23	0

As can be seen from the data, that the electronegativity of X increases the chemical shift, δ increases. This is an effect of the halide atom pulling the electron density away from the methyl group. This exposes the nuclei of both, C and H atoms, "de-shielding" the nuclei and shifting the peak downfield. The effects are cumulative so the presence of more electron withdrawing groups will produce a greater de-shielding and therefore a larger chemical shift, i.e.

Compound	CH_4	CH_3Cl	CH_2Cl_2	$CHCl_3$
δ (ppm)	0.23	3.05	5.30	7.27

These inductive effects are not only felt by the immediately adjacent atoms, but the de-shielding can occur further down the chain, i.e.

NMR signal	$-CH_2-CH_2-CH_2Br$
δ (ppm)	1.25 1.69 3.30

Magnetic Anisotropy: π Electron Effects

The π electrons in a compound, when placed in a magnetic field, will move and generate their own magnetic field. The new magnetic field will have an effect on the shielding of atoms within the field. The best example of this is benzene.

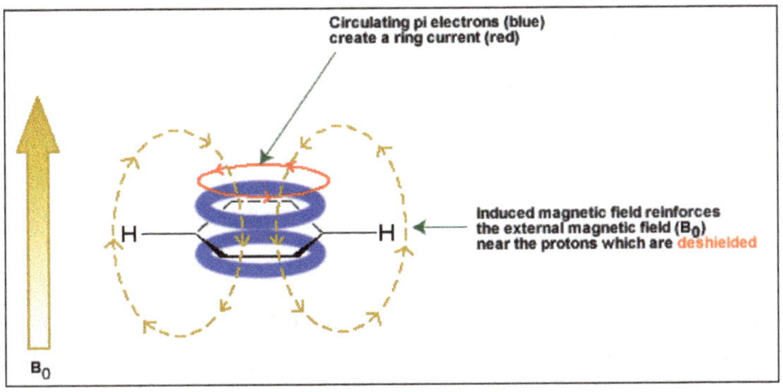

This effect is common for any atoms near a π bond, i.e.

Proton Type	Effect	Chemical shift (ppm)
C_6H_5-H	highly de-shielded	6.5 - 8
C=C-H	De-shielded	4.5 - 6
C≡C-H	shielded*	~2.5
O=C-H	very highly deshielded	9 - 10
* the acetylene H is shielded due to its location relative to the π system		

Hydrogen Bonding

Protons that are involved in hydrogen bonding (i.e.-OH or -NH) are usually observed over a wide range of chemical shifts. This is due to the de-shielding that occurs in the hydrogen bond. Since hydrogen bonds are dynamic, constantly forming, breaking and forming again, there will be a wide range of hydrogen bonds strengths and consequently a wide range of de-shielding. This as well as solvation affects, acidity, concentration and temperature make it very difficult to predict the chemical shifts for these atoms.

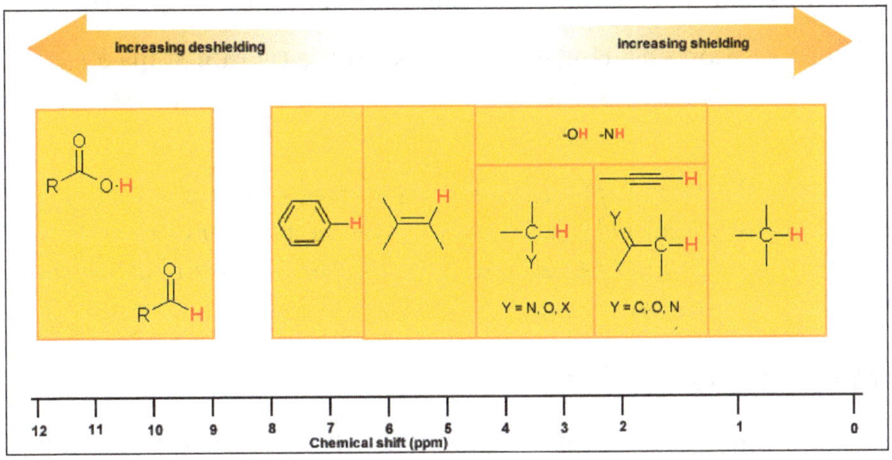

Experimentally -OH and -NH can be identified by carrying out a simple D_2O exchange experiment since these protons are exchangeable.

- Run the normal H-NMR experiment on your sample.
- Add a few drops of D_2O.
- Re-run the H-NMR experiment.
- Compare the two spectra and look for peaks that have "disappeared".

H-NMR Chemical Shifts

^{1}H-NMR Spectra

A ^{1}H-NMR will contain a unique signal for each different type of H atom present in the compound. What do we mean by "type" of hydrogen atom? Since the amount of shielding is dependent on the local chemical environment, the exact chemical shift for H atoms can vary widely. There are three basic methods you can use to determine if H atoms are identical.

Substitution Method

This is the simplest but slowest method. The idea is to replace every H atom, one at a time with another atom i.e. a Cl atom to see if you generate a different compound. Each different product indicates a different type of H atom.

Vebal Description

This requires to you describe each H atom verbally. If you have a different description then the H atoms are different.

For example:

- An -NH is different from a -CH (based on the atom the H is attached to).
- A $-CH_3$ is different to a $-CH_2-$ (based on the number of attached hydrogen atoms).
- A sp^3 C-H is different than an sp^2 C-H which is different to an sp C-H.
- Others include position on a ring or chain, cis / trans relationships etc.

Symmetry

The symmetry method is the most sophisticated but requires a knowledge of molecular symmetry. H atoms that are related by mirror planes, axis of rotation or a center of inversion are equivalent to one another. The first method is the easiest but slowest, the last is the fastest but requires a good knowledge of molecular symmetry.

^{1}H-NMR Spectra: Intensity of Signals

There are several important pieces of information that you can obtain from an ^{1}H-NMR. The first is the chemical shift of the peak. This will aid in identifying the type of H atom that produced the signal. The second is the integration ratios of the peaks. The area under a peak of a ^{1}H-NMR is directly proportional to the number of H atoms that produced the peak. The area is calculated by integrating the area, done automatically for you by the software. On older spectra the integration curve was drawn on the spectra, modern software will produce this as a table attached to the spectra. The example below is for methyl t-butyl ether.

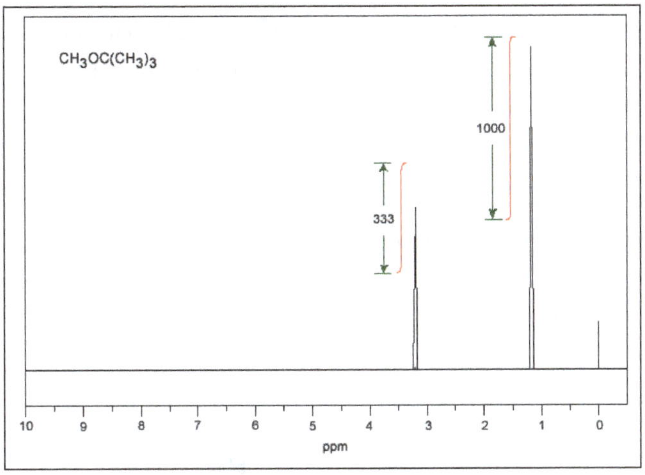

The integrals (shown as the red curves) are in a ratio of 333:1000 or 1:3. This implies that the peak on the left corresponds to the methyl group attached to the oxygen (as expected since the O will deshield the H atoms). The peak on the right is the three methyl groups of the t-butyl group, less deshielded as it is further away from the O atom. Note that we get the simplest ratio of H atom types (1:3, methyl: t-butyl) not the true ratio of 3:9.

Spin-spin Splitting

The NMR above has absorptions which are called singlets, a single sharp peak. However, in most cases, this is not the norm. Absorptions are split into groups of peaks due to coupling between adjacent protons in the molecule.

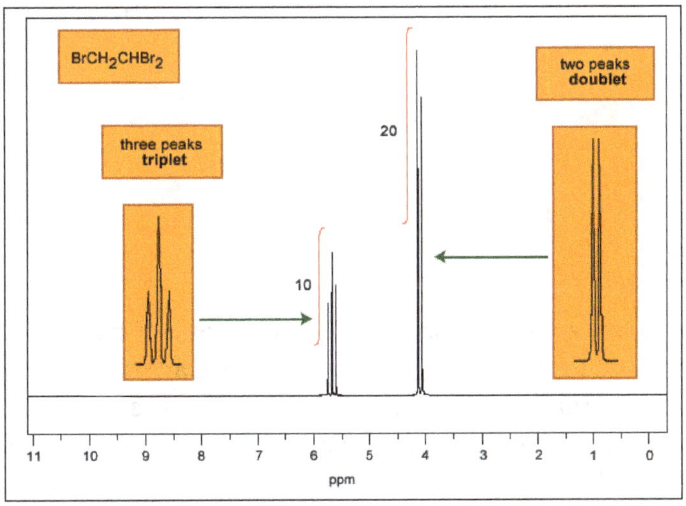

The ^{1}H-NMR of 1,1,2-tribromoethane is shown above (integration ratios of 10:20 or 1:2). Note that the peaks are now split into a group of two peaks on the right, a doublet and into three peaks on the left, a triplet. The relative areas under the peaks are 1:1 for the doublet, and 1:2:1 for the triplet. The triplet corresponds to the -$CHBr_2$ (most de-shielded) proton and the doublet to the -CH_2Br (most shielded) protons. This also agrees with the integration.

Spin-spin splitting occurs between unique types of H atoms on the same or adjacent (vicinal) carbon atoms. This occurs because the magnetic fields of each H atom can interact with the magnetic field of other H atoms. The interaction is only important (i.e. leading to spin-spin splitting) when the H atoms are "chemically different" from one another.

Examples of Spin-spin Splitting

Splitting Patterns in 1, 1, 2-Tribromoethane

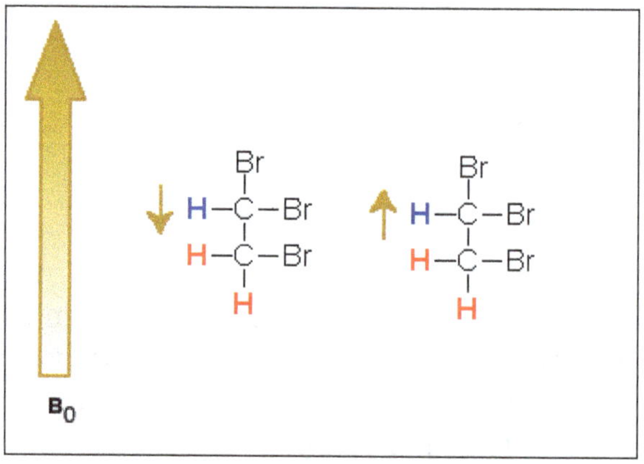

The methine (-CH-) hydrogen can assume two magnetic spin orientations, with or against the external field. As a result the peak for the adjacent methylene group is split into two lines of equal intensity, a doublet.

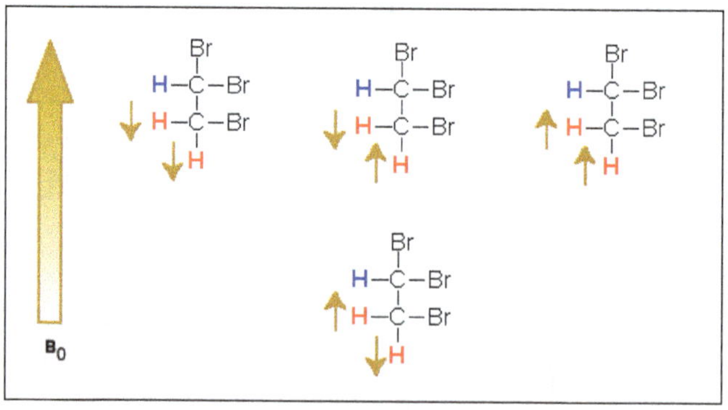

The methylene (-CH_2-) hydrogen atoms can again assume one of two magnetic spin orientations, with or against the external field. However, in this case these are two H atoms that are identical so there are three possible combinations of their two spins. The first (shown on the left) is both are spin down, the next is both spin up (shown on the right). There are two possible ways for the 1 spin up and 1 spin down combination depending on which specific H atom is spin up or down (shown in the center). The combination produces a triplet with relative intensities of 1:2:1 for the adjacent methine group.

Splitting Patterns in Bromoethane

The ^1H-NMR for bromoethane is shown below (integration ratios of 20:30 or 2:3). Note the peaks are now split into a group of three peaks on the right, a triplet and into four peaks on the left, a quartet. The relative areas under the peaks are 1:2:1 for the triplet and 1:3:3:1 for the quartet. The quartet corresponds to the $-CH_2Br$ (most deshielded) protons and the triplet corresponds to the $-CH_3$ most sheilded) protons. This also agrees with the integration.

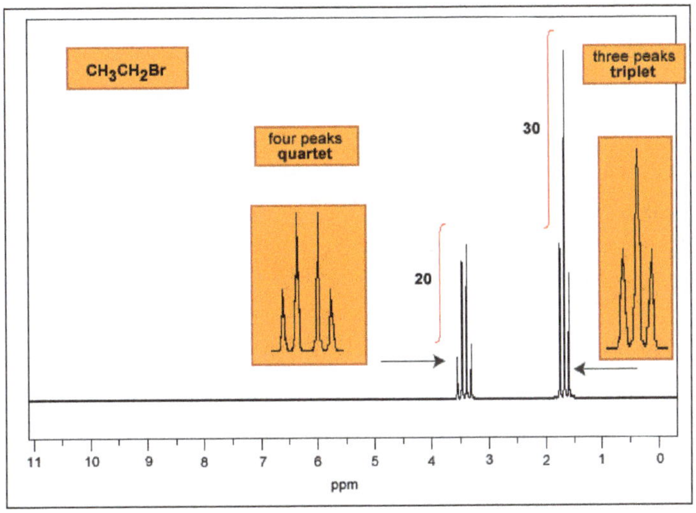

The methylene ($-CH_2-$) hydrogen atoms can assume one of two magnetic spin orientations, with or against the external field. However, in this case these are two H atoms that are identical so there are three possible combinations of their two spins. The first (shown on the left) is both are spin down, the next is both spin up (shown on the right). There are two possible ways for the 1 spin up and 1 spin down combination depending on which specific H atom is spin up or down (shown in the center). The combination produces a triplet with relative intensities of 1:2:1 for the adjacent methyl group.

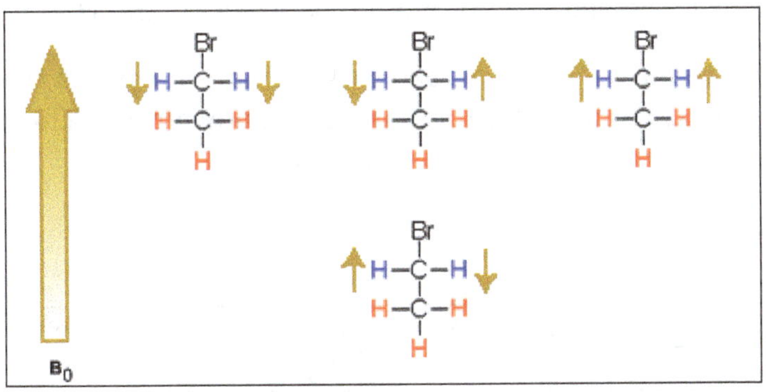

The methyl ($-CH_3$) hydrogen atoms can again assume one of two magnetic spin orientations, with or against the external field. However in this case there are three protons that are identical so there are four possible combinations of their spins. The most de-shielded peak will occur when all three spins are aligned with the field (left most image). The most shielded peak occurs when all three spins are aligned against the field (the right most image). There are now two possibilities left, the case where two spins are with and one against field or two spins against and one with the field. The first case will be most de-shielded than the second case and so downfield form it. This combination will split the adjacent peak for the methylene protons into a quartet with relative intensities of 1:3:3:1.

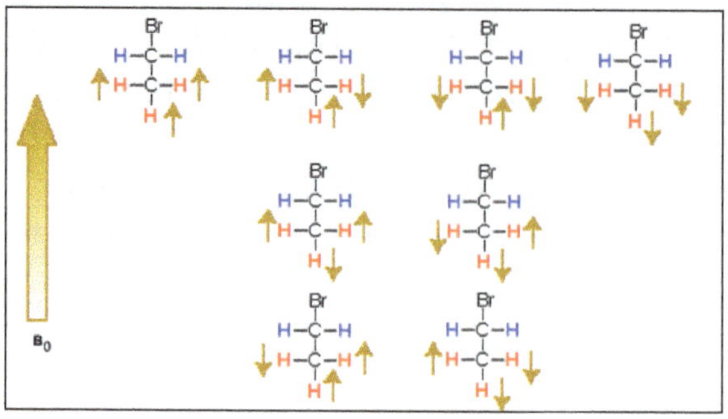

Coupling Constant, J

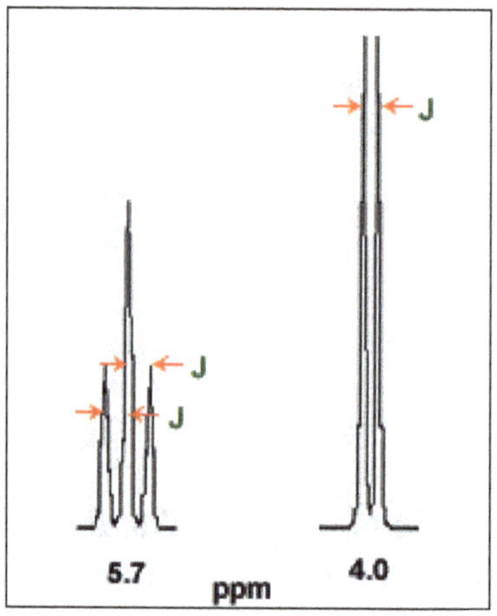

The coupling constant, (symbol J, units Hz) is a measure of the interaction between a pair of protons. In a vicinal type of system, H_a-C-C-H_b the coupling between protons H_a and H_b is J_{ab}. The coupling of H_b and H_a must be equal to the coupling between H_a and H_b, of $J_{ab} = J_{ba}$.

The implications of this are that the spacing between the lines for peaks that are coupled must be the same. Compare the peaks from 1,1,2-tribromoethane shown right. This means we can identify which peaks are coupled by the spacing between the lines.

Common Spin Multiplicities

Number of equivalent adjacent protons	Multiplicity	Ratio of Intensities
0	singlet	1
1	doublet	1:1
2	triplet	1:2:1
3	quartet	1:3:3:1
4	quintet	1:4:6:4:1
5	hextet	1:5:10:10:5:1
6	septet	1:6:15:20:15:6:1

^{1}H NMR Chemical Shifts for Common Functional Groups

It is necessary to know about the chemical shift values for different types of protons because it helps us to determine the structure of various organic compounds. The chemical shift of proton varies with its electronic environment. The proton attached to carbon, oxygen, nitrogen or sulphur appears at different position due to the difference in electronegativity. The chemical shift values also depends on the type of carbon to which hydrogen is attached, for example protons attached to singly bonded, doubly bonded or triply bonded carbons appear at different positions. The anisotropy effect caused by the circulation of the π-electrons also affects the chemical shift position of a proton. In addition we also need to know about the coupling behaviour of the concerned proton with other neighbour protons to assign the structure of a compound completely.

Chemical Shift Values for Various Functional Groups

Alkanes

In alkanes (aliphatic and saturated hydrocarbons), CH absorption occur in the range about 0.7 to 1.7 ppm. The hydrogens in a methyl group are the most shielded followed by methylene and then methine protons. Thus methine hydrogens (CH) have a larger chemical shift than those in methylene or methyl groups. The chemical shift values range for methyl protons is 0.7-1.33 ppm and that for methylene protons is 1.2-1.2 ppm and for methine hydrogens at 1.4-1.7 ppm.

In long hydrocarbon chains or in larger rings, methylene and methine hydrogens may overlap in an unresolvable group. Signal due to the methyl protons is usually separated from other types of hydrogens, being found at lower chemical shift values (higher field). However, even when methyl hydrogens appear within an unresolved cluster of peaks, the methyl peaks can often be recognized as strong singlets, doublets or triplets depending upon the number of hydrogens on the adjacent carbon atom. In the hydrocarbon chains, hydrogens on the adjacent carbons generally couple to each other following the N+1 rule for spin-spin splitting. The coupling constant values are approximately 7- 8 Hz.

Alkenes and Conjugated Systems

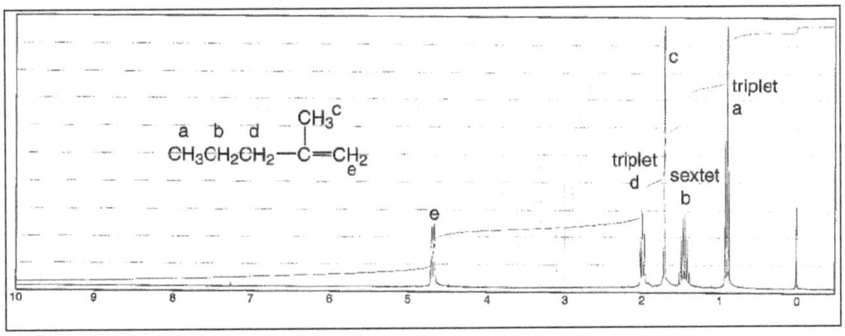

^{1}H NMR spectrum of 2-methyl-1-pentene.

There are two types of hydrogens in alkenes. Hydrogens directly attached to the double bond are called vinyl hydrogens, whereas those located on the carbon atom attached to the double bond are called allylic hydrogens. Each type of hydrogen resonates in a particular chemical shift region. The chemical shift values for vinylic hydrogens (C=C-H) are in the range 4.5-6.5 ppm, whereas for allylic hydrogen (C=C-C-H) it is 1.6-2.6 ppm. Both types of the hydrogens are deshielded due to the anisotropic effect of the π-electrons of the C=C double bond. However, the effect is smaller for the allylic hydrogens because they are more distant from the double bond. If the system is conjugated (C=C-C=C-H), the chemical shift is further moved to the downfield region. Figure shows the ^1H NMR spectrum of 2-methyl-1-pentene. Note that the vinyl hydrogens (H_e) appear at 4.7 ppm and the allylic methyl group at 1.7 ppm and allylic methylene at 1.9 ppm.

There may be three different types of spin-spin interactions in alkenes. The splitting patterns of both vinyl and allylic hydrogens is quite complex. The hydrogen atoms attached to a double bond are rarely equivalent and hence they couple to each other forming a complex coupling system. Further complications may arise due to the allylic coupling. When allylic hydrogens are present in an alkene they also show long-range allylic coupling with hydrogens located on the far double bond carbon as well as the usual splitting due to the hydrogen on the adjacent carbon atom.

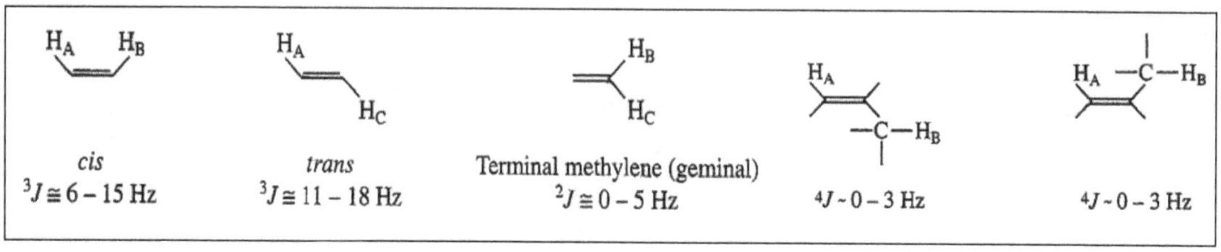

Aromatic Compounds

In general, aromatic compounds have two types of hydrogens; one type of hydrogens which are directly attached to benzene ring and the other type are benzylic hydrogens which are attached to carbon atom attached to benzene ring. Hydrogens attached to the benzene ring have large chemical shift usually in the range 6.5-8.0 ppm. The hydrogens on an aromatic ring are highly de-shielded than those attached to a normal double bonds due to the large anisotropic effect generated by the circulation of the π-electrons of the benzene ring. Benzylic hydrogens are also de-shielded by the anisotropic effect of the π-system but the effect is smaller due to the more distance from the ring. Benzylic hydrogens appear in the range 2.3-2.7 ppm.

If the benzene ring is substituted with electron-withdrawing groups such as $-NO_2$, chemical shift of the ring hydrogens is shifted to downfield region i.e. towards high chemical shift values. These groups de-shield the attached hydrogens by withdrawing electron density from the ring through resonance interaction. On the other hand, the electron-donating groups like methoxy ($-OCH_3$) increase the shielding of ring hydrogens causing them to shift towards up-field region.

In aromatic rings, coupling usually occurs beyond the adjacent carbon atoms (more than three bonds). The ring hydrogens which are chemically as well magnetically non-equivalent interact with one another to produce spin-spin splitting patterns. The extent of interaction depends on the distance between them. Ortho hydrogens ($^3J \approx$ 7-10 Hz) couple more strongly than meta hydrogens ($^4J \approx$ 1-3 Hz), which in turn couple more strongly than para hydrogens ($^5J \approx$ 0-1 Hz). We can determine the substitution pattern of the ring from these splitting patterns and the magnitudes of the coupling constants. The para-disubstituted aromatic rings usually show two doublets. The para-substitution NMR aromatic region pattern usually looks quite different than the patterns for both ortho and meta substituted aromatic rings.

Alkynes

The acetylenic hydrogens (hydrogen attached to the triple bond) have chemical shift values in the range 1.7-2.7 ppm. The chemical shift value is moved towards upfield region than might be expected, or shielded, due to the anisotropic effect of the π-electrons system. The hydrogens on the carbon next to the triple bond are also affected (shielded) by the π-system and these protons resonate at approximately 1.6-2.6 ppm. The figure given below shows the ^1H NMR spectrum of 1-pentyne. The acetylenic hydrogen appeared at 1.95 ppm and the allylic hydrogens at 2.2 ppm.

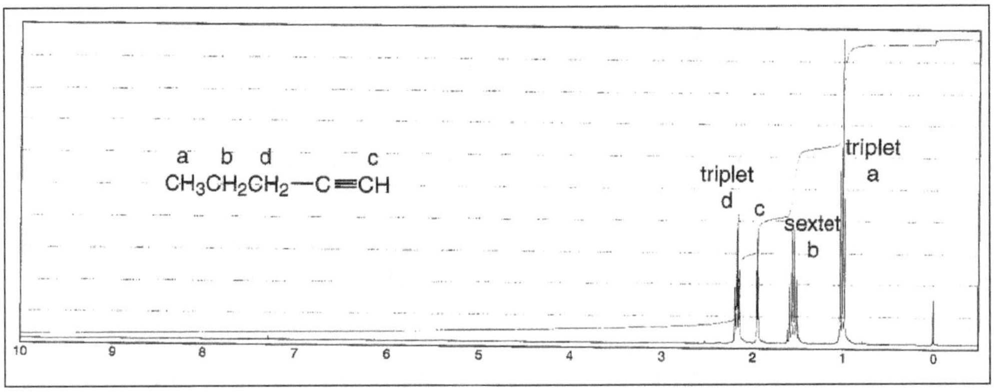

Figure: ^1H NMR spectrum of 1-pentyne

Sometimes, the acetylenic hydrogen may not show a sharp singlet due to the allylic coupling and may be split into a triplet but the coupling constant is quite small (J = 2-3 Hz)

Alkyl Halides

In alkyl halides, hydrogen atoms attached to the carbon, to which halogen atom is attached, are de-shielded due to the electronegativity of halogen atom. The amount of deshielding increases as the electronegativity of the halogen increases. The chemical shift values range in R-CH_2-F, RCH_2-Cl, R-CH_2-Br and R-CH_2-I are 4.2-4.8 ppm, 3.1-4.1 ppm, 2.7-4.1 ppm and 2.0-4.0 ppm.

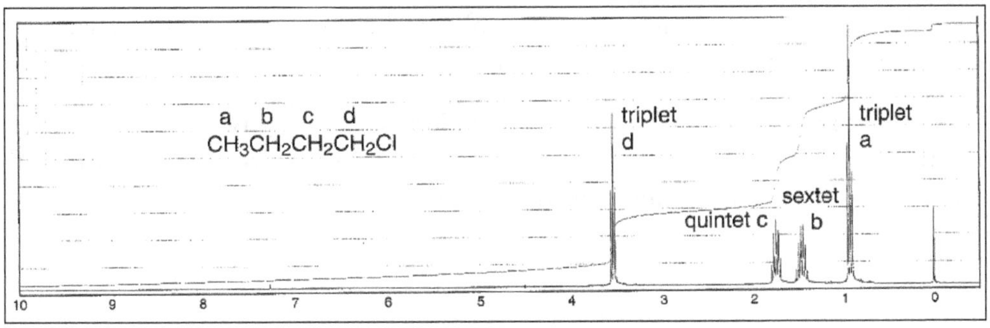

^1H NMR spectrum of 1-chlorobutane.

Compounds containing fluorine will show spin-spin splitting due to coupling between the fluorine and the hydrogens on either the same or the adjacent carbon atom (-CH-F; ^2J = 50 Hz and -CH-CF-; ^3J = 20 Hz). Since the spin of fluorine is ½, the n +1 Rule can be used to predict the multiplicities of the attached hydrogens. The other halogens (I, Cl, Br) do not show any coupling. The ^1H NMR spectrum of 1-chlorobutane is shown in the figure. The hydrogen atoms nearest to chlorine resonate at the highest chemical shift value.

Alcohols

The hydrogen of the hydroxyl group appears in the range 0.5-5.5 ppm. The chemical shift of the – OH hydrogen is variable and its value depends on concentration, solvent, temperature, and presence of water or of acidic or basic impurities. The variability of OH signal is due to the OH proton exchange and the extent of hydrogen bonding in the solution. The peak may be broadened at its base by the same set of factors.

The –OH hydrogen is usually not split by hydrogens on the adjacent carbon (-CH-OH) because rapid exchange decouples this interaction. Exchange is promoted by increased temperature, small amounts of acid impurities, or the presence of water in the solution. In ultrapure alcohol sample – CH-OH coupling is observed. A freshly purified and distilled alcohol may show this coupling.

$$-CH-OH+HA \rightarrow -CH-OH+HA\,(\text{no coupling if exchange is rapid})$$

For detecting the protons exchangeable with D_2O, either the spectrum is recorded in D_2O or a few drops of D_2O are added to the sample and shake it properly before recording the NMR. The proton is exchanged by,

We can use this exchange reaction of an alcohol as a method for identifying the –OH absorption. For this, a small amount of D_2O is added in the NMR tube containing the alcohol solution. After shaking for some time, the –OH hydrogen is replaced by deuterium. Hence the peak due to the alcoholic hydrogen will disappear because deuterium absorbs at different field strength. Therefore there will not be any coupling/ splitting due to the proton which has been exchanged with deuterium.

$$-CH-OH + D_2O \rightarrow -CH-OD+HOD\ (\text{deuterium exchange})$$

The hydrogen on the carbon adjacent to the hydroxyl group (CH-OH) appears in the range 3.2-4.0 ppm due to the de-shielding by the electronegative oxygen atom. If deuterium exchange has taken place, this hydrogen will also not show any coupling with the –OH hydrogen, but may show coupling to other hydrogens on the adjacent carbon atom. If deuterium exchange is not there then the splitting pattern of this hydrogen may be complicated by different coupling with the proton of hydroxyl group and the protons of β-carbon atom.

The phenolic OH is usually found in more de-shielded region 4-7 ppm. The hydroxyl proton in enols appears in the range 15-17 ppm. These chemical shift values are also variable and depend on the temperature, solvent and the concentration of the solution.

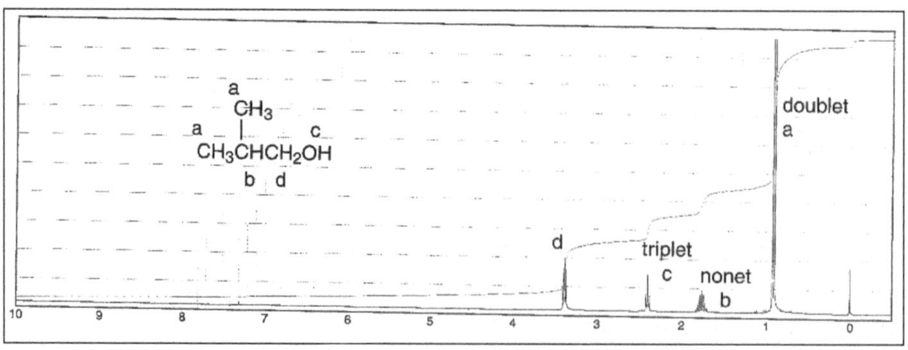

^{1}H NMR spectrum of 2-methyl-1-propanol.

The ¹H NMR spectrum of 2-methyl-1-propanol is shown in figure. Protons attached to the αcarbon appeared in the most downfield region (3.4 ppm). The hydroxyl group appeared as triplet at 2.4 ppm. Here some coupling is observed for hydroxyl proton due to the hydrogens on the adjacent carbon.

Ethers

In ethers, hydrogens on the α-carbons are de-shielded due to the electronegativity of the oxygen atom. These hydrogens appear in the range 3.2-3.8 ppm. Methoxy group can easily be identified as the methyl hydrogens of it appear as a strong singlet. Epoxides are cyclic ethers and in epoxides the de-shielding is not as strong as in alky ethers due to ring strain. The methylene hydrogens in the ring appear in the range 2.5-3.5 ppm.

Figure shows the ¹H NMR spectrum of butyl methyl ether. The absorption of the methyl and methylene hydrogens next to the oxygen is both seen at about 3.4 ppm. The methoxy peak is observed as a sharp singlet. The methylene hydrogens are split into a triplet by the hydrogens on the adjacent carbon in the alkyl chain.

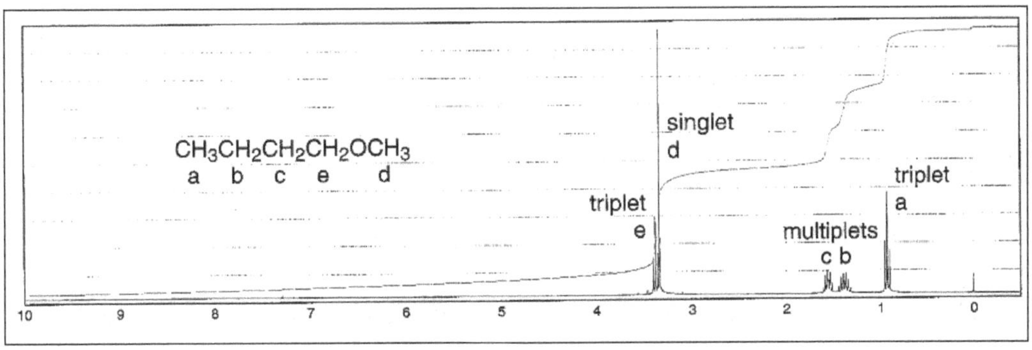

¹H NMR spectrum of butyl methyl ether.

Amines

The amine protons appear as a broad peak (but not as broad as a carboxylic acid proton peak) in the range 0.5–4.0 ppm and the range is extended in aromatic amines. The NH chemical shift is variable and is affected by temperature, acidity, amount of hydrogen bonding, and solvent. Moreover, the –NH protons show often broad and weak signal and rarely show coupling to the hydrogens on the adjacent carbon atom. This condition can be caused by chemical exchange of the –NH proton, or due to quadrupole broadening. The amino hydrogens will exchange with D_2O causing the peak to disappear.

$$-N-H + D_2O \rightarrow -N-D + DOH$$

The –NH peaks are strong in aromatic amines where delocalization of the lone pair of electron on nitrogen atom appears to strengthen the NH bond by changing the hybridization. The hydrogens at α-carbon are slightly de-shielded by the presence of the electronegative nitrogen atom, and they appear in the range 2.2-2.9 ppm. In aromatic amines the α-hydrogen is de-shielded due to the anisotropy of the ring and also due to the resonance that decreases the electron density on nitrogen atom and changes its hybridization.

Figure below shows the 1H NMR spectrum of propylamine. The NH proton appears as weak and broad singlet at 1.8 ppm. Note that the NH proton does not show any coupling with the hydrogens on the adjacent carbon atom.

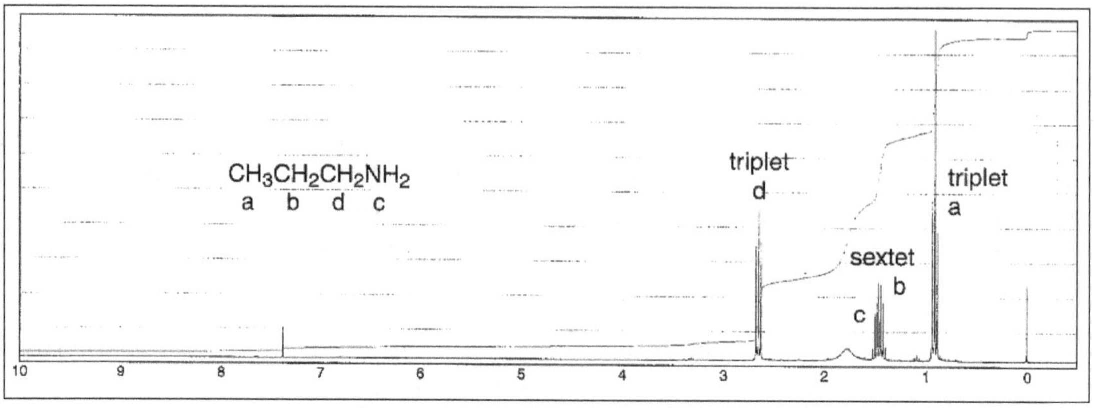

^{1}H NMR spectrum of propylamine.

Nitriles

In nitriles, hydrogens on the α-carbon (-CH-CN) are slightly de-shielded by the anisotropic field of the π-electrons in cyano group. These protons appear in the range 2.1-3.0 ppm.

Nitro Alkanes

In nitroalkanes, hydrogens which are attached to the carbon atom next to the nitro group are highly deshielded and appear in the range 4.1-4.3 ppm. The deshielding effect of nitro group is due to the electronegativity of the attached nitrogen and the positive formal charge assigned to the nitrogen.

Aldehydes

The aldehydic proton (-CHO) appear in the range 9-10 ppm. The high value of chemical shift of this hydrogen is due to the anisotropic effect of the carbonyl group (C=O). The peak in this region is highly particular for aldehydic proton because no other type of hydrogen appears in this region. Hydrogens attached to α-carbon (R-CH-CH=O) are also deshielded due to the anisotropy by carbonyl group but the effect is smaller because of larger distance. These hydrogens appear in the range 2.1-2.4 ppm.

The aldehyde hydrogen (-CHO) couples weakly (^3J = 3Hz) with the hydrogens present on the αcarbon atom. Thus we can determine the number of hydrogen atoms at α-carbon just by looking at the multiplicity of the –CHO proton.

The signal due to the hydrogens on the α-carbon is split by the aldehydic hydrogen (^3J = 3 Hz) as well as by the hydrogens on the next adjacent carbon in the alkyl chain (^3J = 8 Hz). All the carbonyl functional groups such as ketones, aldehydes, esters, amides, and carboxylic acids have this type of α-hydrogens (CH-C=O) hence they all will give NMR peaks in this area (2.1-2.4 ppm). Therefore, it is necessary to look for an aldehydic hydrogen peak at 9-10 ppm to confirm the compound as an aldehyde.

Figure below shows the ^{1}H NMR spectrum of 2-methylbutyraldehyde. The aldehydic proton appears at 9.6 ppm. Notice that this peak is split into a fine doublet by the hydrogen on the carbon adjacent to the carbonyl group. The hydrogen next to the carbonyl group appears at 2.3 ppm.

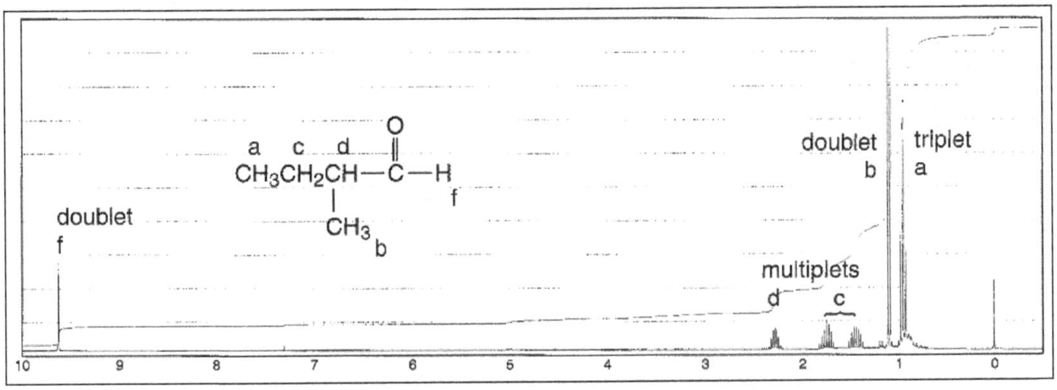

^{1}H NMR spectrum of 2-methylbutyraldehyde.

Ketones

In ketones (RCOR), the hydrogens on the α-carbon atom appear in the range 2.1-2.4 ppm. The other protons in the alkyl group will appear in the range 0-2 ppm. The α-Hydrogens show coupling with hydrogens on the adjacent carbon atom along the alkyl chain. Methyl ketones can easily be distinguished as they show a sharp three-proton singlet at approximately 2.1 ppm.

All the carbonyl functional groups viz. ketones, aldehydes, esters, amides and carboxylic acids will give rise to NMR absorptions in this same area, hence it is necessary to look for the absence of other absorptions (-CHO, -OH, $-NH_2$, $-OCH_2R$, etc.) to confirm the compound as a ketone. Figure shows a ^{1}H NMR spectrum of 3-methyl-2-pentanone. Notice that a sharp singlet at 2.1 ppm is a characteristic peak of a methyl ketone.

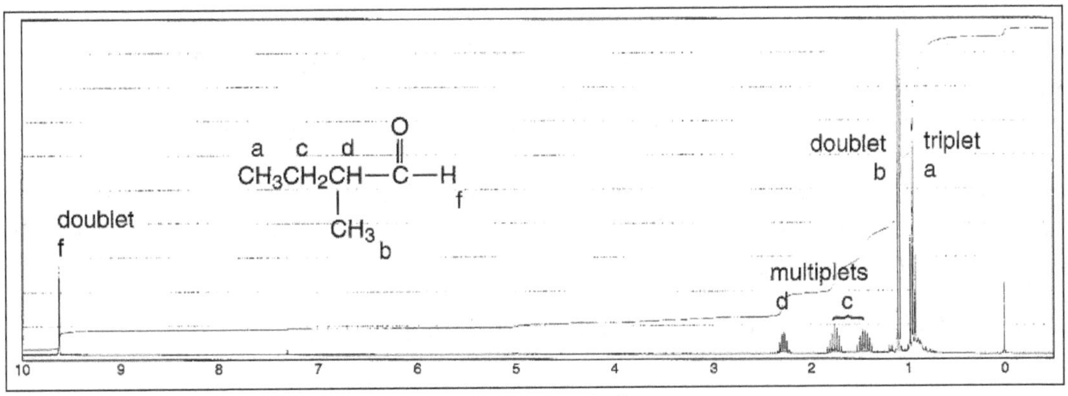

^{1}H NMR spectrum of 3-methyl-2-pentanone.

Esters

There are two different types of hydrogens present in esters. Hydrogens on the carbon attached to the oxygen (R-COO-CH-) are more de-shielded due to the electronegativity of oxygen and they appear in the range 3.5-4.8 ppm. The other types of hydrogens are α-hydrogens (R-CH-COOR) and they resonate in the range 2.1-2.5 ppm. These protons are de-shielded by the anisotropy of the carbonyl (C=O) group. Again all type of carbonyl functional groups give rise to NMR absorptions in the range 2.1-2.5 ppm due to α-hydrogens. Therefore, an ester group in a compound can be identified by the NMR absorptions in the 3.4-4.8 ppm region. Both types of hydrogens can show coupling (normal vicinal coupling) with the protons on the adjacent carbon atoms.

A spectrum of isobutyl acetate is shown in figure. A sharp singlet at 2.1 ppm is due to the protons of CH_3CO group. The proton on carbon attached to oxygen atom appears at 3.8 ppm as doublet due to the coupling with the methine proton.

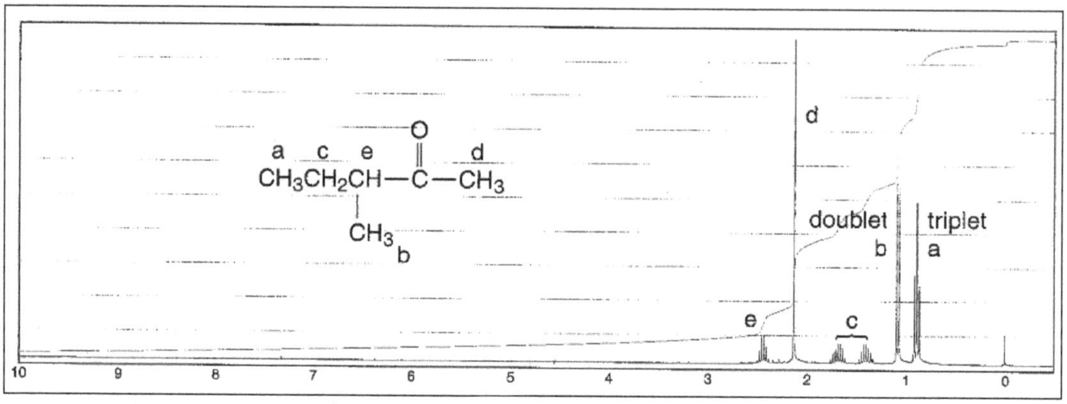

^{1}H NMR spectrum of isobutyl acetate.

Carboxylic Acid

The carboxylic acid proton (-COOH) appear in the range 10.0-13.0 ppm as a broad singlet. Sometimes it is so broad that it is not even seen. With the exception of some special case of hydrogen in an enolic OH group that has strong internal hydrogen bonding, no other type of hydrogen appears in this region. Thus a peak in this region (10.0-13.0 ppm) is a strong indication of a carboxylic acid functional group. Since the carboxylic acid proton has no neighboring proton hence it is usually not split. In dilute and non-polar solutions such as $CDCl_3$, carboxylic acids exist as stable hydrogen-bonded dimers and due to this, the carboxylic acid proton appear as a very broad singlet. In such conditions the integral value cannot be used to identify the number of protons with accuracy as it may cause an overestimation of the number of other types of hydrogens in the molecule.

The carboxyl proton is acidic and may exchange with water and D_2O. In D_2O proton exchange, the –COOH group will convert to –COOD and hence peak near 12.0 ppm will disappear and a new peak appear due to the hydrogen of DOH.

$$R-COOH + D_2O \rightarrow R-COOD + DOH(\text{deuterium exchange})$$

Carboxylic acids are often insoluble in CDCl3 and hence their spectra are used to determine in D_2O to which a small amount of sodium metal is added. This basic solution (NaOD, D_2O) will remove the acidic proton forming a soluble sodium salt of the acid and thus the –COOH absorption will disappear and a new peak around 4.5-5.0 ppm will appear due to hydrogen of DOH.

$$R-COOH\ (\text{insoluble}) + NaOD = R-COONa\ (\text{soluble}) + DOH$$

Figure below shows the ^{1}H NMR spectrum of pentanoic acid. The OH proton of carboxyl group appears at 12 ppm as broad singlet near12 ppm. Notice that this peak is not obtained as broad singlet. The methylene hydrogens next to the carboxylic acid group appear as triplet at 2.1 ppm. This is the most downfield $-CH_2$ signal because it is closest to the electron withdrawing carboxyl group.

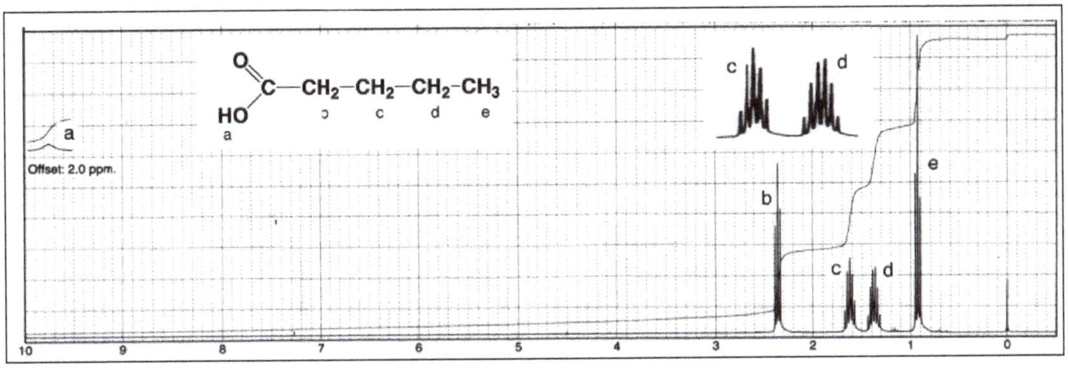

^{1}H NMR spectrum of pentanoic acid.

Amides

Amides have three different types of hydrogens: those attached to nitrogen R(CO)-NH, hydrogens attached to the α-carbon atom -CH-CONH-, and the hydrogens attached to a carbon atom which is also attached to the nitrogen atom R(CO)-N-CH. The chemical shift values for the protons attached to nitrogen atom are in the range 5.0-9.0 ppm. The α-hydrogens absorb in the same range (2.1-2.5) as other acyl (CH_3-C=O) hydrogens. They are slightly de-shielded by the carbonyl group. Hydrogens on the carbon attached to the nitrogen atom are slightly de-shielded (2.2-2.9 ppm) due to the electronegativity of the nitrogen.

It is important to note here that the proton or group of protons attached to the amide nitrogen often exhibit different chemical shifts. For example the 1H NMR spectrum of dimethylformamide (DMF) shows two distinct methyl peaks. In general one can expect the two identical groups attached to nitrogen to be chemically equivalent because of free rotation around the C-N bond. However the rotation is restricted to some extent because of the some double bond character between carbon and nitrogen in the amide bond. The double bond character arises due to the resonance between the unshared pairs on nitrogen and the carbonyl group.

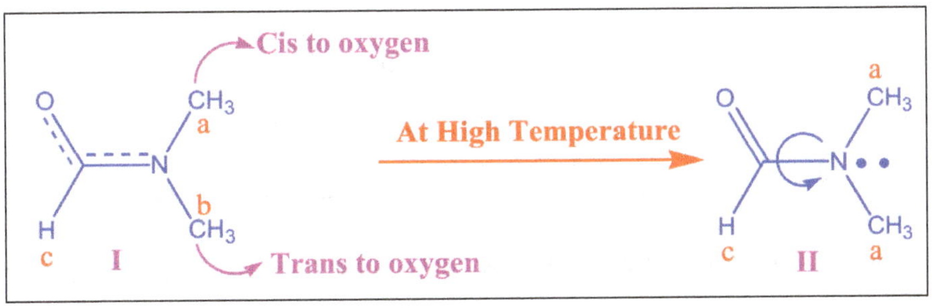

Thus if free rotational is slowed down, the hydrogens or group of hydrogens attached to the nitrogen in an amide are not equivalent and two different absorptions peaks will be observed, one for each hydrogen. Nitrogen atoms also have a quadrupole moment and its magnitude depends on the particular molecular environment. If the nitrogen atom has a large quadrupole moment, the attached hydrogens will show broad peaks with reduced peak intensity.

Figure: shows a ^{1}H NMR spectrum of butyramide. Notice that the two NH2 protons appear at different chemical shift (6.6 and 7.2 ppm). This occurs due to restricted rotation about the C-N bond. The methylene hydrogens next to the C=O group appear at 2.1 ppm.

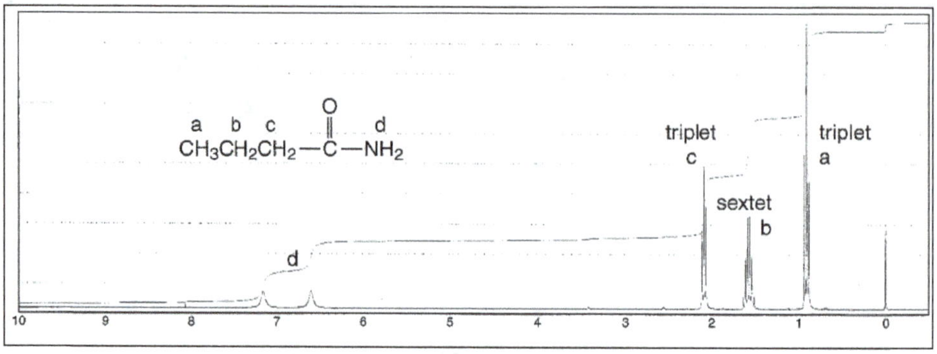

^{1}H NMR spectrum of butyramide

NMR Spectroscopy of N, P and F-atoms

^{15}N NMR spectroscopy is employed to determine the structure of nitrogen containing compounds. Nitrogen has two isotopes i.e. ^{14}N and ^{15}N and both are NMR active. ^{14}N has a nuclear spin of 1, whereas ^{15}N has a value of ½. The natural abundance of ^{14}N (99.63%) is much higher than the ^{15}N (0.37%). Since ^{14}N has an integer nuclear spin (I = 1) and thus it possess quadrupole moment and its signals are usually significantly broadened by quadrupole relaxation mechanism. Sometimes the broadening of the signal is to the extent that they are not observable on a high resolution NMR spectrometer. The other isotope, ^{15}N has a spin of 1/2 and no quadrupole moment, but it is only 0.37 percent abundant and hence is very insensitive. Sensitivity is made worse by its low gyromagnetic ratio ($\gamma \approx -27.126 \times 10^6\ T^{-1}\ s^{-1}$), which is 10.14% that of 1H. The signal to noise ratio for ^{1}H is about 300 fold greater than 15N at the same magnetic field.

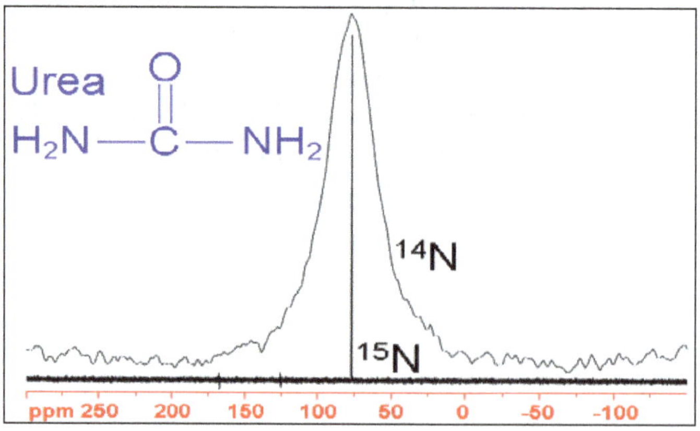

Comparison of ^{14}N and ^{15}N spectra of urea. The ^{14}N signal is broad due to quadrupolar moment.

The extent of broadening depends on the molecule structure. Larger the molecule and the more asymmetric the nitrogen's environment, broader is the signal. Hence, the small and highly symmetric aqueous ammonium ion gives a very sharp line (< 1 Hz widening). Liquid ammonia, being less symmetric, is 16 Hz wide. Urea is larger and asymmetric so the line width is approximately 1 KHz. Molecules that are significantly larger than urea yield signals too broad to be observed.

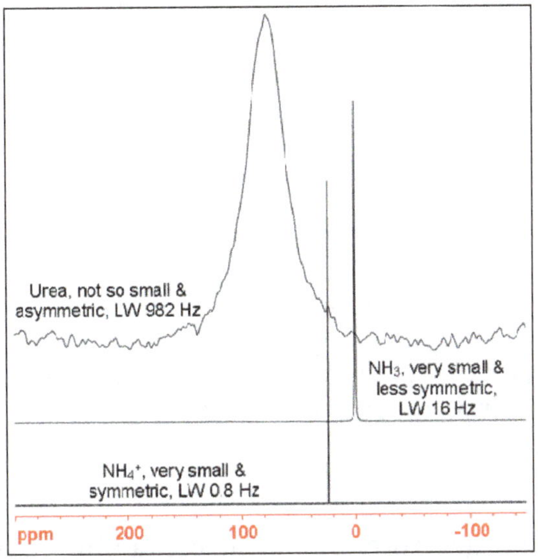

Comparison of line widths of ^{14}N signals. Larger and less symmetric molecule gives wider signal.

With highly sensitive instruments and making maximum use of various signal enhancement methods such as DEPT, NOE, natural abundance spectra can be recorded for ^{15}N. The advent of new techniques, such as pulse sequences and polarization transfer, in conjunction with the use of high-field magnets and large-sample probe heads largely solved the detection problem.

Chemical shifts of nitrogen nuclei cover a wide spectral range of about 900 ppm for most organic and inorganic compounds. There is no measurable isotopic substitution effect on the chemical shift value, hence ^{15}N and ^{14}N shifts may be used interchangeably. Various standards, internal or external, have so far been employed as references for nitrogen chemical shifts. There are two different chemical shift scales in ^{15}N NMR. IUPAC (international union of pure and applied chemistry) recommends nitromethane (CH_3NO_2) for the chemical shift reference (0 ppm). However some people use liquid ammonia. For ^{15}N, chemical shifts referenced with liquid NH3 are 380.5 ppm upfield from CH_3NO_2 (δ: NH_3 = δ: CH_3NO_2 + 380.5 ppm).

The heteronuclear coupling with ^{14}N is rarely observed because of its quadrupolar broadening. It can be observed only when the nitrogen atom is quaternary (ammonium ions) at room temperature and liquid ammonia at low temperatures. Two bond coupling (^{14}N-CH2-) are not observed even when the nitrogen is quaternary, presumably because the coupling is very small.

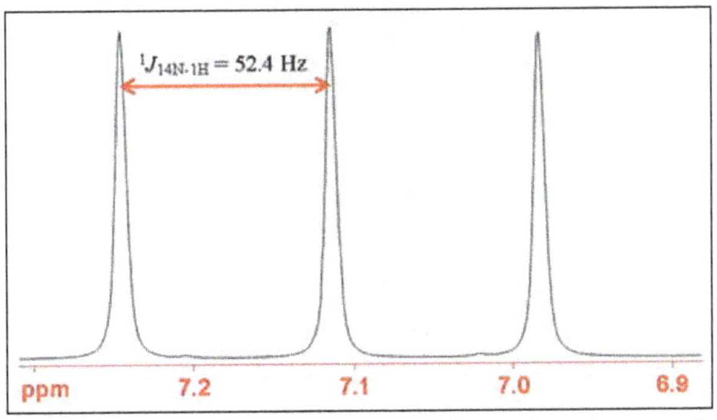

1H spectrum of natural abundance NH4Cl (1.5 M) in 1 M HCl/H_2O showing coupling to ^{14}N (I = 1) as a triplet.

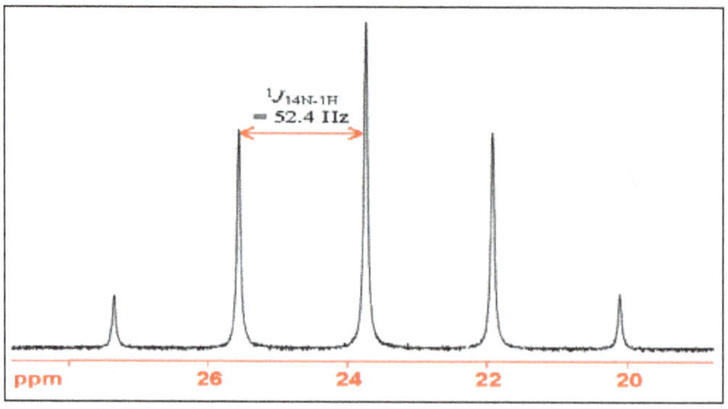

^{14}N spectrum of NH_4Cl (1.5 M) in 1 M HCl/H_2O showing coupling to ^{1}H as a quintet (four protons present at nitrogen atom).

^{19}F NMR Spectroscopy

^{19}F NMR is used to analyze the structure of fluorine containing compounds. ^{19}F is the naturally occurring isotope of fluorine with 100% abundance. It has a nuclear spin of 1/2 and a high gyromagnetic ratio, which means that this isotope is highly sensitive to NMR spectroscopy. Because of its favourable nuclear properties and high abundance, ^{19}F NMR spectroscopy is highly sensitive and fast as compared to ^{1}H NMR spectroscopy. Indeed the ^{19}F nucleus is the second most receptive NMR nucleus after the ^{1}H nucleus.

The ^{19}F NMR chemical shift range is very wide 0-200 ppm as compared to ^{1}H NMR (0-15 ppm). The reference compound for ^{19}F NMR is trichlorofluoromethane $CFCl_3$. It is generally inert, volatile and gives rise to a single ^{19}F peak at δ value of zero. When measurements are done at the same radio frequency, the signals from ^{19}F absorptions occur at considerably different magnetic field strength than that of proton. Thus one does not see the peaks due to ^{19}F absorptions in ^{1}H NMR spectrum. However the peaks due to proton can show splitting due to spin-spin couplings between protons and ^{19}F nuclei.

The figure given below shows the chemical shift range for the fluoro organic compounds. The most commonly encountered signals arising from organofluorine compounds lie between 50 to 70 ppm (for CF_3 groups) and between 200 to 220 ppm (for CH_2F groups). The very wide spectral range can cause problems in recording spectra, such as poor data resolution and inaccurate integration.

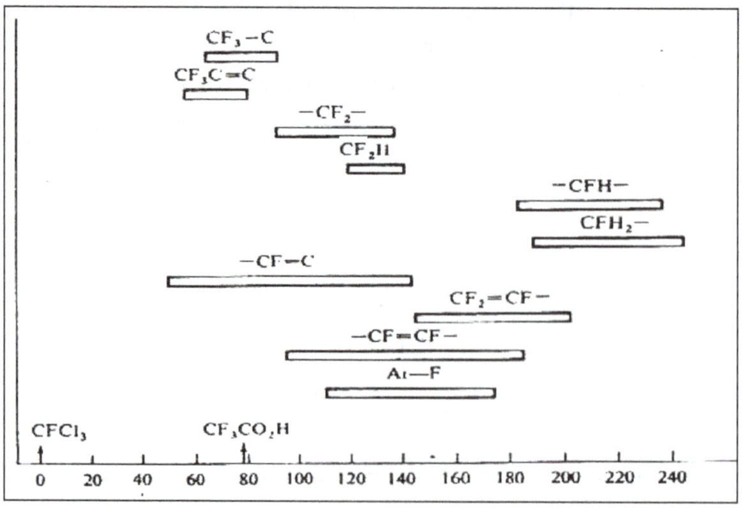

The multiplicity of the peaks gives the information about the neighbouring atoms to the fluorine. The coupling constants between fluorine nuclei are generally larger than those for H. Geminal FF coupling ranges from 40 to 370 Hz and vicinal F-F from 0-45 Hz. The cis vicinal coupling ranges from 0-60 Hz and trans vicinal coupling ranges from 100-150 Hz. Long range coupling more than five bonds is having very low value about 0-18 Hz.

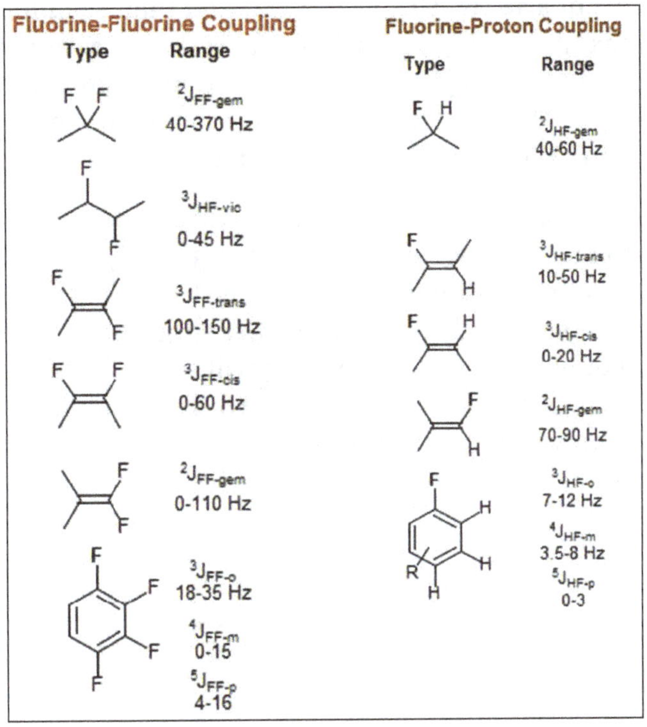

Coupling between H and F nuclei is also strong. Geminal coupling ranges from 40-60 Hz and vicinal coupling from 1.0-30 Hz. The cis H-F coupling constant ranges from 0-20 Hz and trans coupling constant value range is from 10-50 Hz. Fluorine attached to benzene also couples with protons on the ring. The coupling constant values for ortho, meta and para ranges from 7.0-12 Hz, 3.5-8.0 Hz and 0-3.0 Hz, respectively.

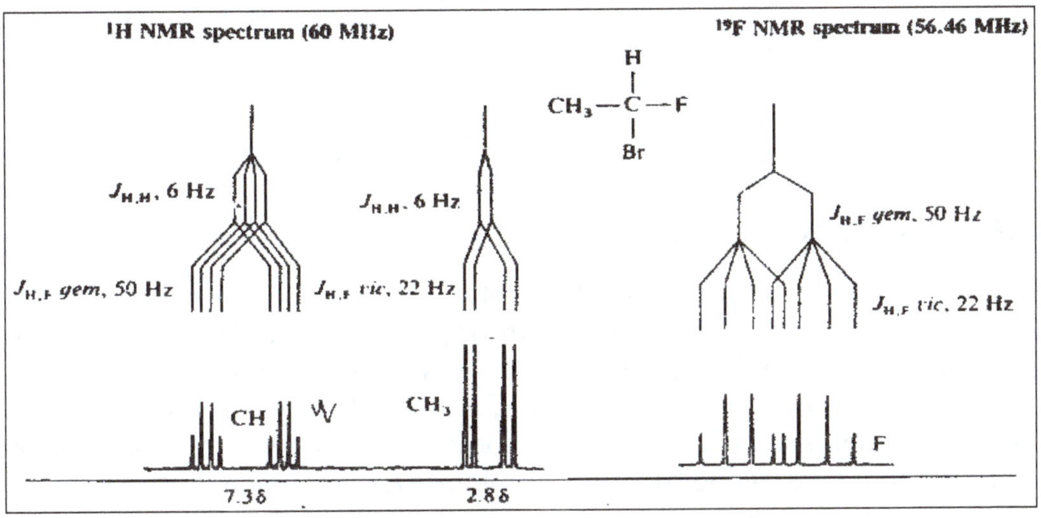

^1H and ^{19}F NMR spectra of 1-Bromo-1-Fluoroethane

The figure given below shows the ^{1}H and ^{19}F NMR spectrum of 1-bromo-1-fluoroethane (CH_3CH-FBr). Each spectrum is recorded using a 1.4 T magnet. In the ^{1}H spectrum, the H-H coupling between CH_3 and CH protons should give a quartet and doublet but the H-F coupling complicates the splitting pattern. The CH_3 signal is further split by ^{19}F, each line of the doublet is split into two by the large vicinal H-F coupling (J = 22 Hz), the CH_3 signal is therefore appears as a double doublet. The CH signal is split into a quartet by CH_3 (J = 6 Hz) and then each line is split into two by the large germinal H-F coupling (J = 50 Hz), the CH signal is therefore appears as a doublet of quartets. In the ^{19}F spectrum the fluorine signal is split into a doublet by the germinal methine proton (J = 50 Hz) and each line is further split into four by the vicinal methyl protons (J = 22 Hz). Thus the peak due to fluorine is an overlapping doublet of quartets.

The figure given below shows the ^{19}F NMR spectrum of 1-bromo-3,4,5-trifluorobenzene. There are two sets of fluorine atoms. The fluorine atoms at the 3 and 5 positions are chemically different from the fluorine atom at 4-position. Consider either F-3 or F-5, both the nuclei couple with the adjacent F-4 nucleus to give a doublet which further split into a doublet by the neighbouring proton to give a doublet of doublets. The fluorine, F-4 will couple with the two equivalent fluorine nuclei (F3 and F5) to give a triplet, and also with the two chemically equivalent protons therefore signal due to this fluorine appears as a triplet of triplets.

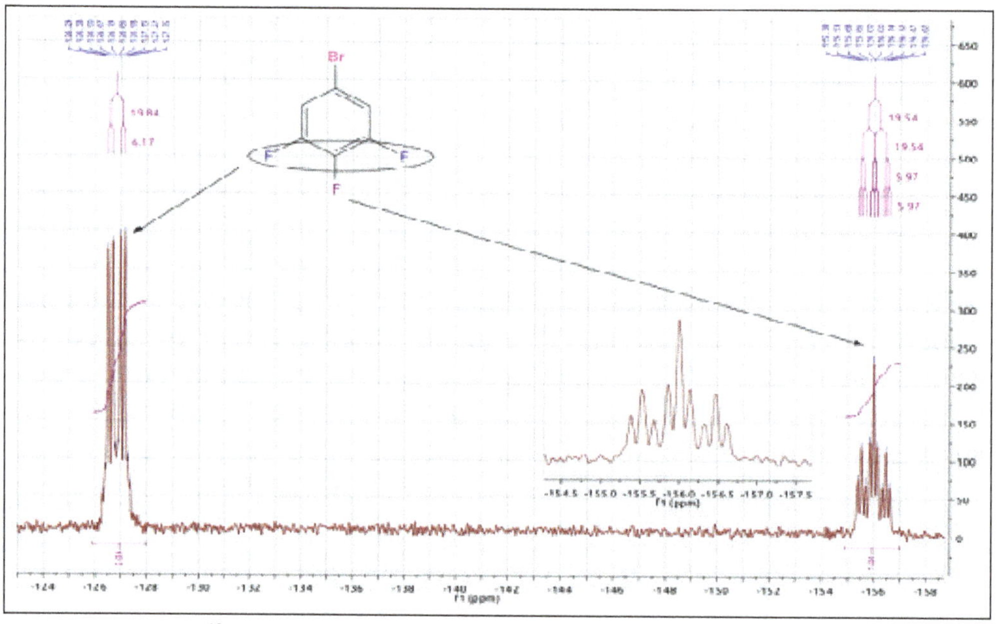

^{19}F NMR spectrum of 1-bromo-3,4,5-trifluorobenzene.

^{31}P NMR Spectroscopy

Phosphorus-31 NMR spectroscopy is used to identify the phosphorous containing organic and inorganic compounds. It is commonly found in organic compounds and coordination complexes (as phosphines, phosphites, phosphonium salts), nucleic acids (phosphate esters), and also other important bio-molecules such as ADP, ATP etc. It is one of the more routine NMR techniques as ^{31}P has an isotopic abundance of 100% and a relatively high gyromagnetic ratio (40.5% of that for ^{1}H). Like ^{1}H and ^{19}F, it has a spin of 1/2 thus making spectra relatively easy to interpret.

85% phosphoric acid (H_3PO_4) is conventionally used as an external standard, which is assigned the chemical shift of 0. The chemical shifts range for phosphorous (0-250 ppm) is much wider than typical for ^1H NMR.

^{31}P shows an interesting pattern in coupling with the neighbouring atoms. The size of the coupling normally decreases dramatically with the number of intervening bonds. A proton directly attached to phosphorous (PH_3) is split by a large coupling constant (one bond coupling) of about 600 Hz. Thus the two components of a doublet will be far apart from each other that it would normally not be accepted as a coupling. The two bond coupling Ph_3P-CH_2- are comparatively smaller (≈ 20-30 Hz), while three bond coupling Ph_3P-CH_2-CH_2- (5-10 Hz) and four bond coupling Ph_3P-CH_2-CH_2-CH_2- are very small (<1 Hz) and rarely observed.

Like the ^{13}C spectra, ^{31}P spectra are recorded with proton decoupling. In ^{31}P NMR spectrum of PPh_3 shows a sharp singlet. If the de-coupler is turned off, the ^{31}P peak will be observed as multiplet or broad due to the coupling with the ortho, meta and para protons of the phenyl rings.

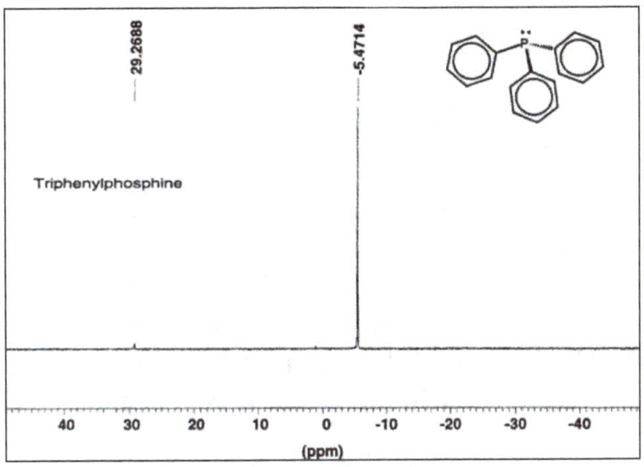

^{31}P NMR spectrum of PPh_3.

Figure represents the ^{31}P NMR spectrum of diethyl phosphonate with ^1H decoupling and showing one-bond and three bond coupling between ^{31}P and ^1H.

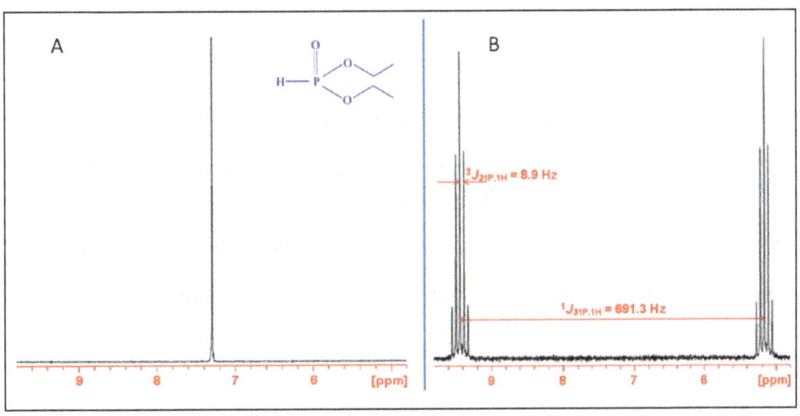

^{31}P NMR spectrum of diethyl phosphonate; A) with ^1H decoupling; B) with onebond and three bond coupling to ^1H.

Figure below shows the ^{31}P NMR and ^1H NMR spectrum of a square planar complex [$MH(PPh_3)_3$]. There are two different sets of triphenylphosphine group in this complex. The two PPh3 groups shown in blue colour are equivalent and the one shown in red colour is different. Hence in the 31P NMR spectrum two sets of peaks will be observed. The red phosphorous nucleus will couple to the two equivalent blue coloured phosphorous nuclei and will give a 1:2:1 triplet. Similarly the blue coloured phosphorous will couple to the single red phosphorous to give a doublet. The coupling constant, J_{P-P} will be same in both the cases. It is important to note here that the coupling between P and H in this spectrum has been removed by the de-coupler and all the coupling is P-P coupling only.

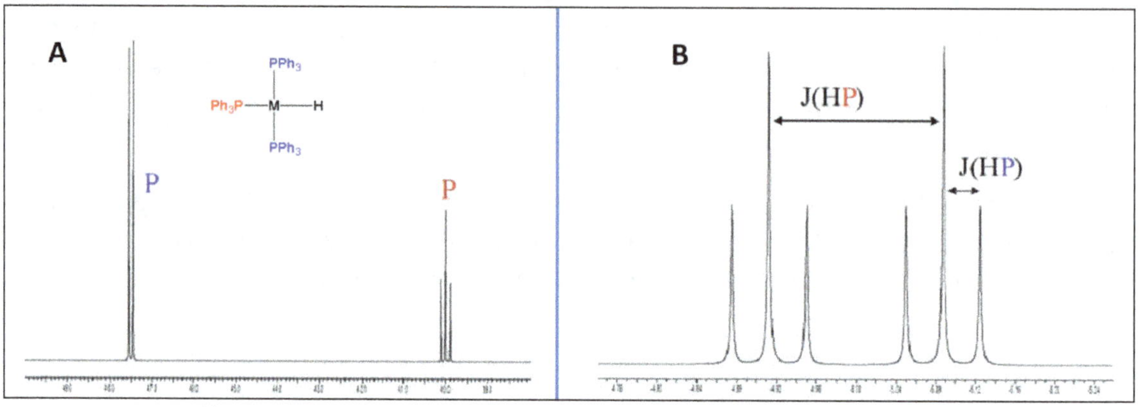

A) ^{31}P NMR B) B) 1H NMR spectrum of square planar complex (signals due to aromatic protons have been removed in the ^{1}H spectrum)

Now consider the ^{1}H NMR spectrum. The hydrogen (in actual hydride attached to metal) will couple to the red PPh_3 to give a doublet. This hydride will also couple to the two equivalent blue PPh_3 to give a 1:2:1 triplet. The first (doublet) splitting would be larger than the second (triplet) splitting (coupling between trans-oriented nuclei is usually larger than the cis-ones). Thus the signal for the proton will be observed as a doublet of triplets. The phenyl protons in the molecule are far away from the hydride atom, hence coupling will not be observed with the aromatic protons.

The hydride atoms attached to metals appear at the extreme right i.e. negative ppm values in the proton spectrum (in the present case from −4.76 to −5.24 ppm). There will, of course, be some signals for the phenyl protons near 7 ppm in the proton spectrum, with an integral of 45 protons (three PPh3 = 9 phenyl rings each with 5 protons).

References

- Optical-spectroscopy, infrared: ru.nl, Retrieved 03, April 2020
- Introductory-chapter-infrared-spectroscopy-a-synopsis-of-the-fundamentals-and-applications, infrared-spectroscopy-principles-advances-and-applications: intechopen.com, Retrieved 27, August 2020
- Analytical-aspects-of-infra-red-spectroscopy, pharma-analysis: pharmatutor.org, Retrieved 30, March 2020
- Nuclear-Magnetic-Resonance-(NMR)-Spectroscopy, A-Organic-Chemistry: libretexts.org, Retrieved 02, June 2020

4

Atomic and Molecular Spectroscopy

Atomic and molecular spectroscopy is a widely used set of analytical techniques. It involves the study of the structure of atoms and molecules. All the diverse principles and theories of atomic and molecular spectroscopy such as Pauli Exclusion Principle and molecular orbital theory have been carefully analyzed in this chapter.

Atomic spectroscopy includes a number of analytical techniques used to determine the elemental composition of a sample (it can be gas, liquid, or solid) by observing its electromagnetic spectrum or its mass spectrum. Element concentrations of a millionth (ppm) or one billionth part (ppb) of the sample can be detected. There are different variations of atomic spectroscopy, emission, absorption, fluorescence, and mass spectroscopy. Determination of an appropriate technique requires a basic understanding of each technique since each has its individual strengths and limitations. Molecular spectroscopy involves the interaction of electromagnetic radiation with materials in order to produce an absorption pattern (i.e. a spectrum) from which structural or compositional information can be deduced. Electromagnetic radiation is conventionally, and rather arbitrarily, divided into a number of distinct regions, each of which covers a range of energies that corresponds to a different type of molecular process.

The interaction of photon of the electromagnetic radiation with matter has given birth to a very important field called spectroscopy. One can go back to the Newton's observation of the spectrum when sun light pass through a prism and light get dispersed in to colour components. The spectrum is basically a series of radiant energy arranged in order of wavelength/frequency.

The spectra can be interpreted for obtaining fundamental information on energy levels, transition probabilities for atoms and molecules. The information so obtained can be used in various kind of analysis like absorption and emission of light, line widths and profile of spectral line and finally came a device named as LASER which became a very important tool as spectroscopy light source. Now a days all types of spectroscopy is done with laser like non-linear spectroscopy, laser.

Raman spectroscopy, time resoled spectroscopy, coherent spectroscopy which further is being used for various applications in chemistry, environmental, biology and medicine. To cite a few, isotope separation with lasers, single molecules detection, atmospheric measurement with LIDAR, spectroscopic detection of water pollution, energy transfer in complexes, cancer diagnostic and therapy and so on.

Neil's Bohr has given an atomic model for hydrogen molecule for explaining the observed spectrum. The model incorporates the idea which was proposed by Rutherford and the concept of light photon given by Einstein (Photoelectric effect) the postulates of the Bohr Theory are:

- An electron in an atom moves in circular orbit around the nucleus under the influence of coulomb field of force.

- Electron in these orbit is in stable state and do not radiate in spite of constant acceleration.
- Out of the infinity number of the orbits, electron exist in the orbit for which the orbital angular momentum $L = n\,h/2\pi$.

The radiation is emitted or absorbed by a transition of the electron from one orbit to another and the frequency is equal to $E_i - E_f$. For hydrogen atom the total energy of electron is given by,

$$E = -\frac{2\pi^2 m_e e^4 z^2}{(4\pi\varepsilon_o)n^2h^2}$$

That is,

$$E = -\frac{13.6(eV)z^2}{n^2}$$

for Hydrogen atom.

In this way from the expression,

$$V = R\infty\left(\frac{1}{n_f^2} - \frac{1}{n_i^2}\right)$$

The spectral lines like Lyman, Balmer, Paschen, Bracket , Pfund are found. Bohr Theory has not taken in the account the fine structure (splitting of spectral lines into several component) of the spectra of hydrogen atom.

Sommerfeld extended the Bohr model and included elliptical orbit. In case of electron in a hydrogen atom, the total energy is negative and hence the path of electron is considered as an ellipse as it is clear from the figure that the radius vector R varies periodically. The Sommerfeld postulated:

- The electron in atoms moves in elliptical path with nucleus at one of the foci.
- Out of the infinite number of the elliptical orbits, only certain discrete orbit following certain conditions occurs.
- The radiation emitted and absorbed when electron undergoes a transition from an orbit of a energy E_i to another orbit of energy E_f and the total energy e is written as exactly the same as derived by Bohr's model of circular orbit.

An electron in an elliptical orbit has velocity that is different at different position, speeding up when the electron is near the nucleus and slow down when it is far away and also precession rate is different for different elliptical orbit so the fine structure splitting of energy levels of the hydrogen atom (n=2 and n=3) are shown in the figure where in the solid lines are observed while the dotted lines are not seen.

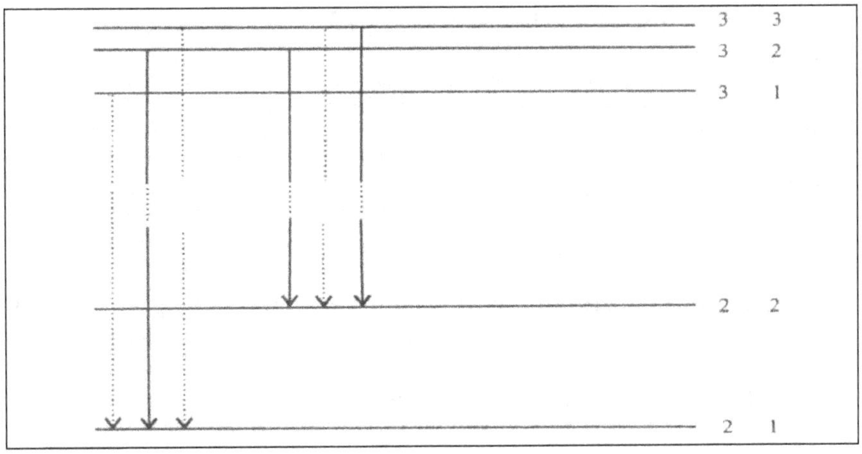

Though the Bohr-Sommerfeld theory has successfully explained the various aspects of atomic spectra of hydrogen like atoms but enable to explain the energy level of atoms with two or more electrons but the theory doesn't explain relative intensities of the spectral lines. These were explained by considering non- relativistic Schrödinger wave equation and the quantum theory of hydrogen atoms. The atomic orbit differs from orbital as:

- The orbits are characterized by principle quantum number n. an orbital doesn't indicate the exact location of an electron. The square of wave function gives the probability finding the electron at a given point in space and hence is characterized by quantum number n, l and m.
- The representation of orbit is two dimensional while that of the orbital is three dimensional and hence spherical Polar coordinates (r, θ and ϕ) are used.
- An orbit having a principal quantum number n can accommodate $2n^2$ electrons (with spin taken into account). The size of the orbital part is determined by the radial part and their shape by angular part the wavefunction.

Notation:

1. The orbitals with l quantum numbers (l=0, 1, 2, 3.....) are denoted by s, p, d, f, g....

2. The principle quantum number n is placed before these letters.

3. The magnetic quantum number m is placed as subscript. For n = 1, l = 0 we have 1s orbital. n = 3, l = 2 we have 3d orbital.

Further we have done with the solution of Schrodinger wave equation and finding $\psi_{nlm}(r,\theta,\phi)$ that contains radial as well as angular parts.

1. The radial part means, the dependence on the radius only describing the values and distribution of electron density in atomic orbitals. It depends on n and l and not on m. this means for 2l+1 functions for a given value of l, there is same radial distribution of electron density.

2. The angular part $\psi_{lm}(\theta,\phi)$ depends on l and m quantum numbers and not on n. Therefore all the orbital with same l have the same form.

3. The quantum no. m determines the orbital orientation.

4. 2l + 1 gives the possible orientation as for:

- s orbital is one.
- p orbital is three.
- d orbital is five.

5. All the possible orientations are equivalent and have the same energy value, so these orbitals are called degenerate.

6. s orbitals are spherically symmetric ; as these depend on radius and are independent of θ and ϕ.

7. The p orbitals exist with n=2.

l = 1, m can take values m= 1, 0,-1.

Three p orbitals are p_1, p_0, $p_{-1.}$

8. The d orbitals exist beginning with n=3 and now for l=2, m =2, 1, 0, -1,-2 five d orbitals are d_2, d_1, d_0, d_1, d_2.

Where $\psi_{nlm}(r,\theta,\phi)=0$ i.e. either $\psi_{lm}(\theta,\phi)=0$ or $R_{nl}(r)=0$: there is some space called nodes or nodal surface.

Electron Spin

Though many aspects could be understood by the introduction of the Schrodinger theory for hydrogen like atom but certain details in the spectrum of hydrogen and of similar atoms were to be understood as the quantum numbers n, l and m alone could not explain certain experimental observations.

- Many spectral lines actually consist of two separate lines that are very close together. For example theory predicts a single line of wavelength 656.3 nm for the first line of Balmer series ($H_\alpha$line). However, two lines 0.14 nm apart were observed for H_α.
- The number of transitions observed when the atoms were placed in an external magnetic field could not be accounted for.

In 1992, Stern and Gerlach measured the possible values of magnetic dipole moments of silver atoms. A beam of silver atoms was used in a region in which there are strong gradient of the magnetic field $\frac{\partial B}{\partial z}$ in z direction (transverse to the beam). If the atom has a magnetic moment μ the there is a force on the atom of magnetic $\mu_z \frac{\partial B}{\partial z}$ and transverse deflection of the beam may be observed. Stren and Gerlach found two traces symmetrically displaced on either side of the beam axis with no un-deflected trace. The magnetic moment of the electron due to orbital motion is $-\beta_e g_1$ L $2\pi/h$ The z component of magnetic moment is proportional to m_1 (magnetic quantum no.) and can take 2l + 1 is odd. If the magnetic moment of an atom is associated only with the orbital motion, there would be an odd number of discrete traces including the undeflected one because of $m_1 = 0$.

This all suggest that there is requirement of another quantum number for which the number of projections on z axis (quantization axis) is even. To explain these observations Uhlenbeck and Goudsmit in 1925 assigned a new angular momentum to the electron. They suggested that the electron in any state spins about its own mechanical axis and therefore the electron has intrinsic angular momentum denoted by s. This spin angular momentum is in addition to the orbital angular momentum. The experimental observations were explained by assigning the electron, a spin angular momentum $\hbar/2$. For spin angular momentum, there are only two orientation of the angular momentum. The idea of electron spin s proved to be successful in explaining the various atomic effects.

The magnitude of s and z components s_z of the spin angular momentum are related to two quantum numbers s and m_s that are identical to those for orbital angular momentum.

$$s = (h/2\pi)\left[s(s+1)\right]^{1/2}$$
$$s = m_s (h/2\pi)$$

Where m_s = ½ and is called spin magnetic quantum number. Now we have four quantum numbers n, l, m_l and m_s to describe each possible state of an atomic electron.

Spin Orbit Interaction

Consider an electron of charge –e moving in Bohr's orbit around the nucleus of charge +Ze. Considering the velocity of the electron relative to the nucleus be v and its position w.r.t. the nucleus is r. As the motion is relative therefore consider the electron to be at rest, while the nucleus is going round the electron in a circle of radius r and velocity –v.

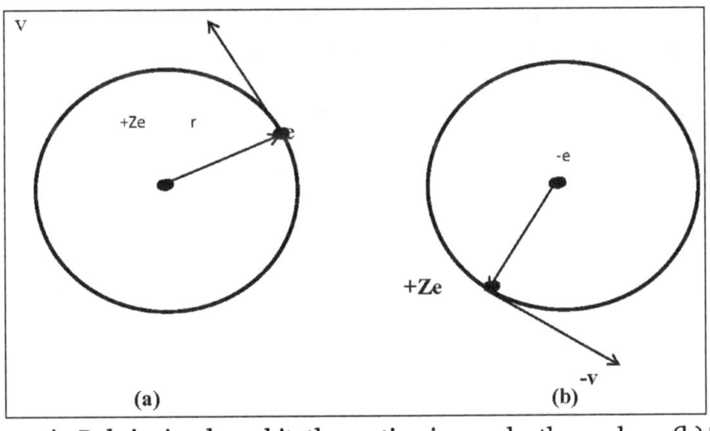

(a) An electron moves in Bohr's circular orbit, the motion is seen by the nucleus. (b) The same motion, but as seen by the electron. From the point view of the view of the electron, the nucleus moves around it.

The motion of the nucleus constitutes a current loop, which produces a magnetic field, and electron is located inside this current loop. The current element j because of motion of nucleus is:

$$j = -Zev$$

According to Ampere's law this produces a magnetic field B that at a distance r is:

$$B = \frac{\mu_o}{4\pi}\frac{j \times r}{r^3} = -\frac{Z_e\mu_o}{4\pi}\frac{v \times r}{r^3}$$

According to Coulomb's law the force between electron of charge –e and nucleus of charge + Ze, The electric field is:

$$E = \frac{Zer}{4\pi\varepsilon_o r^3}$$

$$B = -\varepsilon_o\mu_o v \times E = -\frac{v \times E}{C^2}$$

Here, $c^2 = 1/\varepsilon_o\mu_o$.

The quantity B is magnetic field strength experienced by the electron when it is moving with velocity v w.r.t. the nucleus, and therefore through the electric field of strength E that the nucleus exert on it. The spin – orbit interaction energy ΔE results from the interaction between field B and spin magnetic moment,

$$\mu_s = -\frac{g_{ses}}{2m_e} = -\frac{2\pi g_s \beta_e s}{h}$$

The orientational potential energy of spin magnetic moment in this magnetic field is:

$$\Delta E - \mu_s.B = g_{s\beta_e}\frac{2\pi s.B}{h}$$

But energy has been evaluated in a frame of reference in which the electron is at rest, but the interest is to find its value in the frame of reference in which the nucleus is at rest. Transforming back to frame of references in which nucleus is at rest results in a reduction of the orientational potential energy by a factor of two. Thus, spin - orbit interaction energy (As above) changes to:

$$\Delta E = g_{s\beta_e}\frac{\pi s.B}{h}$$

Using F = – eE,

$$B = \frac{v \times F}{ec^2}$$

and as,

$$F = -\frac{dV}{dr}\frac{r}{r}$$

$$B = -\frac{v \times r}{ec^2 r}\frac{dV}{dr}$$

Multiply and dividing by electron mass m_e,

$$B = \frac{r \times p}{m_e ec^2 r}\frac{dV}{dr}$$

With L = r × p, the above equation reduces to:

$$B = \frac{L}{m_e ec^2 r}\frac{dV}{dr}$$

Magnetic field experienced by the electron because of its motion about the nucleus with orbital angular momentum Lis proportional to the magnetic of L and also that B is in the same direction as L. The spin orbit interaction energy now is:

$$\Delta E = \frac{2\pi\beta_e}{m_e ehc^2 r}\frac{dV}{dr}L.s$$

Energy (wave numbers) is thus,

$$\Gamma = \frac{\Delta E}{hc} = \frac{1}{2m_e hc^3 r}\frac{dV}{dr}L.s = a\ L.s$$

Where a is constant (except for its dependence on r) given by:

$$\frac{1}{2m_e hc^3 r}\frac{dV}{dr}$$

The Vector Model of Atoms

Vector model describes the relationship between the angular momenta and their components in atoms. In quantum mechanics we have the commutation relation among the components of angular momentum L as:

$$[L_x, L_y] = i(h/2\pi)L_z; [L_y, L_z] = i(h/2\pi)L_x; [L_z, L_x] = i(h/2\pi)L_y$$

So, L_x, L and L_z do not commute. If the magnitude of L is considered:

$$L^2 = L_x^2 + L_y^2 + L_z^2$$

Then L^2 commute with L_x, L_y and L_z. Thus L^2 and one of its components can be precisely measured simultaneously. In classical mechanics the angular momentum L is a vector that can have fixed components L_x, L_y and L_z. According to quantum mechanical commutation rules, only one of three components can be fixed. The other two components must vary with time so that L_x and L_y cannot be quantized if L_z is quantized. The vector model is the summary of the general result for angular momentum. The vector model diagram of a single angular momentum vector L is shown in figure. where L appears to precess around z-axis, if it did not precess L_x and are quantized and thus would violate commutation rules. The vector L has a length given by $h/2\pi[L(L+1)]^{1/2}$ where L is an integer and the projection m1 of L along the z-axis can never be perfectly aligned along z-axis.

The figure depicts the precession of orbital angular momentum L about the axis of quantization z. If the angular momentum is only due to spin, then s having a magnitude $h/2\pi[L(L+1)]^{1/2}$ will precess about the z-axis. The projection of s along the z-axis is given by m_s.

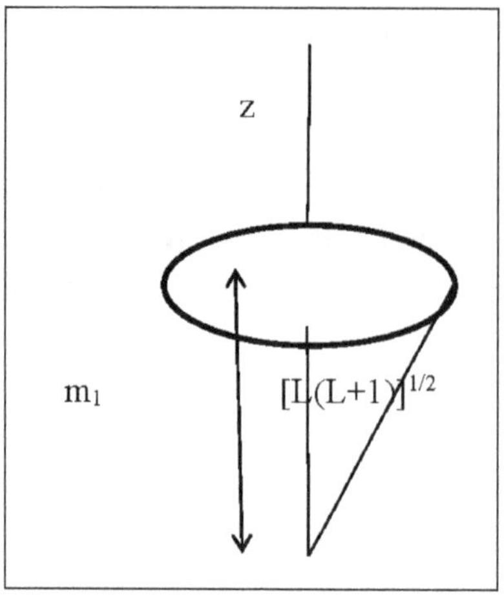

Now considering the two angular momentum L and s that satisfy the commutation rules. If there is no coupling between L and s:

$$m_s\left[s(s+1)\right]^{1/2}$$

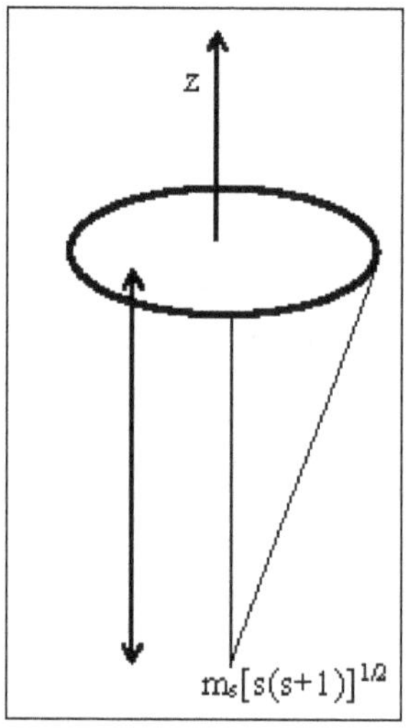

Precession of spin angular momentum s about the axis of quantization z. The vector corresponding to L and s precess independently about z-axis their projection m_l and m_s respectively on the z-axis shown in figure. L and s could precess together and add to form a resultant angular momentum j. The only projection along z-axis in this case would be m_j as shown in above figure.

Pauli Exclusion Principle

The Pauli Exclusion Principle states that, in an atom or molecule, no two electrons can have the same four electronic quantum numbers. As an orbital can contain a maximum of only two electrons, the two electrons must have opposing spins. This means if one is assigned an up-spin (+1/2), the other must be down-spin (-1/2).

Electrons in the same orbital have the same first three quantum numbers, e.g., $n = 1$, $l = 0$, $m_l = 0$ for the 1s subshell. Only two electrons can have these numbers, so that their spin moments must be either $m_s = -1/2$ or $m_s = +1/2$. If the 1s orbital contains only one electron, we have one m_s value and the electron configuration is written as $1s^1$ (corresponding to hydrogen). If it is fully occupied, we have two m_s values, and the electron configuration is $1s^2$ (corresponding to helium). Visually these two cases can be represented as:

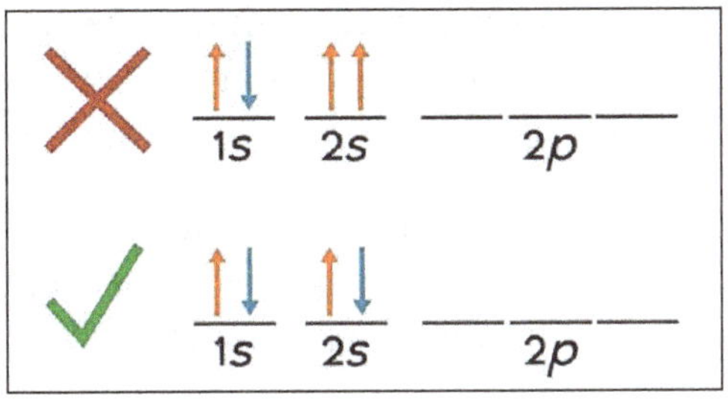

As you can see, the 1s and 2s subshells for beryllium atoms can hold only two electrons and when filled, the electrons must have opposite spins. Otherwise they will have teh same four quantum numbers, in violation of the Pauli Exclusion Principle.

Pauli Exclusion Principle through Quantum Mechanics

The Pauli Exclusion Principle with a single-valued many-particle wave function is equivalent to requiring the wave function to be anti- symmetric. An anti-symmetric two-particle state is represented as a sum of states in which one particle is in one state and the other in other state, and is given by: Anti-symmetry (when the particles are exchanged) means that $A(r, s) = -A(s, r)$. This means $A(r, s) = 0$ when $r = s$ that is, Pauli exclusion.

Quantum mechanically, Pauli's Exclusion Principle dictates that the 'total eigen function of a system of several electrons must be anti-symmetric with respect to exchange of the coordinates (position and spin) of two electrons'.

Considering a system of two non-interacting electrons, one with potential energy, V_1 & the eigen function $\psi(1)$ and the second with values V_2 & $\psi(2)$. As the electrons are non-interacting, total potential energy of the system is $V_1 + V_2$. Now, if it is assumed that the electron '1' is in spin quantum state α and '2' in spin quantum state β, total eigen function of the system is $\Psi_{T=\psi\alpha}(1)_{\psi\beta}(2)$ and on exchange of the particles.

$$\Psi_{T=\psi\beta}(1)_{\psi\alpha}(2)$$

Though the eigenvalues corresponding to Ψ_T and $\Psi_T{}'$ are same yet these functions are not same due to doubly degenerate states, but the probability density functions $\Psi_T * \Psi_T$ and $\Psi_T' * \Psi_T'$ change on the exchange of indistinguishable electrons '1' & '2'. On the other hand, it must not change on exchange of indistinguishable particles.

The total eigen function can be constructed by taking the linear combination of Ψ_T and Ψ_T'. This satisfies time-independent Schrödinger equation and also, the probability density does not change on exchange of particles.

$$\Psi_S = \frac{1}{\sqrt{2}}\left[\psi_\alpha(1)\psi_\beta(2)+\psi_\beta(1)\psi_\alpha(2)\right] \text{ Symmetric wave function.}$$

$$\Psi_A = \frac{1}{\sqrt{2}}\left[\psi_\alpha(1)\psi_\beta(2)+\psi_\beta(1)\psi_\alpha(2)\right] \text{ Anti-symmetric wave function.}$$

Any measurable quantity that can be obtained from Ψ_S and/ or Ψ_A will not be affected by the exchange of particles. If the two electrons are in the same space and spin states, Ψ_A is identically equal to zero. It implies that when the two particles are in a state with the same space & spin quantum numbers, the system can't be described by Ψ_A.

In other words two particles must have different sets of four quantum numbers. For a system of several electrons, the principle is expressed as: A system consisting of several electrons (Fermions) must be described by an Anti-symmetric total eigen function.

Atoms and the Pauli Principle

One of the important consequences of the principle is the elaboration of electron shell structure of atoms and the way atoms share electrons thereby explaining the variety of chemical elements and their combinations. It is known that an electrically neutral atom contains bound electrons equal in number to the protons in the nucleus. Electrons, being fermions, cannot occupy the same quantum state as other electrons, so electrons have different spins (within an atom) at the same electron orbital.

Considering the case of neutral helium atom that has two bound electrons, both can occupy the lowest-energy (1s) states by acquiring opposite spin; as spin is part of the quantum state of the electron, the two electrons are in different quantum states and do not violate the Pauli principle.

But the spin can take only two different values (eigenvalues). Further extending the concept to lithium atom, with three bound electrons, the third electron cannot reside in 1s state, and therefore must occupy one of the higher energy 2s states. The number of electrons in the outermost shell decides the chemical properties. The idea can be extended to successively larger elements where shells of successively higher energy are occupied.

Solid State Properties and the Pauli Principle

In conductors and semiconductors, there are very large numbers of orbital that effectively form a continuous band structure of energy levels. In metal conductors, electrons lie in degenerate states that they cannot contribute much to the thermal capacity of a metal. All most all types of mechanical, electrical, magnetic, optical and chemical properties of solids are the direct consequences of Pauli Exclusion Principle.

Stability of Matter with Pauli Exclusion Principle

The stability of the electrons in an atom is not directly related to the exclusion principle, but the quantum theory of the atom describes it very well. The Heisenberg uncertainty principle explains the idea that close approach of an electron to the nucleus of the atom necessarily increases its kinetic energy. The stability of large systems with many electrons and many nucleons is also explained by the Pauli Exclusion Principle.

The Pauli Exclusion Principle suggest that the bulk matter is stable and occupies volume, as Paul Ehrenfest has pointed out that the electrons of each atom cannot all fall into the lowest-energy orbital and hence must occupy successively larger shells. Atoms cannot be squeezed too closely together and therefore occupy a volume.

Freeman Dyson and Andrew Lenard, considered the balance of attractive; electron–nuclear and repulsive such as electron–electron and nuclear– nuclear forces and showed that ordinary matter would collapse and occupy a much smaller volume without the Pauli principle.

In conclusion, the electrons of the same spin are kept apart by a repulsive exchange interaction that is a short-range effect, acting simultaneously with the long-range electrostatic or Coulombic force. This effect is partly observed in the macroscopic world that two solid objects cannot be in the same place at the same time.

Spin and the Pauli Exclusion Principle

Spin is considered as a rotation of particles around their own axis. The question is whether something as small as an electron is spinning at all. In general, though, spin obeys the same mathematical laws of angular momentum as objects do spinning in classical physics (such as the Earth revolving around its own axis), and there are two important aspects: the speed of rotation and the direction of the axis it rotates about ('up' and 'down').

The concept of spin was first discovered in 1922 by Otto Stern and Walther Gerlach, and they inferred from the experiments that the intrinsic angular momentum, or spin, of a particle such as an electron is quantized which implies that it could only take certain discrete values. The spin of composite particles like protons, neutrons and atomic nuclei, is the sum of the spins and orbital angular momentum of the constituent particles, and is therefore follow the same quantization conditions. Spin is therefore a completely quantum mechanical property of a particle.

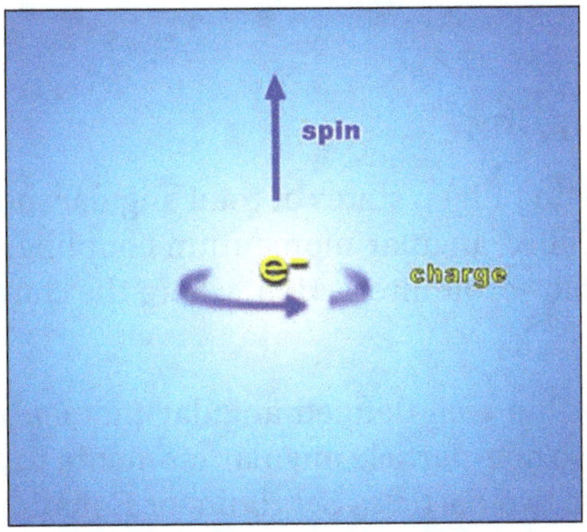

Now the difference between bosons and fermions is understood as: bosons have symmetric wave functions while fermions have anti-symmetric wave functions. Just to understand the concept of spin with half-integer spin that an electron has to spin around TWICE before it presents the same face as earlier. While the probability waves of bosons "flip" or invert before they interfere with each other, Fermions, however, do not flip their probability waves. Therefore, the spins of particles need to be added together using special rules for addition of angular momentum.

Another way of stating the principle is that no two fermions in a quantum system can have the same values of all four quantum numbers at any given time. A very high density white dwarf stars, is explained and also the very existence of different types of atoms in the universe and the large-scale stability and bulk of matter.

By recognizing is the Pauli Exclusion Principle that no two electrons can occupy the same quantum state simultaneously, it effectively stops electrons from "piling up" on top of each other. This explains why matter occupies space exclusively for itself and does not allow other material objects to pass through it and at the same time allows radiations to pass.

The existence of the different atoms in the periodic table of elements and the sheer variety of the universe around us is also understood by this principle. In case an atom gains a new electron, it always goes into the lowest energy state available of course, the outermost shell. Two atoms with 'closed' shells cannot form a chemical bond with each other as the electrons in one atom does not find available quantum states in the other that it can occupy. In this way, the arrangement of electrons in the outermost shell affects the chemical properties of an element and how atoms are able to bond and combine with other atoms. Same way, the molecules interact to form gases, liquids or solids, and how they aggregate themselves in living organisms. The Pauli Exclusion Principle explains the degenerate white dwarf (a type of star) as if two identical particles are forced to have the same quantum numbers, they respond with a repelling outward force, known as "degeneracy pressure" or "Pauli repulsion".

The Pauli exclusion principle is one of the most important principles because the three types of particles (electrons, protons and neutrons) from which ordinary matter is made up concludes that all material particles exhibit space occupying behavior.

Coupling

Angular Momentum Coupling

As the name suggests, constructing Eigen states of total angular momentum out of eigen-states of separate angular momenta is called angular momentum coupling. The orbit and spin of a single particle can interact through spin–orbit interaction, where the complete physical picture includes spin-orbit coupling.

Two charged particles, each with a well-defined angular momentum, may interact by Coulomb forces, where coupling of the two one-particle angular momenta to a total angular momentum is a useful step in the solution of the two particles Schrödinger Equation.

In both the cases the separate angular momenta are no longer constants of motion, but the sum of the two angular momenta usually is still. Angular momentum coupling in atoms is of importance in atomic spectroscopy while angular momentum coupling of electron spins is of importance in quantum chemistry. Likewise, angular momentum coupling in the nuclear shell model is ubiquitous.

In astronomy, spin-orbit coupling reflects the general law of conservation of angular momentum that also holds for celestial systems. The direction of the angular momentum vector is neglected in simple cases. The spin-orbit coupling is the ratio between the frequencies with which a planet or other celestial body spins about its own axis to that with which it orbits another body. This is commonly known as orbital resonance.

Angular Momentum Conservation

The total angular momentum of a system has a constant magnitude and direction if the system is subjected to no external torque. Angular momentum is considered as a property of a physical system:

- The system experiences a spherically symmetric potential field.
- The system moves in isotropic space.

In both the cases the angular momentum operator commutes with the Hamiltonian of the system. Thus the angular momentum and the energy (eigenvalue of the Hamiltonian) can be measured at the same time by using Heisenberg's Uncertainty principle.

An example of the first situation is: An atom whose electrons only experience the Coulomb force of its atomic nucleus. If the electron-electron interaction (and other small interactions such as spin orbit coupling) is ignored, the orbital angular momentum l of each electron commutes with the total Hamiltonian. In this model the atomic Hamiltonian is a sum of kinetic energies of the electrons and the spherically symmetric electron-nucleus interactions. The individual electron angular momenta l_i commute with this Hamiltonian. That is, they are conserved properties of this approximate model of the atom.

An example of the second situation is a rigid rotator moving in field-free space. A rigid rotor has a well-defined, time-independent, angular momentum.

These two situations originate in classical mechanics. The third kind of conserved angular momentum, associated with spin, does not have a classical counterpart. However, all rules of angular momentum coupling apply to spin as well.

In general the conservation of angular momentum implies full rotational symmetry and, conversely, spherical symmetry implies conservation of angular momentum. If two or more physical systems have conserved angular momenta, it can be useful to combine these momenta to a total angular momentum of the combined system—a conserved property of the total system. The building of eigen-states of the total conserved angular momentum from the angular momentum eigen-states of the individual subsystems is referred to as angular momentum coupling.

Application of angular momentum coupling is useful when there is an interaction between subsystems that, without interaction, would have conserved angular momentum. By the very interaction the spherical symmetry of the subsystems is broken, but the angular momentum of the total system remains a constant of motion. Use of the latter fact is helpful in the solution of the Schrödinger equation.

As an example we consider two electrons, 1 and 2, in an atom (Helium atom). If there is no electron-electron interaction, but only electron-nucleus interaction, the two electrons can be rotated around the nucleus independently of each other; nothing happens to their energy. Both operators, l_1 and l_2, are conserved. However, if we switch on the electron-electron interaction that depends on the distance d (1, 2) between the electrons, then only a simultaneous and equal rotation of the two electrons will leave d(1, 2) invariant. In such a case neither l_1 nor l_2 is a constant of motion in general, but the total orbital angular momentum is $L = l_1 + l_2$. Given the Eigen states of l_1 and l_2, the construction of Eigen states of L (that still is conserved) is the coupling of the angular momenta of electrons 1 and 2.

The total orbital angular momentum quantum number L is restricted to integer values. In quantum mechanics, coupling also exists between angular momenta belonging to different Hilbert spaces of a single object, e.g. its spin and its orbital angular momentum. If the spin has half-integer values, such as 1/2 for an electron, then the total (orbital plus spin) angular momentum will also be restricted to half-integer values.

Further, the quantum states of composed systems (i.e. like two Hydrogen atoms or two electrons) in basis sets which are made of tensor products of quantum states that in turn describe the subsystems individually, can be extended. We assume that the states of the subsystems can be chosen as Eigen states of their angular momentum operators (and of their component along any arbitrary z axis).

The subsystems are therefore correctly described by a set of ℓ, m quantum numbers, then there is interaction among the subsystems, the total Hamiltonian contains terms that do not commute with the angular operators acting on the subsystems only. However, these terms do commute with the total angular momentum operator. Sometimes one refers to the non-commuting interaction terms in the Hamiltonian as angular momentum coupling terms, because they necessitate the angular momentum coupling.

Spin-orbit Coupling

The behavior of atoms and smaller particles is well described by the theory of quantum mechanics, in which each particle has an intrinsic angular momentum called spin and specific configurations e.g. electrons in an atom are described by a set of quantum numbers. Collections of particles also have angular momenta and corresponding quantum numbers, and under different circumstances the angular momenta of the parts couple in different ways to form the angular momentum of the whole. Angular momentum coupling is a category including some of the ways that subatomic particles can interact with each other.

In atomic physics, spin-orbit coupling, also known as spin-pairing, describes a weak magnetic interaction, or coupling, of the particle spin and the orbital motion of this particle, e.g. the electron spin and its motion around an atomic nucleus. One of its effects is to separate the energy of internal states of the atom, e.g. spin-aligned and spin-anti-aligned that would otherwise be identical in energy.

L-S Coupling and J-J Coupling

Hamiltonian of the system depends on:

- Kinetic energy of the electrons.
- Potential energy of the electrons due to the interaction between the nucleus and the electrons for these two factors Hamiltonian act as unperturbed Hamiltonian.

- Energy due to interaction between the electrons.
- Spin-spin correlation energy.
- Spin-orbit interaction for these three factors Hamiltonian act as perturbed Hamiltonian.

For lighter elements the spin orbit interaction energy is much less as compared to the interaction energy between electrons and spin-spin correlation energy and there will be L-S coupling for lighter element. For heavier elements the spin orbit interaction energy is much greater than interaction energy between electrons and spin- spin correlation energy and there will be j-j coupling for heavier elements.

L-S Coupling

In L-S coupling it is assumed that, when several electrons are present in an atom, each with a definite value of orbital angular momentum l_i* and spin angular momentum s_i* such that $l_i^* = l_i(l_i + 1)$ h/2π and $s_i^* = s_i(s_i + 1)$ h/2π.

Due to the domination of interaction energy between electrons the orbital angular momentum of indivisible electrons will intract with each other and the total orbital angular momentum is given by L = L(L + 1) h/2π, where L varies from $| l_1 + l_2 + l_3 ....|$minimum to $| l_1 + l_2 + l_3 ....|$maximum the states with different values of L have fairly large energy difference, the state of largest L of being lowest energy. The different levels are designated as S, P, D, F, G........according as L=0, 1, 2, 3, 4, 5....... Thus:

For 3p 4d electrons: $l_1 = 1, l_2 = 2$.

$\therefore \quad L = |l_1-l_2|, |l_1-l_2|+1, ...(l_1+l_2)$

$= 1, 2, 3$ (P, D, F states).

For 3p, 4d, 5f electrons: $l_1=1, l_2=2, l_3=3$. The electrons which are tightly bound to the nucleus will interact first.

$\therefore \quad L = |l_1-l_2|, |l_1-l_2|+1, ...(l_1 + l_2)$

$= 1, 2, 3$ (P, D, F states).

Now L combining with $l_3 = 3$

$\therefore \quad L = 0, 1, 2, 3, 4, 5, 6$(S, P, D, F, G, H, I states).

Due to the domination of spin-spin correlation energy the spin angular momentum of individual electron will interact with each other and the total spin angular momentum is given by $S=\sqrt{S(S+1)}$ h 2π. The quantum number S can take the values:

$S = |s_1 + s_2 + s_3+.........|$min to $|s_1 + s_2 + s_3+.........|$max. The states with different values of S have considerable energy difference, the state of highest S being of lowest energy. The different levels are designated by their multiplicity, (2S + 1). Thus:

For one electron: S = s = ½.

∴ $(2S + 1) = 2$ (doublet level).

For two electrons: $s_1 = ½$, $s_2 = ½$.

∴ $S = |s_1 - s_2|, |s_1 - s_2| + 1, ...(s_1 + s_2) = 0, 1.$

∴ $(2S + 1) = 2$ (singlet and triplet levels)

For three electrons: $s_1 = ½$, $s_2 = ½$, $s_3 = ½$.

To combine three spins, we first combine two of them to obtain S' = 0, 1, and then combine the third S3= ½ to each of them. Thus, if we couple s3 = ½ to s' =0; we get s= ½; and if we couple s3 = ½ to s' =1, we get s= 1/2, 3/2. Thus for three electrons, we get, S =1/2, 1/2, 3/2 (two sets of doublets, and one set of quartets). The followings branching diagram illustrates the possible total spin quantum numbers which can be obtained by combining several independent electron spins.

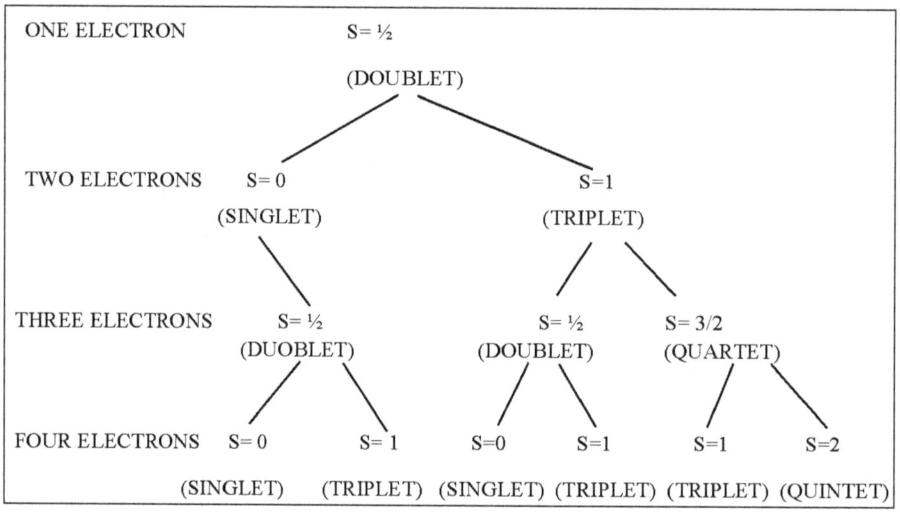

In the general case of N electrons, the possible values of S are:

- S=0,1,2,...........N/2 for even N.
- S= 1/2, 3/2, 5/2.......N/2 for odd N.

The dominant spin-spin correlation and the residual electrostatic interaction having been taken into accounts as a first perturbation, the smaller spin- orbit interaction is included in L-S coupling as an additional perturbation.

As a result of smaller spin – orbit magnetic interaction, the resultant orbital angular momentum vector L and the resultant spin angular momentum vector S are less strongly coupled with each other to form a total angular momentum vector J of the atom:

$$J = L + S$$

The magnitude of J, L and S remaining constant. The magnitude of J is $\sqrt{J(J+1)}\, h/2\pi$, where the quantum number J takes the value:

$$J= |L-S|, |L-S|+1, ...(L-S).$$

J is integral or half – integral according as S is integral or half integral. The number of J values is (2S+1) when L > S, or (2L + 1) when S > L. This mean that due to spin orbit magnet interaction; a level characterized by given values of Land S is further broken up into comparative closer (2S+1) or (2L+1) levels, each characterized by a different J value. The group of these J-levels forms a fine structure multiplet'. The relative spacing of the fine structure levels with in a multiplet is governed by Lande interval rule.

Lande Interval Rule

Under L-S coupling, the spin-orbit interaction energy is of the form:

$$\Delta E_{s,l} = a(L.S),$$

Where a is an interaction constant. Let us write:

$$J = L + S.$$

Taking the scalar self-product we have:

$$J.J = L.L + S.S + 2L.S.$$

$$|J|^2 = |L|^2 + |S|^2 + 2L.S.$$

$$L.S = \frac{1}{2}[|J|^2 - |L|^2 - |S|^2].$$

$$\therefore \quad \Delta E_{s,1} = a/2[J(J+1) - L(L+1) - S(S+1)]h^2/4\pi^2,$$

Because J = J (J+1) h/2π, and so on. We can write it as,

$$\Delta E_{s,1} = A\,[J(J+1) - L(L+1) - S(S+1)],$$

Where, A is another constant.

The various fine-structure levels of a Russell-Saunders multiplet have the same values of L and S, and differ only in the value of J. Hence, the energy difference between two fine-structure levels corresponding to J and J + 1 is:

$$E_{j+1} - E_j = A\,[(J+1)(J+2) - J(J+1)] = 2A(J+1).$$

Thus, the energy interval (spacing) between consecutive levels J and J+1 of a fine-structure multiplet is proportional to J+1, that is, to the larger of the two J-values involved. This is 'Lande interval rule'.

Normal and Inverted Multiplets

Within a given multiplet, usually the level with smallest J value lies lowest. Such a multiplet is a 'Normal' multiplet. There are, however, multiplets in which the largest J level lies lowest. Such multiplets are 'inverted' multiplets.

It is easy to understand how smallest J level lies lowest. The magnet field B produced by the orbital motion of the electron in the electric field of the nucleus is in the same direction as the orbital angular momentum L. In this field the most stable state, that is, the state of the lowest energy must be one in which the spin magnetic moment μs of the electron lies up in the direction of B. We know that μs is directed opposite to S because the electron is negatively charged.

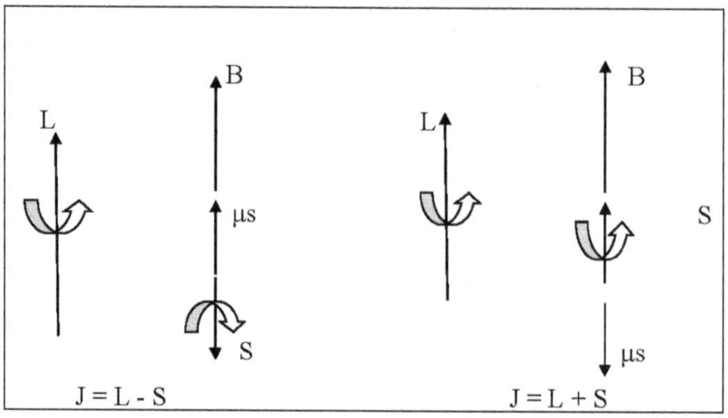

Thus, in the lowest energy state, L and S are in opposite directions and so the value of J is lowest. When L and S are in the same direction, which corresponds to highest J, then μs is opposite to B and the state is least stable. The inverted multiplets arise due to some perturbing influences.

Determination of Spectral Terms for L-S Coupling

(1) Atoms with one optical electron: For hydrogen-like atom, the ground state configuration is 1s.

For this, we have:

$s = ½; 1 = 0,$

So that,

$S = s = ½$; multiplicity $2S+1 = 2$,

$L = l = 0$(S-state),

and,

$J = |L-S|, ... (L+S) = ½.$

Thus, the ground state term of a hydrogen-like atom is:

$^2S_{1/2}$.

The excited state configuration and the corresponding terms for a hydrogen-like atom would be:

2s, 3s, 4s............................. $^2S_{1/2}$

2p, 3p, 4p, ... $^2P_{1/2}$, $^2P_{3/2}$.

The ground state configuration for all alkali atoms is $^2S_{1/2}$.

(2) Atoms with two or more Non-equivalent optical electrons: 4p 4d. For this, we have:

$s_1 = ½, s_2 = ½, l_1 = 1, l_2 = 2.$

The possible values of S and L are:

$$S = |s_1 - s_2|, |s_1 - s_2| + 1 \ldots (s_1 + s_2)$$

$$= 0, 1; \text{multiplicity } (2S + 1) = 1, 3$$

and $L = |l_1 - l_2|, |l_1 - l_2| + 1 \ldots (l_1 + l_2). = 1, 2, 3,$ (P, D, F, state).

Thus, we have in all six terms, three singlet terms and three triplet terms. These terms can be written As:

$^1P, ^1D, ^1F, ^3P, ^3D, ^3F.$

To take into account spin-orbit interaction, let us combine L and S to form J.

$$J = |L-S|, |L-S|+1, \ldots\ldots\ldots\ldots(L + S).$$

For singlet terms, we have:

$S = 0; L = 1; J = 1; ^1P_1.$

$S = 0; L = 2; J = 2; ^1D_2.$

$S = 0; L = 3; J = 3; ^1F_3.$

For triplet terms, we have:

$S = 1; L = 1; J = 0, 1, 2; ^3P_0, ^3P_1, ^3P_2.$

$S = 1; L = 2; J = 1, 2, 3; ^3D_1, ^3D_2, ^3D_3.$

$S = 1; L = 3; J = 2, 3, 4; ^3F_2, ^3F_3, ^3F_4.$

Thus, a single degenerate level of configuration 4p 4d is splitted into 12 levels as shown in figure below.

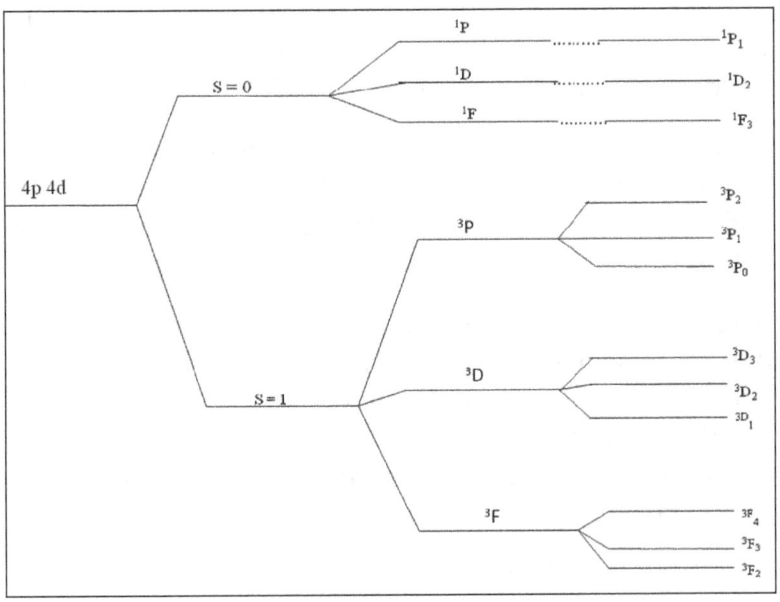

unperturbed level spin-spin correlation energy + residual electrostatic energy + spin-orbit magnetic energy.

Atoms with Two or More Equivalent Electrons

For two equivalent electrons (same n and l values) the values of at least one of the remaining quantum numbers (m_l or m_s) must differ to satisfy Pauli's exclusion principle. Hence terms which were possible for two non-equivalent electrons are now not allowed. Let us now see how to obtain terms from a configuration involving equivalent electrons.

A closed sub-shell, such as s^2, p^6, d^{10}... always forms a $^1s_0$ term only. The closed sub-shell consists of maximum number, 2(2l + 1), of equivalent electrons in antiparallel pairs so that,

$$\Sigma m_l = 0$$
$$\Sigma m_s = 0$$

and this means that,

$$M_l = 0, \ M_s = 0$$

and so,

$$L = 0 \text{ (S - State)},$$

$$S = 0, 2S + 1 = 1 \text{ (Singlet)}$$

and,

$$J = 0.$$

That is, the only possible term is $^1S_0$. Hence, we conclude that when a subshell is completely filled, the only allowed state is one in which the total spin angular momentum, total orbital angular momentum and total angular momentum are all zero. This means that the subshell has no net magnetic dipole moment. The terms of a configuration $(nl)^q$ are the same as the terms of the configuration $(nl)^{r-q}$, where is the maximum number of electrons, that is 2(2l + 1). For example, the terms of p^5 are the same as those of p^1, the terms of p^4 are the same as those of p^2, the terms of d^8 are the same as those of d^2, and so on.

This simplification is based on the fact that a completed subshell like p^6 gives only a $^1s_0$ term (zero angular momentum). This means that the vector addition of the angular momenta of the terms of p^2 to the corresponding quantities for p^4 must give zero. From this it follow that the quantum numbers S and L must be same for p^2 and p^4, that is, the terms of p^2 are the same as those of p^4.

Calculate the spectral terms arising from two equivalent p-electrons (p^2). Let us imagine the atom to be placed in a very strong magnetic field where all the internal coupling are broken down. The individual l and s vectors then precess independently around the magnetic field with quantized components $m_l h/2\pi$ and $m_s h/2\pi$ respectively. The value of l for a p-electron is 1 and hence the values of ml are 1, 0, -1; while those of m_s are +1/2 and -1/2. Now, all the possible combinations of m_l and m_s for a single p-electron are:

$$\begin{array}{llllll} m_l = 1 & 0 & -1 & 1 & 0 & -1 \\ m_s = \tfrac{1}{2} & \tfrac{1}{2} & \tfrac{1}{2} & -\tfrac{1}{2} & -\tfrac{1}{2} & -\tfrac{1}{2} \\ \text{(a)} & \text{(b)} & \text{(c)} & \text{(d)} & \text{(e)} & \text{(f)} \end{array}$$

Thus, there are six possible states (a),(b),(c), (d), (e), (f) in which a single p-electron can exit in an atom. The possible states for two (equivalent) electrons can be obtained by taking all possible combinations of the above six states taken two at a time, with no two alike. There will be 15 such combinations.

ab, ac,ad,ae,af;

bc,bd,be,bf;

cd,ce,cf;

de,df;

ef.

For each of these 15 combinations of very strong field quantum numbers, we add two values of m_l to obtain the strong field values of M_L, and two values of m_s to form M_s [$\sum m_l = M_L$ and $\sum m_s = M_S$].

	Ab	ac	Ad	ae	Af	bc	Bd	Be	bf	cd	ce	cf	De	df	Ef
M_L	1	0	2	1	0	−1	1	0	−1	0	−1	−2	1	0	−1
M_S	1	1	0	0	0	1	0	0	0	0	0	0	−1	−1	−1

The highest value of M_L is 2 which indicates a D term (L = 2). Since this value of M_L occurs only with $M_S = 0$, the term is $^1D(S = 0)$.

$$\left.\begin{array}{lccccc} & ad & ae & af & bf & cf \\ M_L = & 2 & 1 & 0 & -1 & -2 \\ M_S = & 0 & 0 & 0 & 0 & 0 \end{array}\right\}{}^1D.$$

Of the remaining M_L and M_S values, the highest M_L is 1 and the highest M_S is 1. These values must belong to a ^3p term (L=1, S=1).

$$\left.\begin{array}{lccccccccc} & ab & ac & bc & bd & cd & ce & de & df & ef \\ M = & 1 & 0 & -1 & 1 & 0 & -1 & 1 & 1 & -1 \\ M = & 1 & 1 & 1 & 0 & 0 & 0 & -1 & -1 & -1 \end{array}\right\} P.$$

Only one combination be is left for which $M_L = 0$ and $M_S = 0$. It give only ^1S term (L = 0, S = 0).

$$\left.\begin{array}{l} be \\ M_L = 0 \\ M_S = 0 \end{array}\right\}{}^1S$$

Thus, two equivalent p-electrons give rise to ^1D, ^3P and ^1S terms; and no others. The fine-structure levels are $^1D_2$, $^3P_{0,1,2}$ and $^1S_0$.

Orders of Terms and Fine-structure Levels

The relative energies of the various terms and levels which rise from a given electron configuration may be deduced from a set of rules given by Hund. These rules are:

- Of the terms arising from equivalent electrons, those with largest multiplicity lie lowest.
- Of the terms with given multiplicity, and arising from equivalent electrons, that with largest L value lies lowest.
- In the multiplets formed from equivalent electrons in a less half-filled sub-shell, the level with lowest J lies lowest ("normal order").
- In the multiplets formed from equivalent electrons in a more than half-filled subshell, the level with highest J lies lowest ("inverted order").
- Terms arising from half-filled sub-shell show only very slight fine-structure splitting.
- The lowest terms arising from the half-filled sub-shells are the S-terms and are specially stable. These terms are $^2s_{1/2}$ for half-filled s sub-shell, $^4S_{3/2}$ for a half–filled p sub-shell, $^6S_{5/2}$ for a half-filled d sub-shell, and $^8f_{7/2}$ for a half-filled f subshell.

Vector Model for Two-valence Electron Atom under L-S Coupling Scheme

This is a common type of coupling which occurs in most of the lighter atoms. In the vector model for L-S coupling, the individual orbital angular momentum vectors l_1 and l_2 of the two electrons are strongly coupled to each other to form a resultant orbital angular momentum vector L about which both l_1 and l_2 precess rapidly. The corresponding quantum number L can take the values:

$$L = |l_1 - l_2|, \; |l_1 - l_2|+1, \; \ldots\ldots\ldots\ldots (l_1 + l_2).$$

This gives the various terms of the atom. These terms are designated as S, P, D, ...terms accordingly as L = 0, 1, 2,.............................. .

Similarly, the individual spin angular momentum vectors s_1, s_2 of the two electrons are strongly coupled to each other to form a resultant angular momentum vector S about which both s_1 and s_2 precess rapidly. The quantum number S can take the values,

$$S = |s_1 - s_2|, \; |s_1 - s_2| + 1, \; \ldots\ldots\ldots\ldots (s_1 + s_2).$$

Since $s_1 = s_2 = 1/2$, we have,

$$S = 0, 1.$$

Thus, the multiplicity (2S+1) has values 1 and 3; that is, the two (valence) electrons lead to singlet and triplet terms. As a result of spin-orbit interaction, L and S are rather less strongly coupled with each other to form a total angular momentum vector J of the atom, that is,

$$J = L + S.$$

Both L and S precess slowly around J. The quantum number J can take the values,

$$J = |L - S|, \; |L - S|+1, \; \ldots\ldots\ldots\ldots (L + S).$$

Thus, the spin-orbit interaction breaks each level characterized by an L-value in a number of fine-structure levels, each characterized by a j-value. The collection of fine-structure levels is known as a 'multiplet'.

This is a common type of coupling which occurs in most of the lighter atoms. In the vector model for L-S coupling, the individual orbital angular momentum vectors l_1 and l_2 of the two electrons are strongly coupled to each other to form a resultant orbital angular momentum vector L about which both l_1 and l_2 precess rapidly. The corresponding quantum number L can take the values

$$L = |l_1 - l_2|, |l_1 - l_2| + 1, \ldots\ldots (l_1 + l_2).$$

This gives the various terms of the atom. These terms are designated as S, P, D, ...terms accordingly as L = 0, 1, 2,....

Similarly, the individual spin angular momentum vectors s_1 s_2 of the two electrons are strongly coupled to each other to form a resultant angular momentum vector S about which both s1 and s2 precess rapidly. The quantum number S can take the values,

$$S = |s_1 - s_2|, |s_1 - s_2| + 1, \ldots\ldots (s_1 + s_2).$$

Since $s_1 = s_2 = 1/2$, we have,

$$S = 0, 1.$$

Thus, the multiplicity (2S + 1) has values 1 and 3; that is, the two (valence) electrons lead to singlet and triplet terms. As a result of spin-orbit interaction, L and S are rather less strongly coupled with each other to form a total angular momentum vector J of the atom, that is,

$$J = L + S.$$

Both L and S precess slowly around J. The quantum number J can take the values,

$$J = |L-S|, |L-S|+1, \ldots\ldots (L + S).$$

Thus, the spin-orbit interaction breaks each level characterized by an L-value in a number of fine-structure levels, each characterized by a j-value. The collection of fine-structure levels is known as a 'multiplet'.

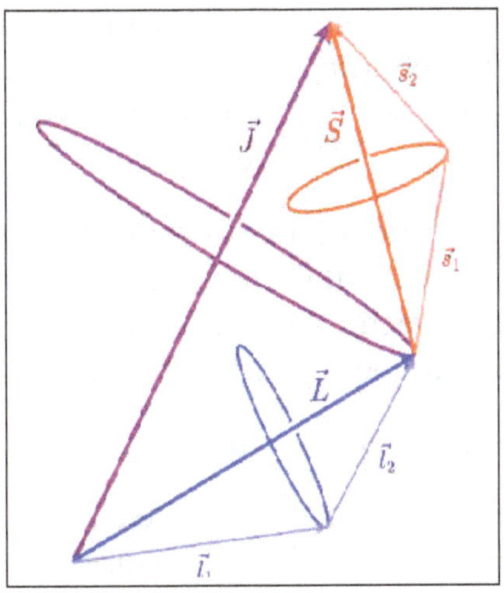

Selection Rules for Multi-electron atoms in L-S Coupling

In the electric-dipole transitions in multi-electron atoms, the selection rules are:

- For one valence electron system:

  $\Delta l = \pm 1$

  $\Delta S = 0$

  $\Delta J = 0, \pm 1$ but $J = 0 \not\leftrightarrow J = 0$.

- For multi-valence electron system:

  $\Delta l = 0, \pm 1$

  $\Delta S = 0$

  $\Delta J = 0, \pm 1$ but $J = 0 \not\leftrightarrow J = 0$.

- There is no restriction on the total quantum number n of either electron.
- Only transition between even and odd terms are allowed for dipole transition.

J-J Coupling

The J-J coupling is an opposite extreme to the ideal L-S coupling and is approached by heavier atoms, for which the spin-orbit (magnetic) interaction term in the Hamiltonian predominates over the residual electrostatic interaction and the spin-spin correlation. This means that the interaction between the orbital and the spin momenta of a single electron is much greater than the interaction between the spin momenta of different electrons. Therefore, in this case the splitting of unperturbed energy level due to the introduction of the various perturbation takes place in the order: (1) spin-orbit interaction, (2) residual electrostatic interaction and spin-spin correlation.

1. Due to strong spin-orbit interaction, the orbital and spin angular momentum vectors of each individual electrons are strongly coupled together to form a resultant angular momentum vector J of magnitude $\sqrt{J(J+1)}h/2\pi$, where $J = l - 1/2$ and $l + 1/2$, that is, J takes half-integral values only. This mean that due to spin-orbit interaction, the unperturbed energy level is splitted into a number of well-spaced levels, each corresponding to a different combination of the possible J-values for the individual optical electrons; the level corresponding to all the electrons having their smaller J value($J = l - 1/2$) being lowest.

2. As a result of the residual electrostatic interaction and spin-spin correlation, the resultant angular momentum vectors J of the individual electrons are less strongly coupled with one another to form the total angular momentum vector J vector of the atom, of magnitude $\sqrt{J(J+1)}h/2\pi$. The total angular momentum quantum number takes the values:

$$J = |J_1 + J_2 + \dots\dots\dots|_{min}, \; |J_1 + J_2 + \dots\dots\dots|_{min+1}, \; \dots\dots\dots\dots (J_1 + J_2 + \dots\dots\dots)$$

This means that each of the above levels is further splitted into a number of levels characterized by different values of J. To understand this coupling let us take some examples,

- For the p-electron: $l_1 = 1$, $s_1 = 1/2$; and so $j_1 = 1/2, 3/2$.
- For the d-electron: $l_2 = 2$, $s_2 = 1/2$; and so $j_2 = 3/2, 5/2$.

This gives four (j_1, j_2) combinations of possible j_1 and j_2 values. These are:

(1/2, 3/2); (1/2, 5/2); (3/2, 3/2) and (3/2, 5/2).

Thus, the spin-orbit interaction splits the unperturbed energy level into four levels, of which (1/2, 3/2) lies lowest and (3/2, 5/2) lies highest.

Each of the above four levels is further spiltted by residual electrostatic interaction and spin-spin correlation into a number of J- levels, equal to the number of integrally spaced values of J that can be formed out of the two J-values. The above four (j_1, j_2) combination give J values as under:

(1/2, 3/2) gives J = 1, 2.

(1/2, 5/2) gives J = 2, 3.

(3/2, 3/2) gives J = 0, 1, 2, 3.

(3/2, 5/2) gives J = 1, 2, 3, 4.

The complete splitting is shown in the following figure.

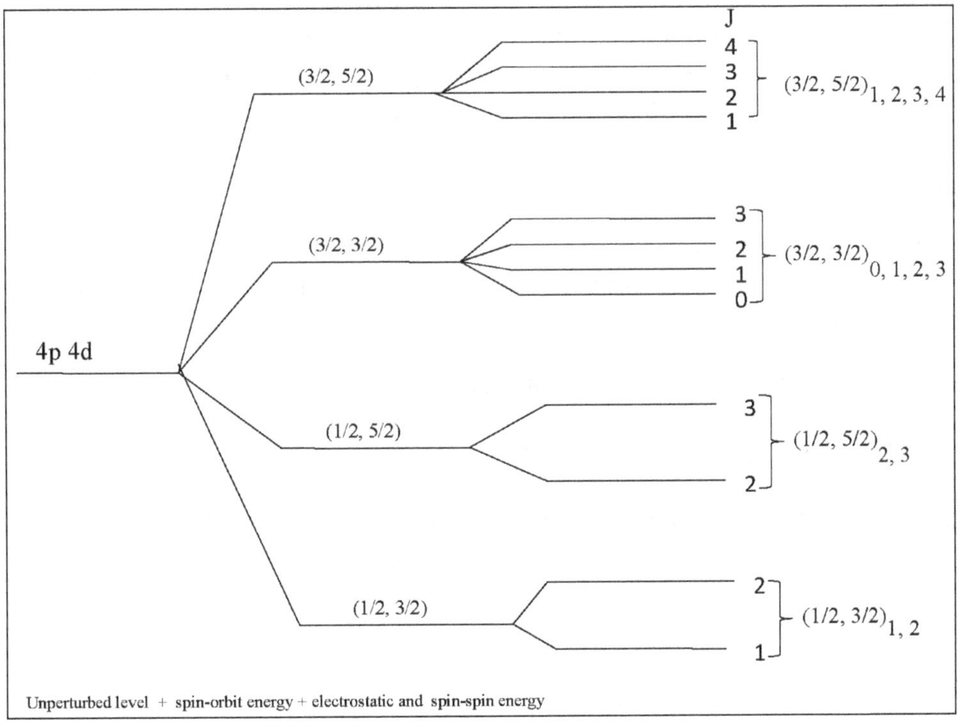

Spin-spin Coupling

Spin-spin coupling is the coupling of the intrinsic angular momentum (spin) of different particles. Such coupling between pairs of nuclear spins is an important feature of nuclear magnetic resonance (NMR) spectroscopy as it can provide detailed information about the structure and conformation of molecules. Spin-spin coupling between nuclear spin and electronic spin is responsible for hyperfine structure in atomic spectra.

Term symbols are used to represent the states and spectral transitions of atoms, they are found from coupling of angular momenta mentioned above. When the state of an atom has been specified with a term symbol, the allowed transitions can be found through selection rules by considering which transitions would conserve angular momentum. A photon has spin 1, and when there is a transition with emission or absorption of a photon the atom will need to change state to conserve angular momentum. The term symbol selection rules are $\Delta S = 0$, $\Delta L = 0, \pm 1$, $\Delta l = \pm 1$, $\Delta J = 0, \pm 1$.

The expression "term symbol" is derived from the "term series" associated with the Rydberg states of an atom and their energy levels. In the Rydberg formula the frequency or wave number of the light emitted by a hydrogen-like atom is proportional to the difference between the two terms of a transition. The series known to early spectroscopy were designated sharp, principal, diffuse and fundamental and consequently the letters S, P, D, and F were used to represent the orbital angular momentum states of an atom.

In atomic nuclei, the spin-orbit interaction is much stronger than for atomic electrons, and is incorporated directly into the nuclear shell model. In addition, unlike atomic-electron term symbols, the lowest energy state is not L − S, but rather, $\ell + s$. All nuclear levels whose ℓ value (orbital angular momentum) is greater than zero are thus split in the shell model to create states designated by $\ell + s$ and $\ell - s$. Due to the nature of the shell model, which assumes an average potential rather than a central Coulombic potential, the nucleons that go into the $\ell + s$ and $\ell - s$ nuclear states are considered degenerate within each orbital (e.g. The 2p3/2 contains four nucleons, all of the same energy. Higher in energy is the 2p1/2 which contains two equal-energy nucleons).

Selection Rules in J-J Coupling

- The parity of the configuration must change in an electric-dipole transition (Laporte rule). This means that if only one electron can jumps in the transition then for this electron we must have $\Delta l = + 1$. If two electrons jump then $\Delta l_1 = + 1$ and $\Delta l_2 = 0, + 2$.
- $\Delta j = 0, + 1$ for the jumping electron, and $\Delta j = 0$ for all the other electrons.
- For the atom as a whole, $\Delta j = 0, + 1$ but $J = 0 \not\leftrightarrow J = 0$.

Vector Model for Two-valence Electron Atom under J-J Coupling Scheme

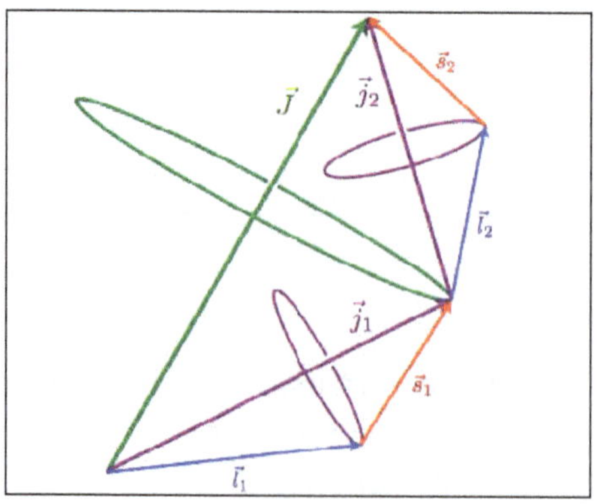

In the vector model of a two-electron atom under j-j coupling scheme, the orbital and the spin vectors l_1 and s_1 of one electron are strongly coupled to each other to form a resultant j_1 about which l_1 and s_1 precess rapidly. Similarly, l_2 and s_2 of the other electron form j_2. The vectors j_1 and j_2 are less strongly coupled to each other and form the total angular momentum vector J of the atom. The vectors j_1 and j_2 precess rather slowly about J shown in figure.

Molecular Orbital Theory

Atomic and Molecular Orbitals

By sharing electron, molecules can form bonds, and it is possible to regard the sharing of two electrons by two atoms as constituting a chemical bond. Atoms can share one, two or three electrons (forming single, double and triple bonds).

A hydrogen atom consists of a nucleus (a proton) and an electron. It is not possible to accurately determine the position of the electron, but it is possible to calculate the probability of finding the electron at any point around the nucleus. With a hydrogen atom the probability distribution is spherical around the nucleus and it is possible to draw a spherical boundary surface, inside which there is a 95% possibility of finding the electron. The electron has a fixed energy and a fixed spatial distribution called an orbital. In the helium atom there are two electrons associated with the helium nucleus. The electrons have the same spatial distribution and energy (i.e. they occupy the same orbital), but they differ in their spin (Pauli exclusion principle). In general: electrons in atomic nuclei occupy orbitals of fixed energy and spatial distribution, and each orbital only contains a maximum of two electrons with anti-parallel spins.

In physics, periodic phenomena are associated with a "wave equation", and in atomic theory the relevant equation is called the "Schrödinger Equation". The wave equation predicts discrete solutions in one dimension for a particle confined to a box with infinite walls, The solutions can be shown as in the figure below:

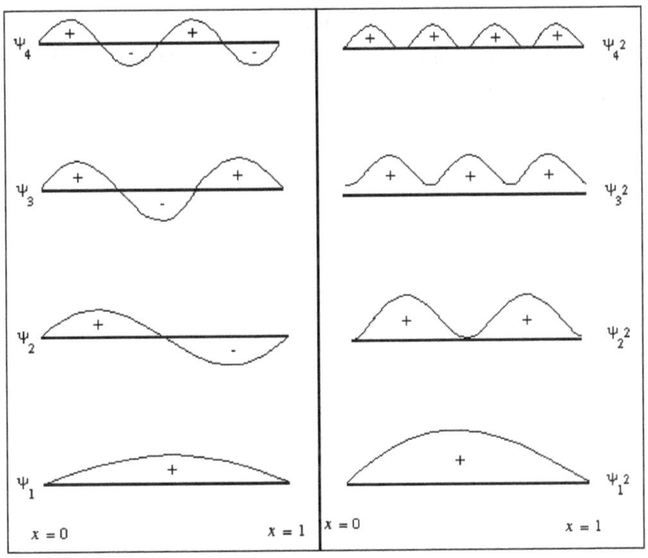

$\psi_1-\psi_4$ represent solutions of increasing energy. In three dimensions, the equation determines the energy and defines the spatial distribution of each electron. Solutions of the wave equations in three-dimensions allows calculation of the "shape" of each orbital. The first five solutions of the wave equation for an electron associated with a proton can be shown in the figure below:

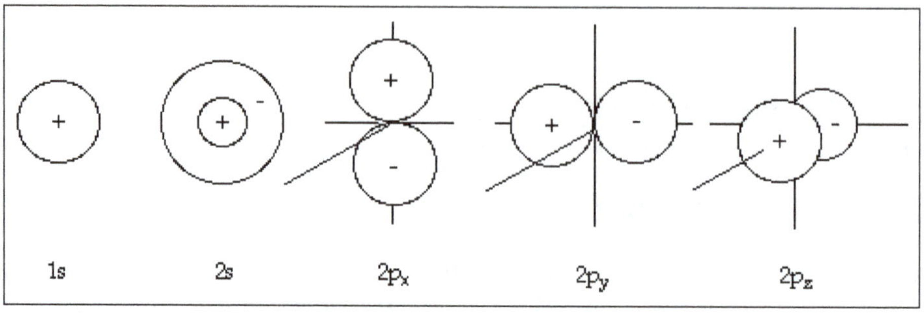

In the hydrogen atom, the 1s atomic orbital has the lowest energy, while the remainder (2s, $2p_x$, $2p_y$ and $2p_z$) are of equal energy (i.e. degenerate), but for all other atoms, the 2s atomic orbital is of lower energy than the $2p_x$, $2p_y$ and $2p_z$ orbitals, which are degenerate.

In atoms, electrons occupy atomic orbitals, but in molecules they occupy similar molecular orbitals which surround the molecule. The simplest molecule is hydrogen, which can be considered to be made up of two separate protons and electrons. There are two molecular orbitals for hydrogen, the lower energy orbital has its greater electron density between the two nuclei. This is the bonding molecular orbital - and is of lower energy than the two 1s atomic orbitals of hydrogen atoms making this orbital more stable than two separated atomic hydrogen orbitals. The upper molecular orbital has a node in the electronic wave function and the electron density is low between the two positively charged nuclei. The energy of the upper orbital is greater than that of the 1s atomic orbital, and such an orbital is called an antibonding molecular orbital.

Normally, the two electrons in hydrogen occupy the bonding molecular orbital, with anti-parallel spins. If molecular hydrogen is irradiated by ultra-violet (UV) light, the molecule may absorb the energy, and promote one electron into its antibonding orbital (s^*), and the atoms will separate. The energy levels in a hydrogen molecule can be represented in a diagram - showing how the two 1s atomic orbitals combine to form two molecular orbitals, one bonding (s) and one antibonding (s^*). This is shown below - by clicking upon either the s or s^* molecular orbital in the diagram - it will show graphically in a window to the right:

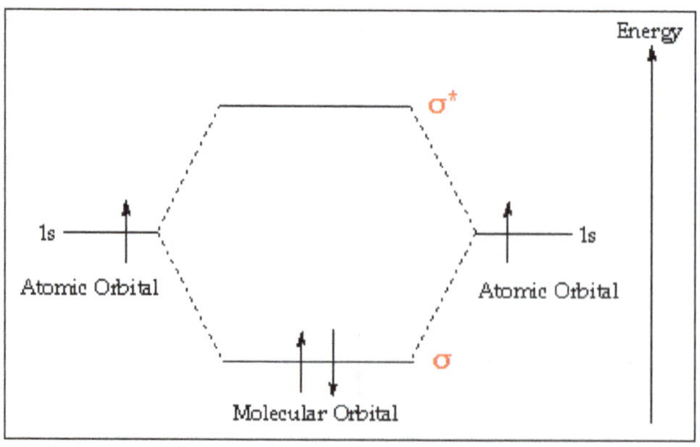

Orbitals for Selected Molecules

Saturated Molecules

These are molecules in which all valence electrons are involved in the formation of single bonds. There are no non-bonded lone pairs. These molecules are generally less reactive than either electron-rich or electron-deficient species, with all occupied orbitals having relatively low energies.

Methane

The valence molecular orbitals of methane are delocalized over the entire nuclear skeleton - that is, it is not easy to assign any one orbital to a particular C-H bond. It is possible to see how complex the orbital structure becomes with the increase in energy. Methane has four valence molecular orbitals (bonding), consisting of one orbital with one nodal plane (lowest occupied) and three degenerate (equal energy) orbitals that do have a nodal plane.

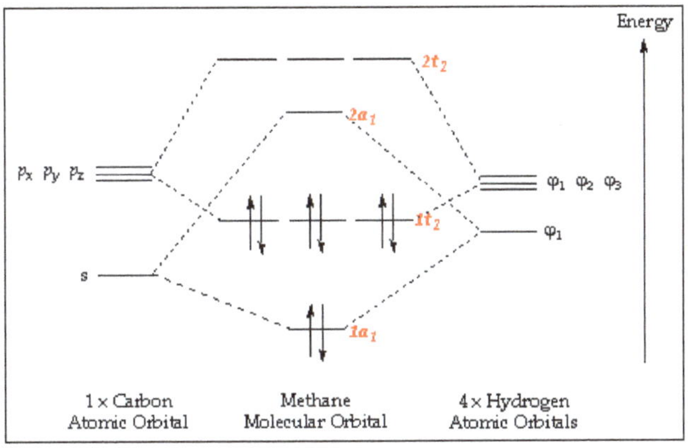

Ethane

The ethane molecule has fourteen valence electrons occupying seven bonding molecular orbitals. As can be seen from the energy diagram - four of the molecular orbitals occur as degenerate pairs. Like in methane - the molecular orbitals of ethane show increasing nodal structure with increasing orbital energy. Energy diagram and pictorial view of the orbitals shown in figure below:

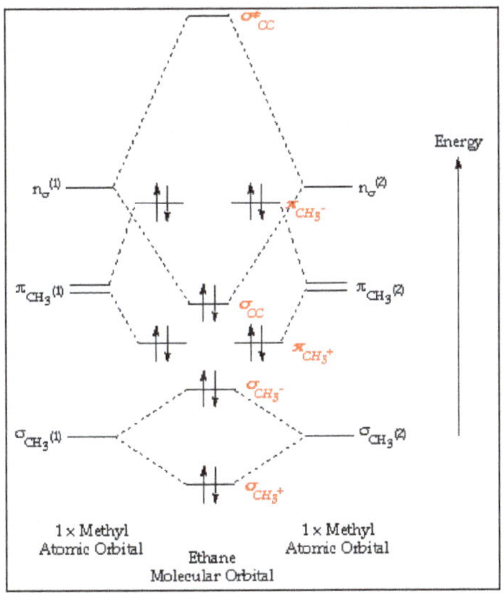

Molecules with Double Bonds

In molecules where the number of bonding electron pairs exceeds the number of unions between atoms, the extra electrons occupy higher energy molecular orbitals than the orbitals found in molecules where the number of bonding electron pairs equals the number of unions between atoms. These are double bonds, and the orbitals have a nodal plane containing the atoms sharing these p π-type orbitals.

Ethene

The simplest alkene is ethene. Its chemistry is dominated by two "frontier orbitals", that is the Highest Occupied Molecular Orbital (HOMO) and the Lowest Unoccupied Molecular Orbital (LUMO). For the ethene orbital energy diagram these are shown as π_{CC} for the HOMO, and π_{CC} for the LUMO. An important property of the ethene molecule, and alkenes in general is the existence of a high barrier to rotation about the C=C which tends to hold the molecule flat. Energy diagram and pictorial view of the orbitals shown below:

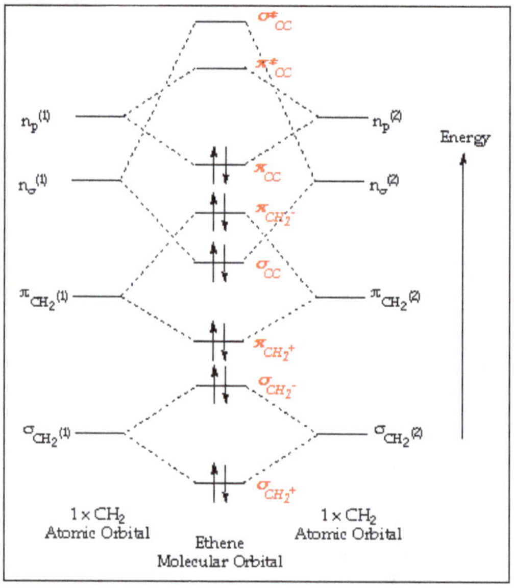

Molecules with Triple Bonds

Ethyne

Energy diagram and pictorial view of the orbitals shown below in figure.

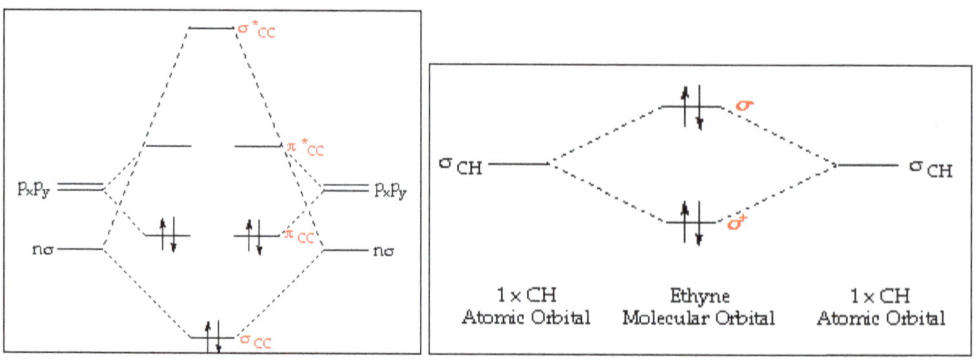

Molecules with Electron Lone Pairs

Hydrogen Fluoride

A simple diatomic molecule is Hydrogen fluoride. There are eight valence electrons which occupy four molecular orbitals. The two highest energy MO's are degenerate, are π-type and have no electron density associated with the hydrogen atom, i.e. they are Non-Bonding Orbitals (NBO) and in Lewis Theory are represented as two "Lone Pairs". Another important difference between Hydrogen Fluoride and previous molecules is that the electron density is not equally distributed about the molecule. There is a much greater electron density around the fluorine atom. This is because fluorine is an extremely electronegative element, and in each bonding molecular orbital, fluorine will take a greater share of the electron density.

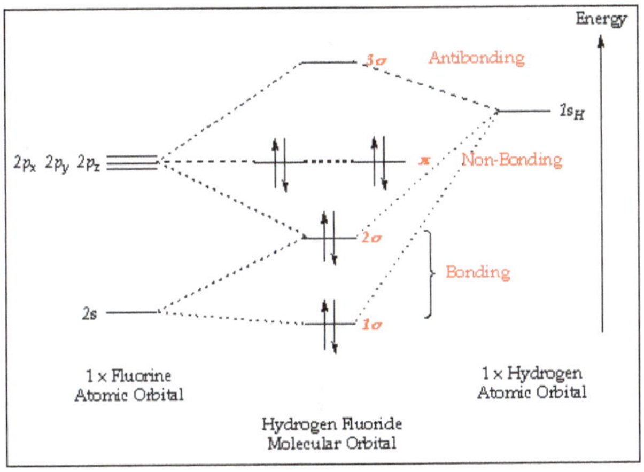

Water

In the water molecule the highest occupied orbital, ($1b_1$) is non-bonding and highly localized on the oxygen atom, similar to the non-bonding orbitals of hydrogen fluoride. The next lowest orbital ($2a_1$) can be thought of as a non-bonding orbital, as it has a lobe pointing away from the two hydrogens. From the lower energy bonding orbitals, it is possible to see that oxygen also takes more than its "fair share" of the total electron density.

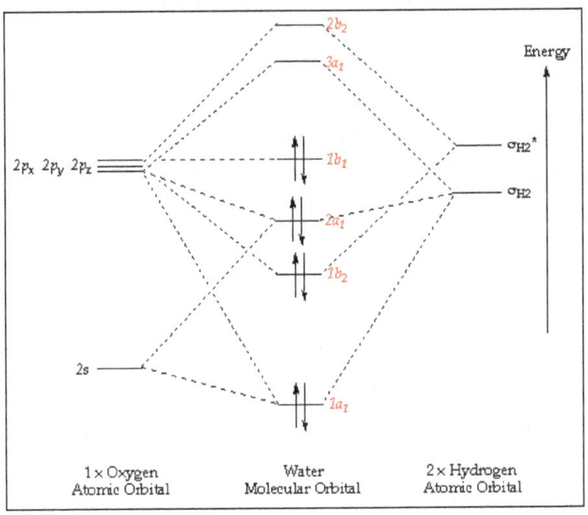

Ammonia

Ammonia has two pairs of degenerate orbitals, one bonding and one antibonding, and like hydrogen fluoride and water has a non-bonding orbital ($2a_1$). This highest occupied orbital has a lobe pointing away from the three hydrogens, and corresponds to a lone pair orbital localized upon the nitrogen, whereas the three lowest energy MO's lead to the description of the three N-H bonds of the Lewis structure. The lone pair is relatively high in energy, and is responsible for the well-known Lewis base properties of ammonia.

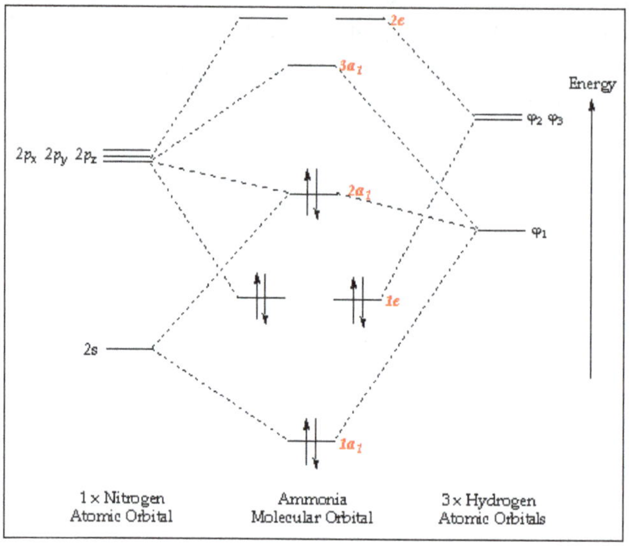

The next molecule in the series HF, H_2O and H_3N, is H_4C (methane) - and unlike the other three molecules has no non-bonding orbitals.

Conjugated and Aromatic Molecules

π bonds in close proximity will often interact. Some of the delocalized molecular orbitals that result will be stabilized, while others will be destabilized. The individual combinations may be polarized, providing an increase in wave function amplitude on some centers at the expense of a decrease in amplitude on others. This gives rise to the possibility of more varied reactivity patterns than are observed for simple alkenes.

Aromatic molecules exhibit a wide range of reactivity patterns toward both electron rich and electron deficient species. These mainly depend on the structures and energies of the frontier π-type molecular orbitals, the HOMO and LUMO. Except for non-bonded lone pairs, the s framework plays little role in the overall reactivity.

trans-1, 3-Butadiene

The energies of the p-molecular orbitals of conjugated molecules like butadiene, - occur in pairs, with their energies equal to $(\alpha \pm x\beta)$, where α and β are constants. For each bonding orbital of and energy $\alpha + x\beta$ there is a corresponding antibonding orbital of energy $\alpha + x\beta$. The π-molecular orbitals are extended over the whole molecule. For butadiene, the p manifold contains four electrons, leading to an electronic configuration of $\pi_1^2\, \pi_2^2$.

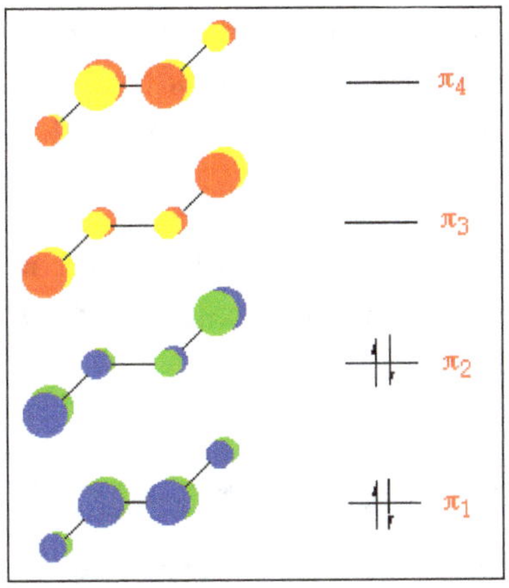

Allyl Radical

The radicals allyl and pentadienyl have the same arrangement of p-orbitals, (i.e. the occur in pairs of energy), but because there is an odd number of carbon atoms in the conjugate chain, there must be a non-bonding orbital with energy x=0. Also, because of the pairing properties of the p-molecular orbitals of conjugated chains, there will be a node at every alternate carbon atom in the non-bonding orbital. This is important for the unpaired electron of allyl, which will occupy this non-bonding orbital. If an electron is added to the allyl radical to form the anion, the negative charge will appear at the terminal carbon atoms. If the unpaired electron is removed forming the cation, the resulting positive charge is also spread over the terminal carbon atoms.

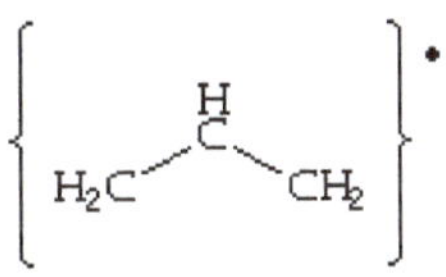

Radicals allyl.

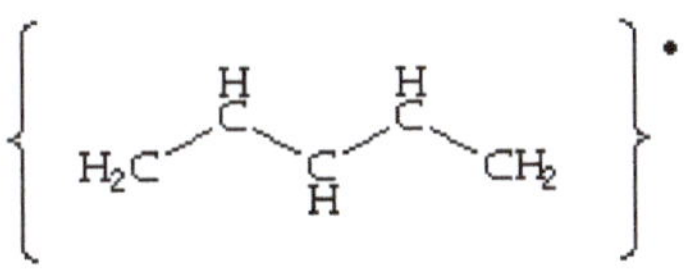

Pentadienyl.

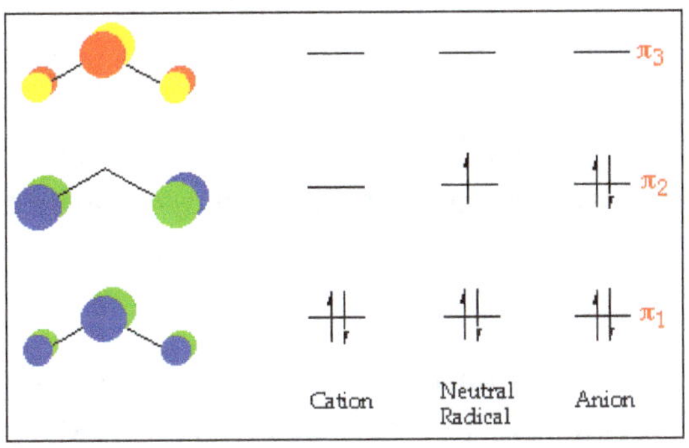

There are three p-molecular orbitals for allyl, the π_1 is bonding, the π_2 orbital is non-bonding and the π_3 is anti-bonding. In the neutral allyl species - there are a total of seventeen valence electrons - of which three fill the π-orbital manifold. A pictorial representation of the energy diagram for the neutral, cationic and anionic allyl species are shown below - (orbitals are shown only for the cationic species).

In the pentadienyl anion, the negative charge is centred on the carbon atoms in the 1, 3 and 5 position - similarly with the positive charge for the cation. These ions are represented in resonance theory as two or three canonical forms:

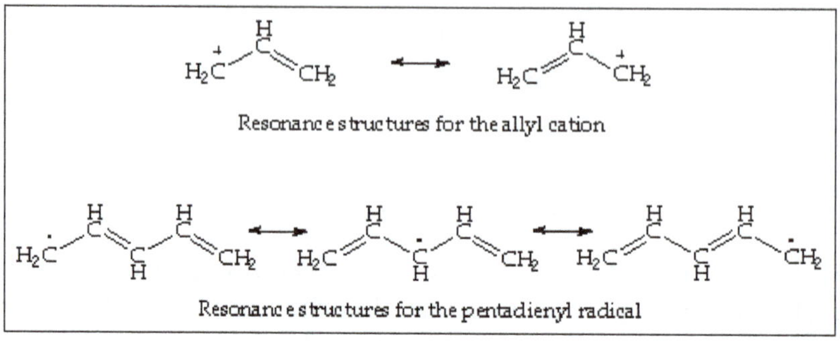

The delocalisation of π -electrons is associated with a lowering of the orbital energy. Therefore the total energy of the occupied π-orbitals of butadiene is lower in energy then two isolated ethene-type double bonds. Further delocalisation of p-electrons occurs in aromatic hydrocarbons. A figure showing the comparative energy levels of the p-orbitals of the cyclic molecules C_nH_n, for n = 3-6, is shown below:

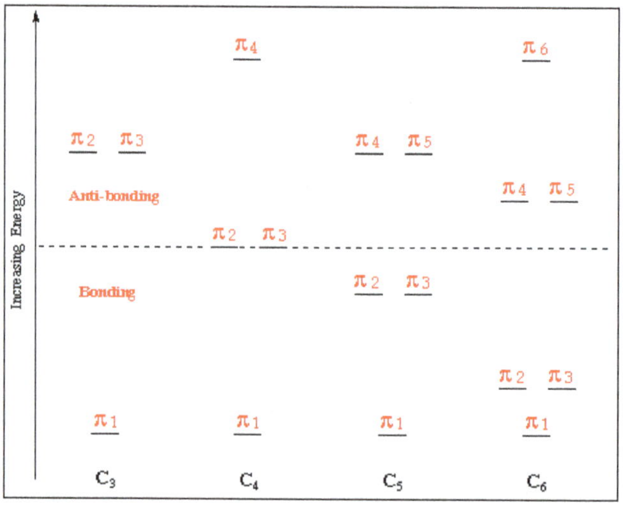

In all cyclic polyenes (C_nH_n), the p-molecular orbitals occur in degenerate pairs, except for the lowest p-orbital, and for the cyclic polyenes with even numbers of carbon atoms, the highest π-orbital.

Cyclobutadiene

From the cyclic polyene diagram - the square molecule cyclobutadiene (C_4H_4) has four p-orbitals, a bonding orbital (π_1), two degenerate non-bonding orbitals $(\pi_2 \text{ and } \pi_3)$ and an anti-bonding orbital (π_4). Four electrons are placed into these four orbitals; twon into the bonding orbital, and one each with parallel spins into the degenerate non-bonding orbitals (Hund's rule).

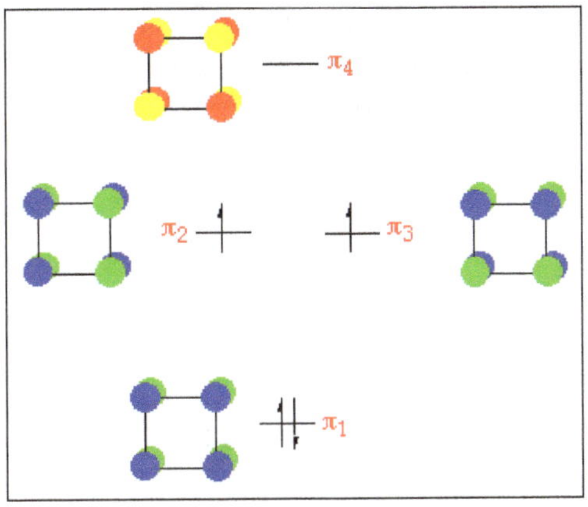

There is no reason to expect cyclobutadiene to be square, theoretically calculations show an oblong with two double bonds structure has lower energy. Experimentally it is shown that cyclobutadiene acts a very strained cyclo-olefin, rather than as a bi-radical species.

Cyclopentadiene

This molecule is a relatively acidic hydrocarbon, and the anion is formed by the treatment of cyclopentadiene with a strong base. From the cyclic polyene diagram it can be seen that cyclopentadiene has three p bonding orbitals which are delocalised over the five carbon atoms. The uppermost π bonding orbitals are a degenerate pair, and are the highest occupied molecular orbitals (HOMO's).

These orbitals are much higher in energy than those in neutral aromatic species such as benzene, indicating that this anion is far more susceptible t attack by electrophiles. The anion is far more capable of coordinating to transition metals with available empty d-orbitals. The anion has six π-electrons, making the system aromatic. The six electrons are arranged as in the diagram below:

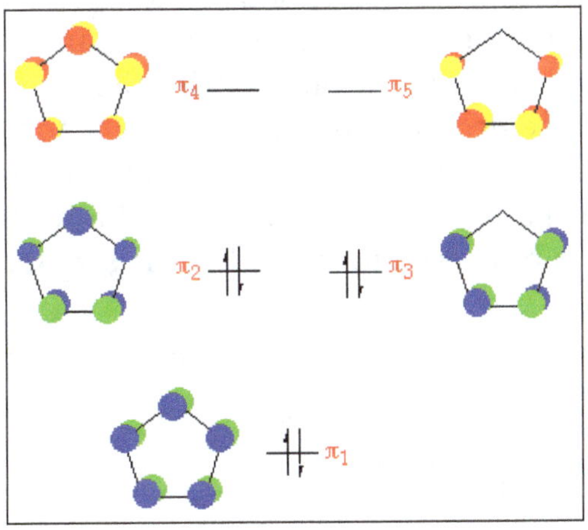

Benzene

Benzene is the archetypal aromatic compound. It has a symmetrical π system and so is not over reactive on any one site. From the cyclic polyene diagram it can be seen that benzene has six π-molecular orbitals, (which contain the six π-electrons), three bonding and three anti-bonding. The upper bonding degenerate pair of orbitals are the HOMO's of benzene. The π orbital manifold, is shown below - but also of interest is that the pattern of the π orbitals is repeated within the σ system. The σ functions - like the π orbitals are delocalized throughout the carbon skeleton. The six π electrons are arranged as in the diagram below:

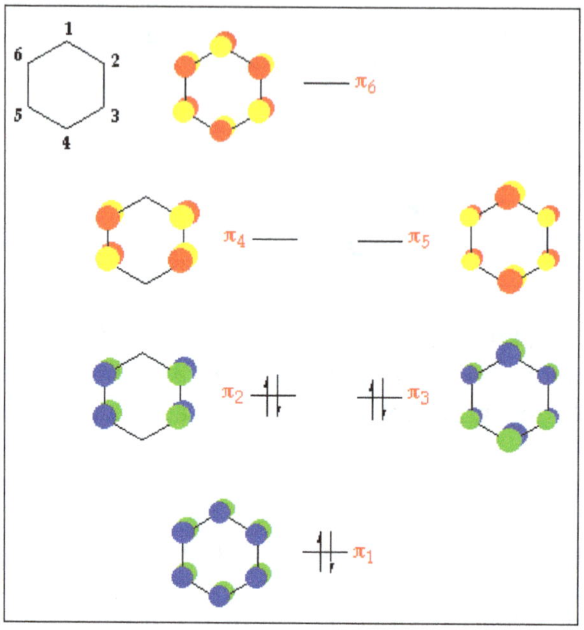

Bonding in Benzene

From the above diagram it can be seen that the lowest lying orbital, π_1, the orbital coefficients are such that the bonding character between each pair of adjacent carbon atoms is equal. In π_2 bonding only occurs between atoms C_2 and C_3 and between C_5 and C_6 since the coefficients on C_1 and C_4 are zero. In π_3, C_1, which is bonded to C_2 and C_6 and C_4 is bonded to C_3 and C_5, there are anti-bonding interactions between C_2 and C_3 and between C_5 and C_6. Therefore if we consider the pair of orbitals π_2 and π_3 the contribution to the C-C p bonding is equal for each bond. Since there are three occupied bonding orbitals and six CC linkages - the p bond order is $^1/_2$. This description is in accord with the two resonating mesomeric forms (or Kekulé structures in a) below in which single and double bond characters alternate around the ring. Conventionally, the diagram in b) is used to show that the six electrons are delocalized around the ring:

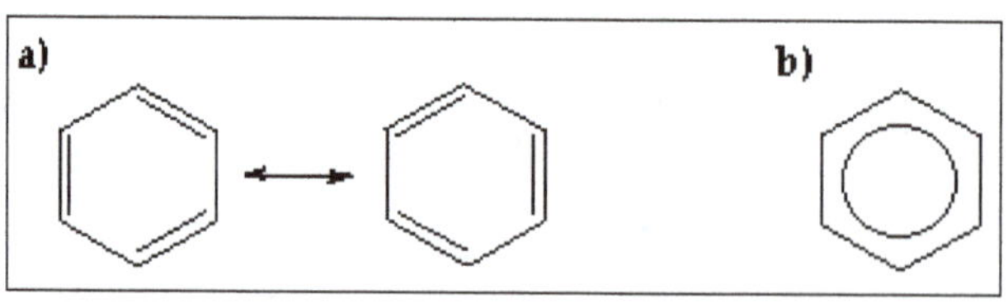

A_2 Molecules

Nitrogen

This molecule has ten electrons. The atomic orbitals combine to produce the following molecular orbital diagram.

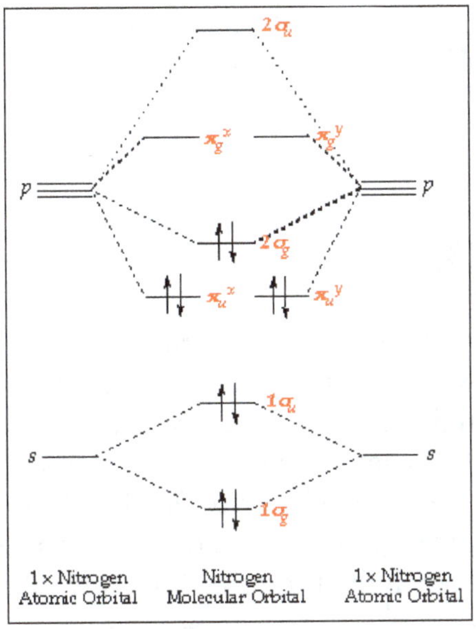

Here the $2\pi_g$ orbital is occupied by two electrons to give a total bond order of three. This corresponds well with the Lewis structure $(N \equiv N)$, although the orbital approach tells us that there is one σ and two π.

Oxygen

This molecule has twelve electrons, two more than nitrogen - and these extra two are placed in a pair of degenerate π_g orbitals. The atomic orbitals combine to produce the following molecular orbital diagram:

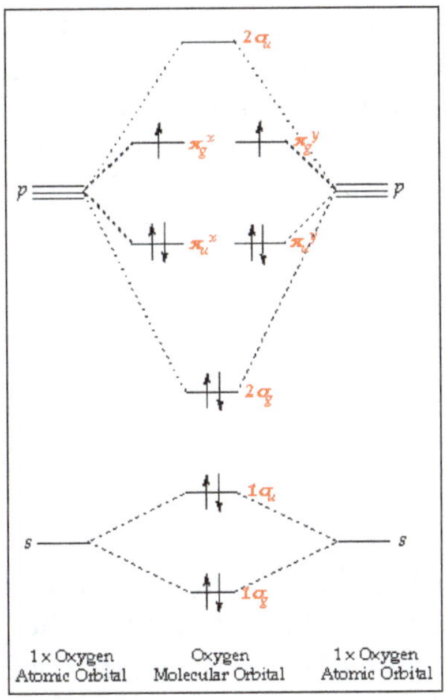

Comparison of the above energy level diagram with that for nitrogen - you can see that the $2\sigma_g$ level lies lower than π_u. Here, we are starting to fill the anti-bonding orbitals originating from the p orbital interactions and so the bond order decreases from three to two.

The lowest energy arrangement (Hund's rule) - has a single electron, each with parallel spins, in each of the π_g^x and π_g^y orbitals. This produces a paramagnetic molecule, with a double bond and has two unpaired electrons.

X–ray Spectroscopy

X–rays are radiations in the electromagnetic spectrum of wavelength between ~.01 to ~10 nm. These are characterized by index of refraction very near to unity for all materials. Electrons are emitted thermally from the heated cathode C and are accelerated towards the anode target A (a metal) by the applied potential V. When accelerated electrons strike the target, X–rays are produced. The face of the target is at an angle relative to the direction of electron beam and the X–rays that leave the anode target pass through the side of the tube. The geometry of the tube is such that it evacuates to allow electrons to reach the target without any interruption.

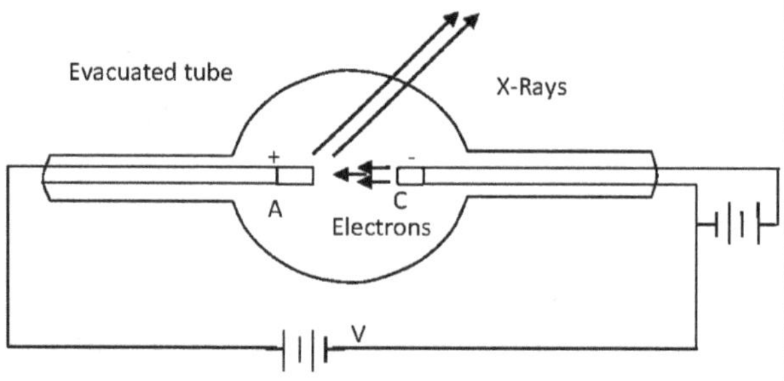

X – Ray tube

The X–rays are produced both by deceleration of electrons in the metal target and by the excitation of the core electrons in the atom of the target. The first process gives a broad continuous spectrum and the second gives sharp lines. When a moving electron is stopped suddenly, all its energy appears as photon of frequency υ of X–rays. The energy of an electron of charge 'e' in dropping through a potential difference V is eV and,

$$E = hv = \frac{hc}{\lambda} = eV$$

Or,

$$\lambda = \frac{hc}{E} = \frac{hc}{eV}$$

An electron does not lose all its energy in this way; it will have a number of glancing collisions with the atoms that it collides and causing them to vibrate. The temperature of the target increases due to this process. The minimum value of λ is possible that accounts for the short wavelength cut off. However, the larger wavelengths are more probable and thus the rapid increase in the intensity. The figure presents the X–ray spectrum that result when Molybdenum target is bombarded by electron at 35 keV. The electron beam on striking the target gets decelerated and also a small fraction of electron of the beam strikes the target and ejects the inner shells electrons. The atom becomes unstable and outer shell electrons in the same atom drops into the hole/vacancy caused by the ejection of the electron. In this process it loses energy and a photon is emitted.

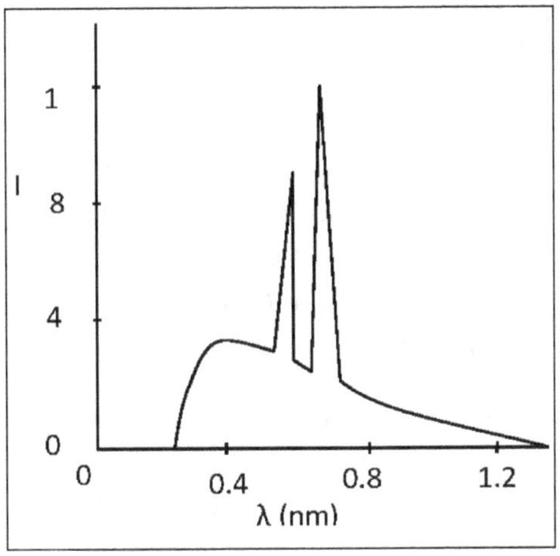

Figure: X–ray spectra from Molybdenum target at 35kV accelerating potential.

As E is a definite quantity associated with the electron energy change in the atom, the wavelength achieved is specific. Several wavelengths are possible and they constitute the characteristic X-ray line spectrum shown as peaks in the figure. The energy of the characteristic X-ray produced is very weakly dependent on the chemical structure in which the atom is bound indicating that non-bonding shells of atoms are the characteristic X-ray source. The resulting characteristic spectrum is superimposed on the continuum. An atom remains ionized for a very short time of the order of ~10^{-14}s. Thus the incident electrons that arrive about every ~10^{-17}s can repeatedly ionize an atom as not all outer incident electrons fall into holes to produce X-rays.

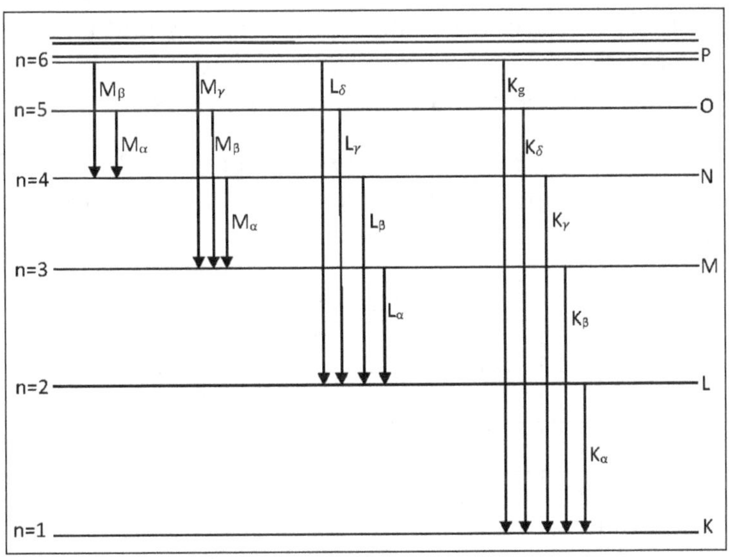

Assuming that an electron from K shell is ejected and a hole is created in the shell, then an electron from the higher shell L, M, N,.... makes a radiative transition filling the hole. The energy of the photon emitted is in the range from a few keV to a few hundred keV and thus lie in the X-ray region. The emission spectrum that results in a line spectrum forms a simple series. The lines originating from a K shell vacancy are called K_α, K_β, K_g lines and corresponding to L→K, M→K, N→K transition. Kα is the strongest and the emission of K series is accompanied by other series. The vacancies are created in L, M, N, ... shells as a result of lines of K series, those are filled by M,N, ... shell electrons. Thus K series may be accompanied by L, M, N, ... and so on. These transitions can be shown on an energy level diagram.

This is different from that used in the atomic spectra of valence electrons in the sense that X-ray diagram gives the energy of the atom when one electron of the quantum number n, l, j is missing. Thus the diagram depicts the energy levels of a hole, with quantum number n, l, j that jumps from one subshell to the next when the atom emits X-ray line spectrum. The absence of an electron of negative energy (representation of a hole), the energy associated with the hole is positive. In the figure, the levels are also represented by capital letters K, L, M, N, corresponding to n = 1, 2, 3, 4, respectively.

X-rays can be classified as hard or soft depending upon the wavelength,. The K X-rays are called hard and L radiation is soft and M, N, O X–rays are very soft. M and L radiation arising from heavy elements are harder and are of shorter wavelengths than K radiation of lighter elements.

Emission Spectra of X–ray

Spin, orbital and total angular momenta for every filled subshell or shell are zero. Suppose that one electron is removed from the closed subshell, then the allowed values of spin, orbital and total angular momenta for the rest of the electrons in the subshell are identical with those of the single electron that can complete the subshell. This infers that the states of np^1, nd^1....... are same as that of np^5, np^9, .. , respectively, that is, the quantum number n, l and j of subshell with one missing electron are identical with those of an electron that completes the subshell. As only one electron is missing from the subshell, therefore s = S = ½ and multiplicity is 2. The figure depicts the energy levels corresponding to various values of n. There are n-1 values of l for each value of n, and for each value of l, there are two j values corresponding to j = 1 + 1/2 and j =1 – ½. As an example, for L shell: n = 2, the l values are 0 and 1 and corresponding j values are ½ corresponding to l = 0 and ½, 3/2 for l = 1. Thus each n level splits up into 2n-1 components.

In X-rays, shells are designated with n from 1 to 7 with letters K, L, M, N, O, P, Q, respectively and levels in each shell with Roman numerals in indices. The Nomenclature of X-ray states is shown in the table for ready reference.

n	l	j	Designation	State
1	0	1/2	K	$1s_{1/2}$
2	0	1/2	L_I	$2s_{1/2}$
2	1	1/2	L_{II}	$2p_{1/2}$
2	1	3/2	L_{III}	$2p_{3/2}$
3	0	1/2	M_I	$3s_{1/2}$
3	1	1/2	M_{II}	$3p_{1/2}$
3	1	3/2	M_{III}	$3p_{3/2}$
3	2	3/2	M_{IV}	$3d_{3/2}$
3	2	5/2	M_V	$3d_{5/2}$

A part of energy level diagram for X-ray emission spectrum Indicating only K and L levels is shown below:

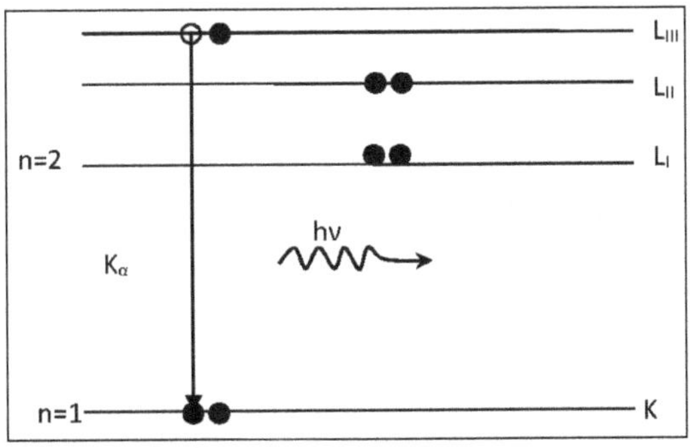

There is a transition from one excited state to another for emission of an X-ray line and it is notable that, both the initial and final states are excited. The emission of K line is due to transition of an electron level L to that of K accompanied by emission of a photon whose energy is equal to the difference in energy between these levels, so the selection rules for the transitions are:

$$\Delta 1 = \pm 1, \ \Delta j = 0, \pm 1$$

All the transitions between K and L levels are not allowed as per the selection rules. As $L_1 \rightarrow K$ is not allowed since $\Delta 1 = 0$ is forbidden. One of the two allowed transitions is shown in the figure. The transition has been are denoted by an arrow pointing in the direction of electron transition with emission of X-ray photon i.e. from level L_{III} to level K.

The two allowed transitions between L and K levels are further given the name K_{a1} and K_{b2}. The K_{a1} is the intense line while weaker is K_{a2}. Similarly, K_{b1} and K_{b2} are designated. A similar system is used for γ lines. The K_{a1} line arises from a K-L_{III} transition ($1_{S1/2} \rightarrow 2_{P3/2}$) for the atom corresponding to the electron from L_{III} filling the vacancy in the K shell. The K_{a2} line corresponds to electron from L_{II} filling the vacancy in the K shell. The statistical factor of L_{III} is 2j + 1 = 4 while that of L_{II} is 2, therefore K_{a1} is to be twice as intense as the K_{a2} line. Similarly, K_{b1} (K-M_{III} transition) is twice as strong as K_{b2} (K-M_{II} transition). It is noteworthy that, that K_{a1} and K_{b1} are not equally intense. The rule of proportionality to the statistical weight is good only for the intensity of lines of very nearly the same energy.

Atomic Number and the Position of Emission Lines

Moseley, in 1913, observed that the wavelength of characteristic X-ray lines shift continuously with the change in the element atomic number Z. He observed that with increasing Z, wavelength of X-ray decreases and hence the frequency of X-ray increased.

A plot of square root of frequency versus Z gives a straight line. This increase in frequency with increase of Z is understood to be due to increasing binding energy of the electron with increase of number of protons in the nucleus. The electron energy E_{nl} for a given n and l is given by,

$$E_{nl} = \frac{hcRZ^{*2}}{n^2}$$

Here $Z^* = Z - \sigma_{nl}$ is the effective charge of the nucleus and σ_{nl} is value of screening constant. The term value is,

$$T = \frac{E_{nl}}{hc} = \frac{RZ^{*2}}{n^2}$$

For an emission line, the frequency (in cm^{-1}),

$$\bar{\nu} = R\left(\frac{Z_f^{*2}}{b_f^2} - \frac{Z_i^{*2}}{n_i^2}\right)$$

Taking $n_i > n_f$ and σ approximately constant for initial and final state, that is, $Z_j^* = Z_f^* = Z^*$.

$$\bar{\nu} = RZ^{*2}\left(\frac{1}{n_f^2} - \frac{1}{n_i^2}\right)$$

$$\bar{\nu} = RZ^{*2}a$$

where, $a = \left(\frac{1}{n_f^2} - \frac{1}{n_i^2}\right)$

$$\sqrt{\frac{\bar{\nu}}{R}} = Z^*\sqrt{a} = \sqrt{a}(Z - \sigma_{nl})$$

This equation is known as Mosley Law. This provided the information that where elements were missing from the periodic table and has led to the discovery of some of the elements. For K_α line, $n_f = 1$ and $n_i = 2$ and thus σ~1 and 7.4 for K and L shells, respectively.

From the above equation wavelength for K series is $\lambda = 1/\bar{\nu}$ for L series it can also be obtained by putting the suitable values Z* and a.

X–ray Emission (Doublet) Spectra

An X-ray doublet is referred to the separation between adjacent X- ray terms of the same principal quantum number. The intervals that correspond to differences between levels with the same n, l and s but different j are called spin doublets also called regular doublets. The frequency interval remains constant all through the spectrum of the same element but increases very slowly with rising value of Z. Examples of spin doublets are L_{II} - L_{III}, M_{II} or M_{IV} - M_V etc.

The irregular doublets or the screening doublets are pair of energy levels having equal n and j but quantum number l differs by 1. The separation between lines remains constant for all the elements. The examples of screening doublets are:

L_I - L_{II} ($2s_{1/2} - 2p_{1/2}$);

$M_{III} - M_{IV}$ ($3p_{3/2} - 3d_{3/2}$);

M_I - M_{II} ($3s_{1/2} - 3p_{1/2}$) ....

The explanation of the doublet separation is the following: The energy of a given level due to the Coulomb interaction of an electron with the nucleus and the energy of a level (in a first approximation) is,

$$E_n = -\frac{hcRZ^{*2}}{n^2} = \frac{hcR(Z-\sigma_1)^2}{n^2}$$

σ_1 being the screening constant depending on the value of l. The position of the energy level is given by a similar expression to that of hydrogen thereby taking into account the correction due to relativistic effect and spin-orbit interaction, given by, $E_{nlj} = \frac{hcR(Z-\sigma_1)^2}{n^2} + \frac{hcR(Z-\sigma_2)^2}{n^3}\left(\frac{3}{4n} - \frac{1}{j+1/2}\right)$ with Z replaced by Z*.

As the screening constant represents an average over ranges of effective Z value, therefore different screening constants for expression with different power of Z are expected. The screening constant $\sigma_2 < \sigma_1$.

The term value is,

$$T_{nlj} = -\frac{E_{nlj}}{hc} = \frac{R(Z-\sigma_1)^2}{n^2} + \frac{R(Z-\sigma_2)^4}{n^3}\left(\frac{3}{4n} - \frac{1}{j+1/2}\right)$$

And the doublet separation,

$$\Delta T = \frac{R\alpha^2 (z-\sigma_2)^4}{n^3 1(1+1)}$$

Thus the energy difference ΔT (cm^{-1}) between regular doublets is proportional to the fourth power of $(Z - \sigma_2)$. The value of σ_2 is independent of Z but depend on the orbit. σ_2 increases with the subshells going farther from the nucleus. It is reported that $\sigma_2 = 3.5$ for L_{II}, L_{III}; 8.5 for M_{II}, M_{III}; 13 for M_{Iv} and M_v etc.

Ignoring the term in σ^2,

$$T = \frac{R(Z-\sigma_1)^2}{n^2}$$

For the two levels a and b,

$$\sqrt{T_a} = \sqrt{R}\frac{(z-\sigma_1)_a}{n}$$

$$\sqrt{T_a} = \sqrt{R}\frac{(z-\sigma_1)_b}{n}$$

σ_1 is different in the two case as a and b involve different orbits,

$$\Delta\sqrt{T} = \sqrt{T_a} - \sqrt{T_b} = \frac{\sqrt{R}}{n}\left[(\sigma_1)_b - (\sigma_1)_a\right] = \frac{\sqrt{R}}{n}\Delta\sigma_1$$

$\Delta\sigma_1$ is the difference in screening constant and the screening constant σ_1 increases with Z.

Satellites

The lines in the X-ray spectra are often found to be accompanied by faint satellites. As these lines do not fit into conventional energy level diagram and are therefore, called non-diagram lines. They occur usually on the short wavelength side of the diagram lines. The origin of satellite is related to transition between the states of double and multiple ionization. Considering that the incident electron beam striking the target of X-ray tube has sufficiently high energy and it ejects an electron from K shell as well as from L shell that results in holes in K and L shells and such a state undergoes a radiative transition into any one of a number of other double ionization like KL→LL (an electron dropping from L shell into K shell). The loss of energy due to emission of line in the transition KL→LL is greater than the normal K→L transition that gives rise toK_α line.

The other important source of satellite line is the auger process. In this process when an electron from an outer orbit make a transition to a hole in the core orbital, a photon is emitted that may be passed to another electron of outer orbit leading to its ionization. Thus a doubly ionized state is produced. For example, an electron from one of the higher shells. This infers that Auger effect produces a doubly ionized state, that give rise to the emission of satellite of L series.

Continuous X–ray Emission

In addition to characteristic spectrum a continuous spectrum of a different nature arises, when X-rays are excited by electron. The continuous part of the spectrum is due to slowed down movement of an electron striking the target in the Coulomb field of the nucleus.

The continuous spectrum has a sharp limit λ_o on the lower wavelength side whose value depends on the X-ray tube voltage and is independent of atomic number Z of the target element. The relationship between wavelength and X-ray tube voltage is already established as:

$$\lambda = \frac{hc}{E} = \frac{hc}{eV}.$$

Below this short wavelength limit no X-ray radiation is observed. Wavelength of maximum intensity λ_m is approximately one and half time λ_o and therefore,

$$\lambda_m V^{1/2} \text{ constant}$$

Intensity of continuous X-ray is nearly proportional to the square of voltage and to the first power of atomic number that is,

$$I = k\, Z\, V^2$$

Here k is constant of proportionality. The position of characteristic spectrum is independent of voltage of X-ray tube. It is to note that by raising voltage the limit of continuous X-ray spectrum shifts towards shorter wavelength side.

X–ray Absorption Spectra

The intensity of rays become weaker because of absorption and scattering, when a beam of X-ray passes through a sample. The absorption of X-ray emission energy occurs as a result of a single process. The X-ray photon knocks out electron of a shell and the energy of the absorbed photon is thus transformed into kinetic energy of the electron plus the potential energy of the excited atom that equals binding energy of the electro.

The X-ray emission line yields information on the difference in binding energies between two electronic states. However, to determine absolute binding energies, absorption of X-ray has to be studied. For such a purpose an X-ray continuum is used and sample absorption as a function of wavelength is measured.

The X-rays of the longest wavelength force out electrons from the outer shell. With increasing energy of X-rays (decreasing wavelength) a small part of it is required to knock out the electron from the given shell. This causes reduced absorption. This decrease continue till the X-ray energy is sufficient to knock out electron from the next deeper lying inner shell and give rise to a sharp increase in absorption as shown in the figure.

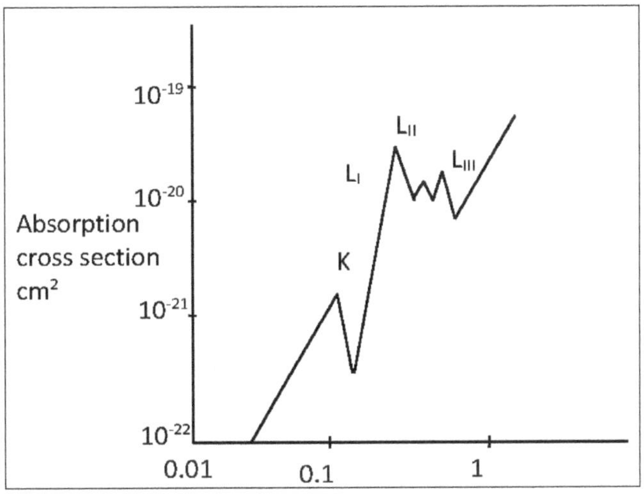

It causes a discontinuous behavior of absorption μ with λ. The discontinuous behavior of absorption and wavelength or frequency is called absorption edge and corresponds to photon critical energy. The absorption edges are labeled as K edge, L edge, M edge etc. from the observed absorption edge an approximate binding energy is obtained.

For increasing energy of X-ray, the threshold for photoemission of deeper shells are reached and additional contributions to the total absorption are obtained resulting in a number of edges. Apart from the edge there is a general fall off in absorption due to λ^3 dependence of the absorption coefficient given by $\mu = k \varrho \lambda^3 Z^3$. The figure depicts the X –ray absorption spectrum with absorption edges. Here k is a constant and ϱ is the density.

It is worth to learn that the soft X-rays are absorbed strongly in comparison with hard X-rays. The wavelength of the absorption edge is defined in terms of the orbital energy,

$$\lambda_k = \frac{hc}{E_k}$$

The absorption edge exhibits fine structure corresponding to the fine structure of the core state. However, K edge also exhibits a structure that is due to discrete levels close to the series limit.

References

- Pauli-Exclusion-Principle, Electronic-Configurations, Physical-and-Theoretical-Chemistry-Textbook-Maps: libretexts.org, Retrieved 18, January 2020
- Introduction-to-Molecular-Orbital-Theory: ch.ic.ac.uk, Retrieved 23, May 2020

5

Understanding Laser Spectroscopy

In laser spectroscopy, lasers are used as light sources. The invention and development of the laser have helped in advancing the field of spectroscopy. Science and technology have undergone rapid developments in the past decade which has resulted in the discovery of significant tools and techniques in the field of laser spectroscopy; which have been extensively detailed in this chapter.

Spectroscopy denotes methods where the interaction of light with matter is utilized. In many cases, lasers are used as light sources for spectroscopy, which is then called laser spectroscopy (or sometimes laser spectrometry). In a wider sense, laser spectroscopy can be understood to use laser-like sources, including not only real lasers, but also optical parametric oscillators (OPOs) or the outputs of other nonlinear frequency conversion devices such as frequency doublers or difference frequency mixers.

Laser is an abbreviation used for light amplification by stimulated emission of radiation. There are three basic components of a laser system:

- An active medium of the atoms or molecules.
- A mechanism for creating population inversion wherein we require a greater population of electrons in an excited state that existed in a lower state.
- A resonator or cavity which has resonant frequency equal to the photon frequency.

Transition Probabilities and Population Inversion

Assuming that there are n atoms in equilibrium with a radiation field and considering electron transition between two quantum states- the upper state being labeled 2 and the lower state labeled 1. Now there are n_2 atoms instantaneously in the upper state and n_1 in the lower state. The population n_1 and n_2 may change in three ways as shown in figure below:

- Stimulated absorption,
- Spontaneous emission,
- Stimulated emission.

The emitted photons have random directions and phases, and the radiation is said to be incoherent. In the stimulated emission process the two photons have the same phase and direction, and a beam of such photons is said to be coherent radiation. All the three processes mentioned above occur in a laser. Now, the transitions cause the populations of the two levels to change with time and the rate of change of n_1, and the corresponding rate of change of energy in the field need to be described.

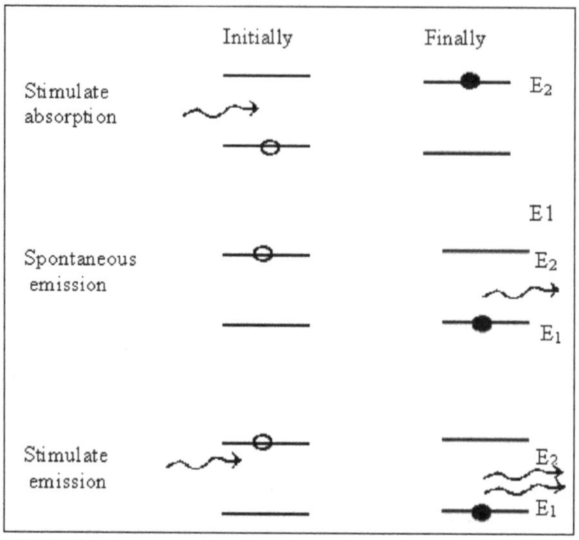

Initial and final electron energies corresponding to stimulated absorption, spontaneous emission, and stimulated emission.

There is certain probability per second of an electron in level 1 absorbing a photon of energy hv= E_2- E_1 making a transition to level 2. This transition probability is proportional to n_1 and the density of photons having frequency v.

It is noteworthy that the transition probability is expressed in terms of the photon density per unit frequency $J/m^3 \cdot s^{-1}$. The constant of proportionality is referred to the Einstein B coefficient (B_{12}) ($m^3 \cdot s^{-2}$ /J):

$$\frac{dn_1}{dt} = -n_1\ B_{12}p(v) \text{ (number of stimulated upward transitions per second),}$$

where the negative sign means that stimulated absorption causes n_1 to decrease with time.

The rate of change of n_1 is proportional to n_2. (If n_2 = 0, i.e. no electrons in level 2, that means spontaneous emission does not occur). The constant of proportionality is the Einstein A coefficient,

$$\frac{dn_1}{dt} = +\ n_2\ A_{21} \text{ (number of spontaneous downward transitions per second).}$$

The rate of change of n_1 due to stimulated emission is proportional to the product n_2 x p(v). (If there are no electrons in level 2 to be stimulated or no photons to stimulate them, then stimulated emission does not occur). The constant of proportionality is the Einstein B coefficient, labeled B_{21} (same units as B_{12}). Thus, we have,

$$\frac{dn_1}{dt} = +n_2B_{21}\,p(v) \text{(number of stimulated downward transitions per second).}$$

Therefore the total rate of change of n_1 is given by,

$$\frac{dn_1}{dt} = n_2A_{21} + (n_2B_{21} - n_1B_{12})p(v) \text{ (number of transitions per second to and from level 1).}$$

Under thermal equilibrium conditions, n_1 and n_2 are related and the Einstein coefficients. If the atomic system is in thermal equilibrium with the radiation at a given temperature T, the relative population of any two energy levels, such as 1 and 2 are given by Boltzmann's equation:

$$\frac{n_2 = e^{-E2/kT}}{n_1 = e^{-E1/kT}} = e - hv / kT \text{ (thermal equilibrium)}$$

where k is Boltzmann's constant. And $n_2 < n_1$.

Also under equilibrium conditions, the net number of downward transitions must be equal to the net number of upward transitions, namely,

$$n_2A_{21} + n_2B_{21}\, p(v) = n_1B_{12}p(v) \text{ (thermal equilibrium).}$$

There applies the principle of detailed balance, that states that for thermal equilibrium every process must be balanced by its exact opposite.

Now by solving for p(v), one obtains,

$$p(v) = \frac{n_2A_{21}}{n_1B_{12} - n_2B_{21}} \left(J / m^3 . s - 1\right)$$

By using above equations,

$$p(v) = \frac{A_{21}}{B_{21}} \frac{I}{(B_{12} / B_{21})e^{hv/kT} - 1}\left(J / m^3 . s - 1\right)$$

we all know Planck's formula for emission of blackbody radiation by a body at temperature T is,

$$p(v)dv = \frac{8\pi v^2}{c^3} \frac{hv}{e^{hv/kT} - 1} dv \left(J / m^3\right)$$

Now in order to agree with Planck's formula, the following equation must hold:

$$B_{12} = B_{12} \left(m^3 . s^{-2} / J\right)$$

$$\frac{A_{21}}{B_{21}} = \frac{8\pi h v^3}{c^3} \left(J / m^3 \cdot s^{-1}\right)$$

It is worth noting that in 1917, Einstein derived these equations but he did not know how to calculate the A and B coefficients. The calculation of these coefficients is a quantum mechanical, time-dependent perturbation theory calculation, and quantum mechanics was not developed until 1926. The 1932 Nobel Prize in Physics was awarded to Werener Heisenberg, Germany. For this creation of quantum mechanics, the application of which has, interred alia, led to the discovery of the allotropic forms of hydrogen. The 1933 Nobel Prize in Physics was awarded jointly to Erwin Schrödinger. For the discovery of new productive forms of atomic theory. Heisenberg developed "matrix mechanics," Schrödinger developed "wave mechanics," and Dirac developed relativistic quantum mechanics.

The development of quantum mechanics lead to the calculations of Einstein coefficients and B_{21}, in the electric dipole approximation, is written as,

$$B_{21} = \frac{2\pi^2}{\varepsilon_0 h^2} \left| \psi * (er) \psi_1 dv \right|^2 \left(m3 \cdot s^{-2} / J \right)$$

where ψ_1 and are the eigenfunctions of levels 1 and 2, er is the electric dipole moment operator, and ε_0 is the permittivity of the vacuum.

Using equations $\frac{\ _{21}}{\ _{21}} = \frac{8\ hv}{} \left(J / m^3 \cdot s^{-1} \right)$ and $B_{21} = \frac{2\pi^2}{\varepsilon_0 h^2} \left| \psi * (er) \psi_1 dv \right|^2 \left(m3 \cdot s^{-2} / J \right)$, equation $\frac{dn_1}{dt} = n_2 A_{21} + (n2B_{21} - n_1 B_{12}) p(v)$ (number of transitions per second to and from level 1) can be expressed in terms of B_{21},

$$\frac{dn_1}{dt} = n_2 \frac{8\pi hv^3}{c^3} B_{21} + (n_2 - n_1) B_{21} (v) \text{ (number of transitions per sec to and from level 1).}$$

First ignore the spontaneous emission term because it gives rise to incoherent radiation and as we are interested in coherent radiation, that is, "Laser light." Then,

$$\frac{dn_1}{dt} = (n_2 - n_1) B_{21} p(v)$$

Consider radiation passing through an optical medium that has atoms in levels E_1 and E_2. The rate of stimulated downward transitions exceeds that of upward transitions if $n_2 > n_1$, and, hence, the coherent radiation is amplified that means an increase in the number of coherent photons. This clearly mentions that the condition required for "lasing," i.e. $n_2 > n_1$, is contrary to the thermal equilibrium distribution given by equation $\underline{n}_2 = \underline{e}^{-E2/kT} = e - hv / kT$. This condition is called population inversion.

The equation $\frac{dn_1}{dt} = n_2 \frac{8\pi hv^3}{c^3} B_{21} + (n_2 - n_1) B_{21} (v)$ (number of transitions per sec to and from level 1. contains the matrix element of the electric dipole moment operator between 1 and 2. In case, a matrix element is zero, the transitions between the levels are called forbidden transitions. Forbidden transitions may occur by means of mechanisms other than electric dipole radiation. Such transitions occur at a much slower rate than the permitted (electric dipole) transitions.

The above description is for an ideal system of two atomic levels but actually more levels are involved in the laser emission. There exists the concept of metastable state. A metastable state is a state from which all transitions to lower energy states are forbidden. A metastable state has a long lifetime, ~10^{-3}s, and atoms making upward transitions due to collisions or radiation absorption tend to accumulate in metastable states. An ordinary state has permitted transitions to lower-energy states; hence, it has a short lifetime ~10^{-8}s.

Relevant Properties of Lasers

Lasers have enormous potentials for applications in spectroscopy in several respects:

- They can exhibit a very large degree of temporal coherence. This is related to a narrow linewidth in extreme cases of precision spectroscopy, sub-hertz linewidths are used. Note that a narrow line with does not only help in terms of measurement resolution, but also for improving sensitivity. For example, the absorption related to a specific absorption line is more easily seen if it applies to the entire probe light and not only to a part of its spectral content.

- The spatial coherence of laser light is usually very high, and this leads to further improvements. For example, one can use multipass gas cells, where a large number of round trips can be realized only with a collimated beam having a low beam divergence.
- Laser wavelengths can often be tuned (→ tunable lasers); this can be used for highly precise scanning of optical transitions. Some lasers can be tuned quite rapidly (→ wavelength-swept lasers) for quick acquisition of spectra.
- Some lasers can provide rather high optical powers; particularly pulsed lasers can generate very high peak powers. This is relevant, for example, for laser-induced breakdown spectroscopy.
- Ultrafast lasers can generate ultrashort pulses, which also have various interesting properties for spectroscopy. For example, they allow for time-resolved spectroscopic measurements with an extremely high temporal resolution. Also, extremely precise frequency measurements can be done based on frequency combs.

Properties of Laser Light

Directionality

The ordinary light sources emit in all directions while Laser light emit only in one direction. The directionality of laser beam is expressed in terms of the 'Full Angle Beam Divergence' that is twice the angle that the edge of the beam makes with the axis of the beam. The outer edge of the beam is defined as the p-point at which the strength of the beam is dropped to 1/e times its value at the centre.

As the beam moves straight and only in one direction, the laser beam can be sharply focused, while light from an ordinary source travels in all directions and in latter case, the energy intensity rapidly decreases as one moves away from the source similar to the fact the sun's intensity diminishes when it finally reaches the earth.

This simply means that a laser beam can give very high energies to a very small area. It is also important to note the spread of the beam. The spread of the beam on the surface of moon is a few kilometres though the moon is ~380000 Kms. Thus the beam has a too little divergence. If a laser beam is sent from the earth to a mirror on the moon, the beam will be reflected to the earth again with no divergence.

Intensity

The intensity of a laser beam is quite high in comparison to any other light source. As the laser gives light into a narrow beam, its energy is concentrated in a very small region and this lead to great intensity. Now we have output intensity per unit area that is highly directional and very large.

Flux densities for focused laser light of 10^{15} W cm^{-2} are achievable. However, an oxyacetylene flame has a flux density of only $10^3 Wcm^{-2}$. An object needs to be heated to a temperature at ~10^{30} K to radiate with an intensity of a typical laser source. It is worth noticing that a tungsten lamp, while radiating is only ~3000 K in temperature and our sun is only about 10^8 K.

Intensity is the power transferred per unit area (Watts per square meter), where the area is an imagined surface that is perpendicular to the direction of propagation of the energy. Intensity can be found by taking the energy density (energy per unit volume) at a point in space and multiplying it by the velocity at which the energy is moving. The resulting vector has the units of power divided by area. If a point source is radiating energy in all directions producing a spherical wave, and is assumed that no energy is absorbed or scattered by the medium, then the intensity decreases in proportion to distance from the object squared because of the inverse-square law.

Applying the law of conservation of energy,

$$P = \int I \cdot dA$$

where P is the power radiated, I is the intensity as a function of position, and dA is a differential element of a closed surface that contains the source.

Integrating over a surface of uniform intensity I, may be over a sphere centered around the point source,

$$P = |I| \cdot A_{surf} = |I| \cdot 4\pi^2$$

where I is the intensity at the surface of the sphere, and r is the radius of the sphere.

$$|I| = \frac{P}{A_{surf}} = \frac{P}{4\pi^2}.$$

For a monochromatic propagating wave of Gaussian beam, if E is the complex amplitude of the electric field then the time-averaged energy density of the wave is given by,

$$\langle U \rangle = \frac{n^2 \epsilon_0}{2} |E|^2$$

and the local intensity is obtained by multiplying this expression by the wave velocity, c/n:

$$I = \frac{cn\epsilon_0}{2} |E|^2$$

where n is the refractive index, c is the speed of light in vacuum and E_0 is the vacuum permittivity.

For non-monochromatic waves, the intensity of different spectral components is to be simply added. The treatment above does not hold for arbitrary electromagnetic fields. In general, an evanescent wave may have finite electrical amplitude and may not transfer any power.

Monochromaticity

A laser source is highly monochromatic means comes with a precise wavelength, while ordinary light source has a broad spectral output. The spread in frequency of a line is characterised by the line width. The laser light has a high degree of monochromaticity. A typical laser emits line of line width as 3A°. It is almost the purest monochromatic light available so far. In fact Lasers can be generated with short time durations from pico second to femto seconds and this can be considered dual property of mono chromaticity. This means the energy concentrated in wavelength.

Coherence

Laser Light has high degree of coherence whereas the ordinary light is not coherent. This is the unique property of laser beam. It has arisen due to stimulated emission process. The emitted photons have a definite phase relation to each other. This coherence is of two types: temporal coherence and spatial coherence, both of which are important. Ordinary light is not coherent because it comes from independent atoms which emit on time scales of about 10^{-8} seconds. Though there is a good degree of coherence in sources like the mercury green line and some other useful spectral sources, but their coherence is not comparable with that of a laser.

It will be of interest to introduce the two types of Coherence: Spatial and Temporal. If we consider two points S_1 and S_2 in such a way that at time $t = 0$, these lie on the same wave front of given Electromagnetic wave, where $E_1(t)$ and $E_2(t)$ be the corresponding electric fields. The difference between phases at these two fields at time $t = 0$ is zero and if it remains zero at $t > 0$, there exist perfect coherence between these points. The wave will be said to be spatially coherent if even otherwise any two points of the wave front are considered. The idea can be understood through Double slit experiment. Spatial coherence implies a fixed phase relationship between the electric fields at different locations across the wave front of beam. Spatial coherence is the essential prerequisite of the strong directionality of laser beams.

To define temporal coherence, consider the electric field of electromagnetic wave at a given point S at times t and $t + \tau$. Now for a delay time of τ, the phase difference for two field values remains the same for any time t, then it is understood that there is temporal coherence over a time τ. If this remains same for any value of τ, there exists perfect time coherence. The idea can be understood through Michelson Interferometer experiment. Temporal coherence implies a strong correlation between the electric fields at one location but different times.

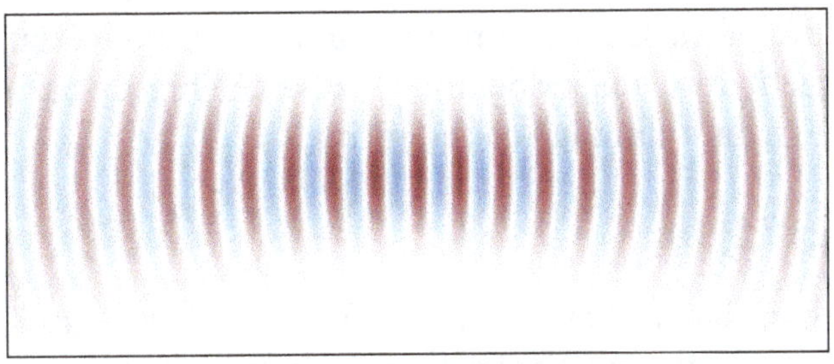

Electric field distribution around the focus of a Gaussian laser beam with perfect spatial and temporal coherence.

Lasers generate the Gaussian beams with very high spatial coherence, and this is the fundamental difference between laser light and radiation from other ordinary light sources. High spatial coherence arises from the existence of resonator modes that define spatially correlated field patterns. The difference between spatial and temporal coherence can be presented by the following figure that depicts a monochromatic Gaussian beam, exhibiting perfect spatial and temporal coherence.

The figure below shows a beam with high spatial coherence, but poor temporal coherence. The wave fronts are formed as above, and the beam quality is also very high, but the amplitude and phase of the beam varies along the propagation direction. It is to observe that both the local amplitude and the spacing of the wave fronts vary to some extent.

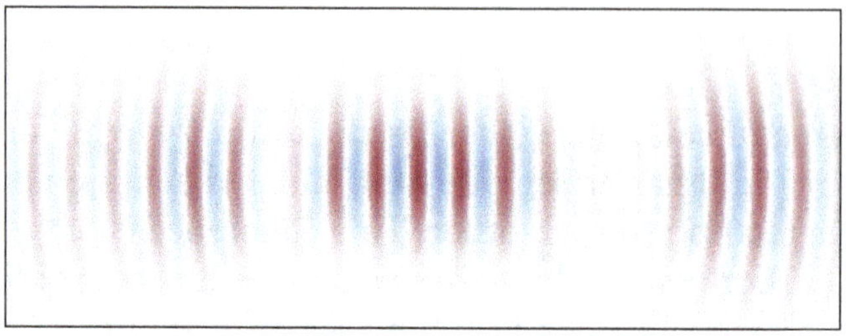

A laser beam with high spatial coherence, but poor temporal coherence.

The figure below depicts a laser beam with poor spatial coherence, but high temporal coherence. The wave fronts are deformed, and this results in a high beam divergence and poor beam quality. On the other hand, the beam is monochromatic, so that the spacing of the deformed wavefronts remains constant.

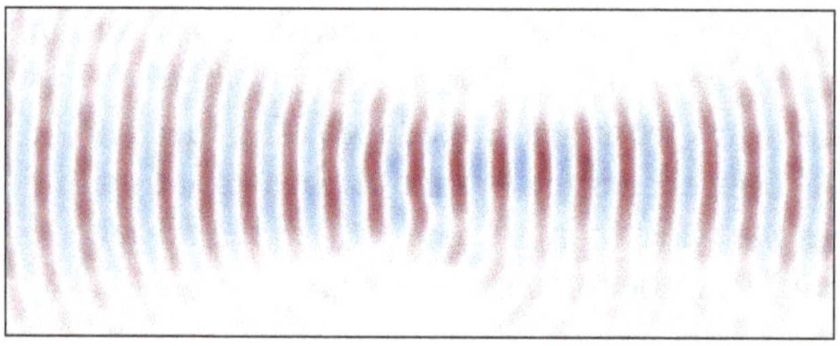

A laser beam with poor spatial coherence, but high temporal coherence.

In conclusion, The high degree of collimation has arisen due to the fact that the cavity of the laser has very nearly parallel front and back mirrors which constrain the final laser beam to a path which is perpendicular to those mirrors. The back mirror is almost perfectly reflecting while the front mirror is ~90% reflecting. But the light has passed back and forth between the mirrors many times in order to gain intensity by the stimulated effect of more photons at the same wavelength. The highly collimated nature of the laser beam contributes both to its danger and to its usefulness. One should never look directly into a laser beam, because the highly parallel beams focused on the retina of the eye will cause instant damage to the retina.

Self-focusing of Laser Light

The laser is known to be a non-linear phenomenon and therefore has a unique property of Self-focusing. It is well known that the refractive index of a material is related to the susceptibility by the relation.

$$\eta = \sqrt{1 + x}$$

As the susceptibility x is a function of the field E, η depends on E. This dependence of the refractive index on the field strength gives rise to a nonlinear effect. It is note that the self-focusing behavior does not alter the frequency of the light waves.

A dielectric medium when placed in an electric field, it gets polarized. Each constituent molecule acts as a dipole, with a dipole moment P_i. The dipole moment vector per unit volume P is given by,

$$P = \sum_i P_i$$

Here, the summation is over the dipoles in the unit volume. It is evident that the orienting effect of the external field on the molecular dipoles depends both on the properties of the medium and on the field strength. Thus,

$$P = \varepsilon_o x E$$

Here x is called the polarizability or dielectric susceptibility of the medium.

The quantity x is a constant only in the sense of being independent of E; its magnitude is a function of the frequency. With sufficiently intense laser radiation the relation does not hold good and has to be generalized to,

$$P = \varepsilon_o \left(x^{(1)} E + x^{(2)} E^2 + x^{(3)} E^3 + \ldots \right)$$

Here $x^{(1)}$ is the same as x, the coefficients $x^{(2)}$, $x^{(3)}$, ... define the degree of nonlinearity and are known as nonlinear susceptibilities. If the field is low, as it is in the case of ordinary light sources, only the first term is retained. The higher order terms become more significant with the increased value of the electric field. Thus, dielectric permittivity, refractive index, etc. that depend upon susceptibility, also become functions of the field strength E, now the medium is called a "nonlinear medium."

Therefore, considering,

$$P^{(1)} = \varepsilon_o \left(x^{(1)} + \frac{3}{4} x^{(2)} E_o^2 \right) E.$$

The expression for the refractive index is,

$$\eta = \sqrt{1 + \left(x^{(1)} + \frac{3}{4} x^{(2)} E_o^2 \right)}$$

or

$$\eta = \sqrt{\varepsilon_l + \varepsilon_{nl}}$$

Here, $\varepsilon_l = 1 + x^{(1)}$

That gives the dielectric permittivity of the linear medium, and

$$\varepsilon_{nl} = \frac{3}{4} x^{(2)} E_o^2$$

is a nonlinear increment in the expression for dielectric permittivity.

$$\eta = \sqrt{\varepsilon_l}\sqrt{1+\frac{\varepsilon_{nl}}{\varepsilon_l}} \approx \sqrt{\varepsilon_l}\left(1+\frac{\varepsilon_{nl}}{2\varepsilon_l}\right)$$

$$(\because \varepsilon_{nl} << \varepsilon_l)$$

$$= \eta_l\left(1+\frac{\varepsilon_{nl}}{2\eta_l^2}\right) = \eta_l\left(1+\frac{3x^{(2)}E_o^2}{8\eta_l^2}\right)$$

or $$\eta = \eta_l\left(1+\eta_{\eta l}E_o^2\right)$$

where $\eta_l = \sqrt{\varepsilon_i}$ is the refractive index of the linear medium, and the nonlinear increment in the expression for the refractive index is,

$$\eta_l \eta_{\eta l} E_o^2 = \frac{\varepsilon_{nl}}{2\sqrt{\varepsilon_l}} = \frac{3x^{(2)}E_o^2}{8\sqrt{1+x^{(1)}}}$$

This implies that the refractive index of a nonlinear medium is proportional to the square of the amplitude of the field, that is, to the intensity. Now the intensity of a laser beam is not constant over its cross-section. It peaks at the axis of the beam and falls off gradually away from the axis. The velocity of the light wave is given by $v = c/\eta$. As η decreases owing to the falling of the intensity of the light beam, plane wave-front incident on material becomes concave as it propagates through the medium and contracts towards the axis and self-focuses.

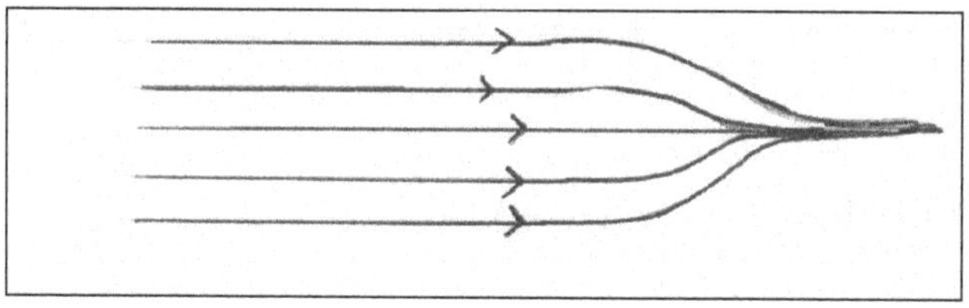

Mode Locking in Laser

Mode locking (sometimes written as mode locking) is a method (or actually a group of methods) to obtain ultrashort pulses from lasers, which are then called mode-locked lasers. Here, the laser resonator contains some kind of mode locking device (mode locker) – either an active element (an optical modulator) or a nonlinear passive element (a saturable absorber), which causes the formation of an ultrashort pulse circulating in the laser resonator. In the steady state, the various effects influencing the circulating pulse are in a balance so that the pulse parameters are unchanged after each completed round trip, or often even nearly constant throughout each round trip. Each time the pulse hits the output coupler mirror, a usable pulse is emitted, so that a regular pulse train leaves the laser. Assuming a single circulating pulse, the pulse repetition period corresponds to the resonator round-trip time (typically several nanoseconds), whereas the pulse duration is much lower: typically between 30 fs and 30 ps, in extreme cases down to ≈ 5 fs. For that reason, the peak power of a mode-locked laser can be orders of magnitude higher than the average power.

Active and Passive Mode Locking

Active mode locking involves the periodic modulation of the resonator losses or of the round-trip phase change, achieved e.g. with an acousto-optic or electro-optic modulator, a Mach–Zehnder integrated-optic modulator, or a semiconductor electro absorption modulator. If the modulation is synchronized with the resonator round trips, this can lead to the generation of ultrashort pulses, usually with picosecond pulse durations. In most but not all cases, the pulse duration achieved is governed by a balance of pulse shortening through the modulator and pulse broadening via other effects, such as the limited gain bandwidth.

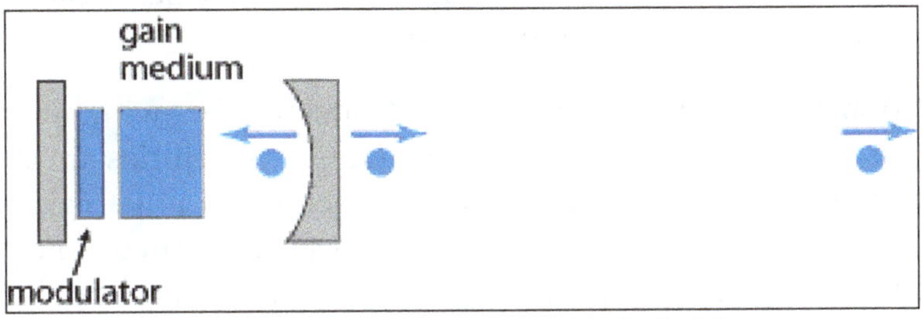

Schematic setup of an actively mode-locked laser.

Passive mode locking (with a saturable absorber) allows the generation of much shorter (femtosecond) pulses, basically because a saturable absorber, driven by already short pulses, can modulate the resonator losses much faster than an electronic modulator: the shorter the pulse becomes, the faster the loss modulation, provided that the absorber has a sufficiently short recovery time. The pulse duration can be even well below the recovery time of the absorber. In some cases, reliable self-starting mode locking is not achieved.

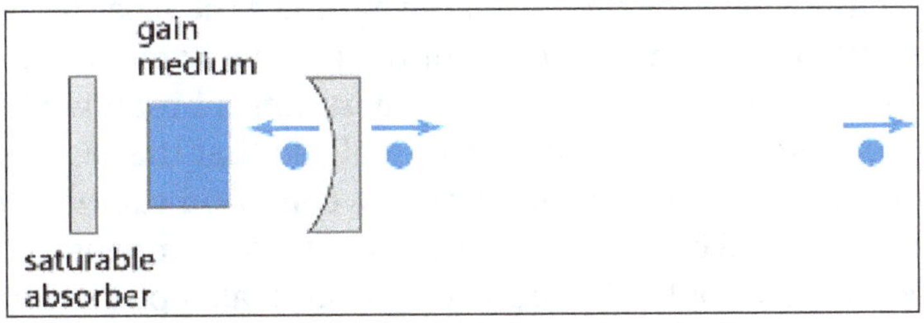

Schematic setup of a passively mode-locked laser.

In some lasers (particularly in mode-locked diode lasers), active and passive mode locking are simultaneously applied. Such hybrid mode-locked lasers combine some key advantages, such as an externally controlled pulse repetition rate and fairly short pulses.

Continuous Versus Synchronous Pumping

Most mode-locked lasers are continuously pumped, often with a laser diode. The pump source then continuously supplies energy to the gain medium, while the circulating pulse extracts energy in regular time intervals. In most cases, the pulse spacing is very short compared with the upper-state lifetime, so that the gain medium can easily store the energy during such times. Gain saturation will then be the same as for continuous laser emission with the same average power.

Even some gain media with a fairly short upper-state lifetime, e.g. in dye lasers or some semiconductor lasers (e.g. VECSELs), continuous pumping is possible, provided that the pulse repetition rate is not too low. If this condition cannot be fulfilled, so that gain saturation by a single pulse becomes strong (which can make the pulse formation unstable), synchronous pumping can be a solution. This, however, requires another mode-locked laser as the pump source.

High and Low Pulse Repetition Rates

High repetition rate pulse trains are sometimes obtained with harmonic mode locking, where multiple pulses are circulating in the laser resonator with a constant temporal spacing. This allows the generation of multi-gigahertz pulse trains even with fiber lasers, which typically have fundamental round-trip frequencies of only some tens of megahertz.

For high pulse repetition rates with fundamental mode locking, i.e., without harmonic mode locking, very short laser resonators are required. If Q-switching instabilities are avoided with a proper design, such lasers can be very simple, stable and compact. Due to the high pulse repetition rate, the pulse energies obtained from mode-locked lasers are fairly limited – normally nanojoules or picojoules. Higher pulse energies combined with lower repetition rates can be obtained with cavity-dumped mode-locked lasers and with regenerative amplifiers.

The Optical Spectrum: Frequency Combs

The optical spectrum of a pulse train, as generated in a mode-locked laser, is not smooth (as for a single pulse), but rather consists of discrete lines with an exactly constant spacing which equals the pulse repetition rate (→ frequency combs). This is true although the resonance frequencies of the resonator modes are usually not exactly equidistant due to the effect of chromatic dispersion, e.g. in the gain medium: the mode-locking mechanism forces the laser to emit frequencies which can to some extent deviate from the frequencies of the resonator modes. These frequency deviations may not be arbitrarily high, and therefore the generation of broadband spectra is usually possible only if the resonator dispersion is sufficiently small, so that the resonator mode frequencies are approximately equidistant. In the time domain, this condition can be understood via the temporal broadening of pulses caused by dispersion, which must be compensated by the mode-locking mechanism. Note, however, that optical nonlinearities often also play an important role, so that considering the "cold cavity modes" misses an important part of the physical mechanisms.

Origin of the Term "Mode Locking"

The term mode locking originates from a description in the frequency domain: a short pulse is formed in the laser resonator when a fixed phase relationship is achieved between its longitudinal modes, or more precisely, between the lines in the spectrum of the laser output. However, the basic mechanisms leading to mode locking can usually be much more easily understood in the time domain. Also, the concept of resonator modes becomes questionable under the influence of strong optical nonlinearities. Strictly, it is not even advisable to use the term modes for the lines in the spectrum of a mode-locked laser, even though these lines are related to the resonator modes.

It is instructive to consider the synthesis of a periodic pulse train by superposition of sinusoidal oscillations with equally spaced frequencies, corresponding to different axial resonator modes in a mode-locked laser. The larger the number of frequency components involved, the shorter can be the duration of the generated pulses.

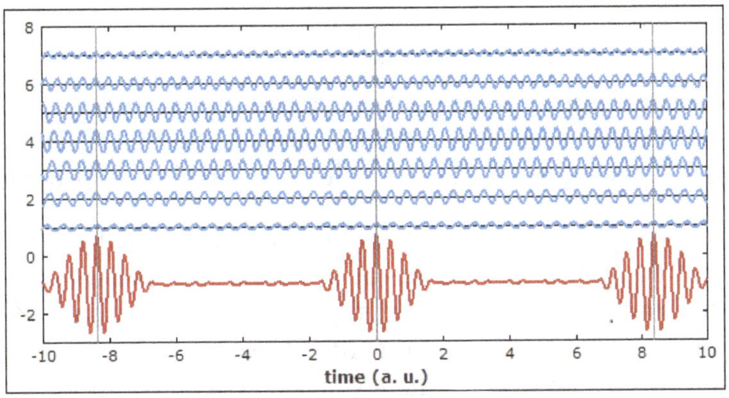

Synthesis of a periodic pulse train (red curve) by adding seven oscillations with slightly different frequencies (blue curves). The vertical lines indicate points in time where all the oscillations add up in phase.

An important aspect is that there must be a fixed phase relationship between these modes. This is illustrated in Figure: the blue curve shows a pulse train with a fixed phase relationship, so that at regular temporal positions (e.g. at t = 0) the electric fields of all frequency components add up to a maximum of the total field strength. The red curve shows the electric field for the same strength of all frequency components, but with random relative phases.

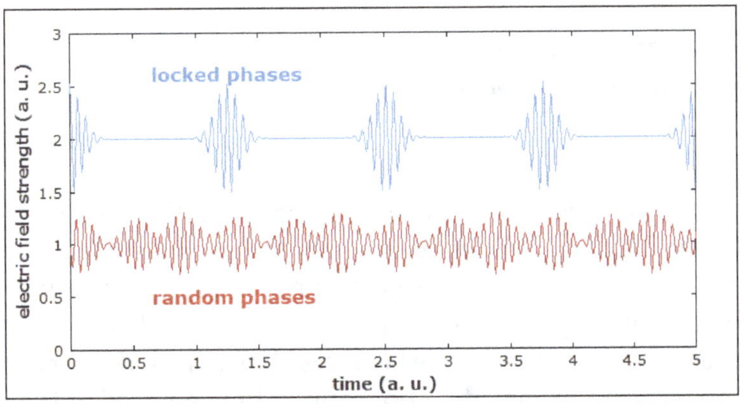

Temporal evolution of the intracavity field in a laser, once with a fixed phase relationship between the modes (mode-locked state), once with random phases.

Due to chromatic dispersion, the axial modes of a laser resonator would normally not have exactly equidistant mode frequencies, and therefore the mentioned phase relationship should normally be lost within a short time. However, that phase relationship can be enforced by a strong enough mode-locking mechanism provided that the de-phasing tendencies are not too strong. The frequencies of the lines in the optical spectrum are then exactly equidistant (disregarding noise), thus slightly deviating from the mode frequencies.

The explanations and diagrams above are actually simplified in an important detail: it has been assumed that all involved optical frequencies are integer multiples of the pulse repetition frequency. In reality, it is generally such that every frequency is in addition shifted by a certain amount, called the carrier–envelope offset frequency. One then still gets a periodic pulse train in terms of intensity, but a change of the optical phase at the pulse maximum which evolves continuously.

Instabilities and Noise of Mode-locked Lasers

The various mechanisms for mode locking can exhibit various kinds of instabilities. For example, a laser may generate bunches of pulses (instead of single pulses), pulses with unstable energy (→ Q-switching instabilities), pulses which fall apart after some time and are later replaced with new pulses, or pulses which are accompanied by a noise background. Different kinds of modeling can be very helpful in tracking down such problems and to find appropriate remedies.

Even without particular instabilities, the output of a mode-locked laser contains various kinds of noise; the timing jitter (random fluctuations of the pulse positions) is often of special interest. Other types of noise include phase noise, intensity noise, and fluctuations of other parameters such as the pulse duration, chirp, and optical center frequency. The phase noise results in a finite width of the lines in the generated frequency comb. There are interesting relations between different types of noise in mode-locked lasers.

Mode Locking Devices

Optimum mode-locked performance depends on a well worked-out laser design, including the choice of a suitable mode locking device, be it some kind of modulator or a saturable absorber.

Lasers for Spectroscopy

Semiconductor Laser

A semiconductor laser (LD) is a device that causes laser oscillation by flowing an electric current to semiconductor. The mechanism of light emission is the same as a light-emitting diode (LED). Light is generated by flowing the forward current to a p-n junction. In forward bias operation, the p-type layer is connected with the positive terminal and the n-type layer is connected with the negative terminal, electrons enter from the n-type layer and holes from the p-type layer. When the two meet at the junction, an electron drops into a hole and light is emitted at the time.

Basic Structure of Semiconductor Lasers

The basic structure of a semiconductor laser is shown in figure. The active layer (light emission layer) sandwiched between the p- and n-type clad layers (double heterostructure) is formed on an n-type substrate, and voltage is applied across the p-n junction from the electrodes. Both edges of the active layer has mirror-like surface. When forward voltage is applied, electrons combine with holes at the p-n junction, and emit the light. This light is not a laser yet; it is confined within the active layer because the refractive index of the clad layers is lower than that of the active layer. In addition, both ends of the active layer act as a reflecting mirror where the light reciprocates in the active layer. Then, the light is amplified by the stimulated emission process and laser oscillation is generated.

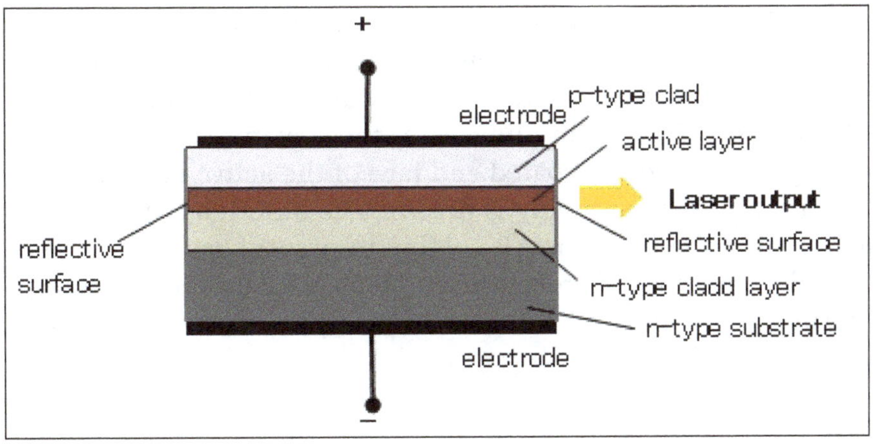

Types of Semiconductor Laser

The center wavelength of a semiconductor laser principally depends on the band gap energy of the active layer semiconductor. However, the details of laser spectra are different depending on the LD types even though the band gap energies are same.

- Fabry-Perot semiconductor laser: This laser has the simplest structure and is used for many applications, including optical pickups for CDs, DVDs, and BDs; laser printers; and excitation of fiber lasers. It is characterized by the use of the cleavage plane of a laser crystal for reflection of the light emitted in the active layer, as shown in Figure. A cleavage plane of a crystal is extremely smooth and can be used as a reflecting mirror. Since Fabry perot-LD has no mechanism to select a specific oscillation wavelength, the multiple laser oscillation occurs within the gain bandwidth as shown in Fig. The oscillation at multiple wavelengths causes pulse broadening in an optical fiber transmission, and therefore it cannot be applied for long-distance communications.

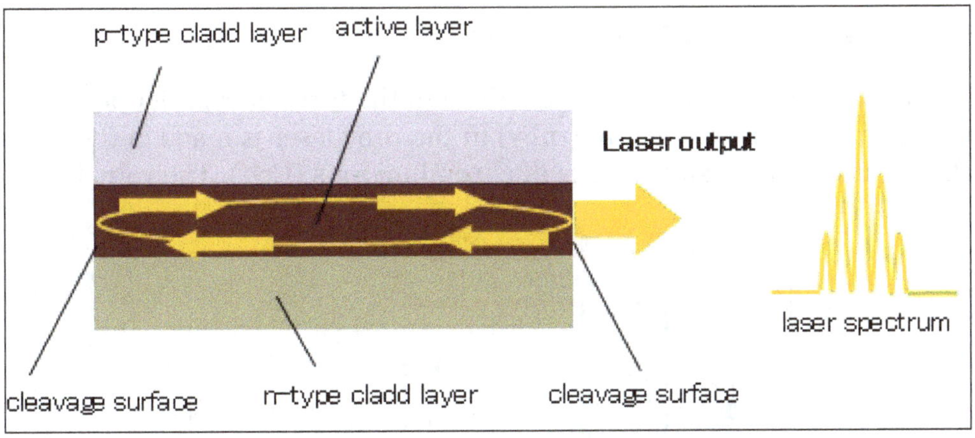

Schematic diagram of Fabry-Perot type semiconductor laser.

- DFB semiconductor laser: The DFB laser (Distributed Feedback Laser) has a grating below or above the active layer, and oscillates at the single wavelength defined by the Bragg wavelength of the grating. It exhibits superior performances, including a narrow spectrum width and low noise, and hence is used as an optical signal source in a long-haul optical communication.

- FBG wavelength stabilized semiconductor laser: Though DFB laser has an excellent performance like a single wavelength oscillation it is expensive for manufacturing difficulty. More economical laser to oscillate a single wavelength is a laser diode stabilizing the wavelength by using FBG. Figure illustrates the structure outline. In this case, the laser cavity is composed of the HR coated end face of the active layer and the low reflectance FBG. In figure, the fiber core only is drawn. The other end face of the active layer and the incident end of the fiber has AR coating to prevent reflecting light. In addition, the incidence end of the fiber is processed to a lens configuration to collect the light. Since the FBG formed in the fiber core only reflects a specific wavelength, a single wavelength oscillation occurs.

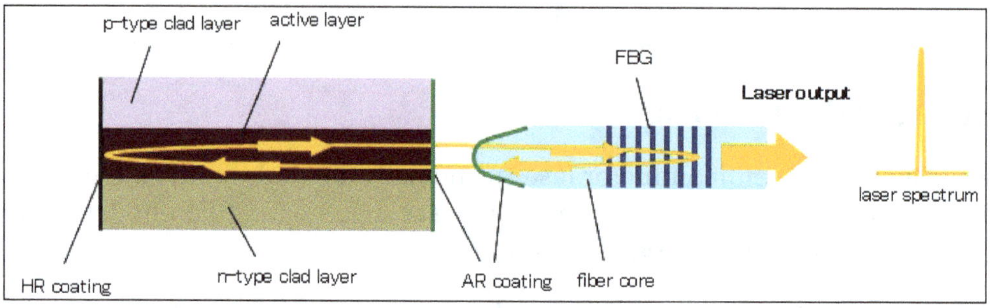

Schematic diagram of FBG wavelength stabilized semiconductor laser.

Ruby Laser

A ruby laser is a solid-state laser that uses the synthetic ruby crystal as its laser medium. Ruby laser is the first successful laser developed by Maiman in 1960. Ruby laser is one of the few solid-state lasers that produce visible light. It emits deep red light of wavelength 694.3 nm. A ruby laser consists of three important elements: laser medium, the pump source, and the optical resonator.

Laser Medium or Gain Medium in Ruby Laser

In a ruby laser, a single crystal of ruby (Al_2O_3: Cr^{3+}) in the form of cylinder acts as a laser medium or active medium. The laser medium (ruby) in the ruby laser is made of the host of sapphire (Al_2O_3) which is doped with small amounts of chromium ions (Cr^{3+}). The ruby has good thermal properties.

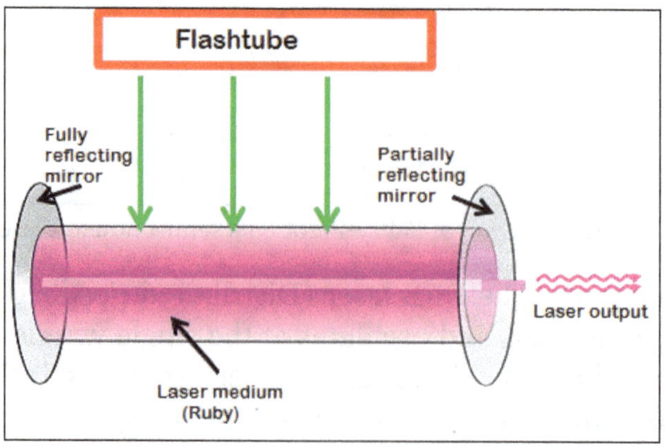

Pump Source or Energy Source in Ruby Laser

The pump source is the element of a ruby laser system that provides energy to the laser medium. In a ruby laser, population inversion is required to achieve laser emission. Population inversion is the process of achieving the greater population of higher energy state than the lower energy state. In order to achieve population inversion, we need to supply energy to the laser medium (ruby). In a ruby laser, we use flashtube as the energy source or pump source. The flashtube supplies energy to the laser medium (ruby). When lower energy state electrons in the laser medium gain sufficient energy from the flashtube, they jump into the higher energy state or excited state.

Optical Resonator

The ends of the cylindrical ruby rod are flat and parallel. The cylindrical ruby rod is placed between two mirrors. The optical coating is applied to both the mirrors. The process of depositing thin layers of metals on glass substrates to make mirror surfaces is called silvering. Each mirror is coated or silvered differently. At one end of the rod, the mirror is fully silvered whereas, at another end, the mirror is partially silvered. The fully silvered mirror will completely reflect the light whereas the partially silvered mirror will reflect most part of the light but allows a small portion of light through it to produce output laser light.

Working of Ruby Laser

The ruby laser is a three level solid-state laser. In a ruby laser, optical pumping technique is used to supply energy to the laser medium. Optical pumping is a technique in which light is used as energy source to raise electrons from lower energy level to the higher energy level. Consider a ruby laser medium consisting of three energy levels E_1, E_2, E_3 with N number of electrons. We assume that the energy levels will be $E_1 < E_2 < E_3$. The energy level E_1 is known as ground state or lower energy state, the energy level E_2 is known as metastable state, and the energy level E_3 is known as pump state.

Let us assume that initially most of the electrons are in the lower energy state (E_1) and only a tiny number of electrons are in the excited states (E_2 and E_3).

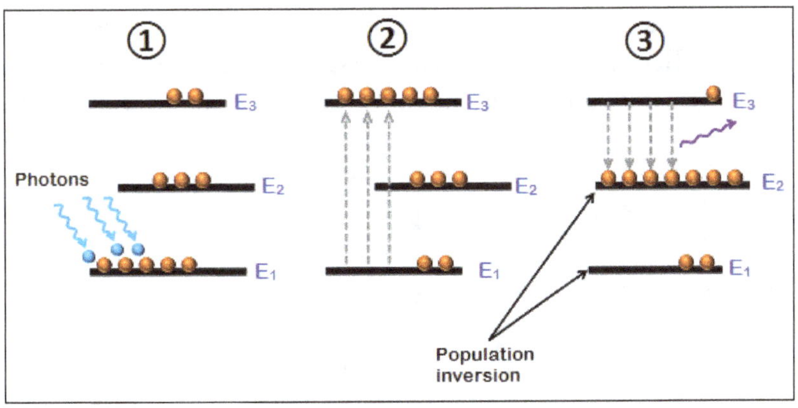

Physics and radio electronics.

When light energy is supplied to the laser medium (ruby), the electrons in the lower energy state or ground state (E_1) gains enough energy and jumps into the pump state (E_3). The lifetime of pump state E_3 is very small (10^{-8} sec) so the electrons in the pump state do not stay for long period. After a short period, they fall into the metastable state E_2 by releasing radiation less energy. The lifetime of metastable state E_2 is 10^{-3} sec which is much greater than the lifetime of pump state E_3. Therefore, the electrons reach E_2 much faster than they leave E_2. This results in an increase in the number of electrons in the metastable state E_2 and hence population inversion is achieved. After some period, the electrons in the metastable state E_2 falls into the lower energy state E_1 by releasing energy in the form of photons. This is called spontaneous emission of radiation.

When the emitted photon interacts with the electron in the metastable state, it forcefully makes that electron fall into the ground state E_1. As a result, two photons are emitted. This is called stimulated emission of radiation. When these emitted photons again interacted with the metastable state electrons, then 4 photons are produced. Because of this continuous interaction with the electrons, millions of photons are produced.

In an active medium (ruby), a process called spontaneous emission produces light. The light produced within the laser medium will bounce back and forth between the two mirrors. This stimulates other electrons to fall into the ground state by releasing light energy. This is called stimulated emission. Likewise, millions of electrons are stimulated to emit light. Thus, the light gain is achieved. The amplified light escapes through the partially reflecting mirror to produce laser light.

Laser System

CO_2 Laser

CO_2 is a linear and symmetric molecule and its ground electronic state is a ${}^1\Sigma$. As CO_2 is a linear molecule, therefore it has $3N - 5 = 4$ internal vibrational degree of freedom where $N = 3$ is the number of nuclei in the molecule. These four vibrations correspond to (i) symmetric stretching (ii) doubly degenerate bending mode and (iii) asymmetric stretching mode and have been shown in the figure.

In the symmetric stretching mode, the oxygen atoms oscillate along the axis of the molecule simultaneously departing or approaching the carbon atom that is stationary. In the bending mode, that is doubly degenerate, the molecule ceases to be exactly linear. It occurs both in the plane of the figure and the plane perpendicular to it. In this mode all the three atoms in the molecule undergo vibrational motion perpendicular to the molecular axis. In antisymmetric stretching all the three atoms oscillate but both the oxygen atoms move in one direction and carbon atom moves in the opposite direction.

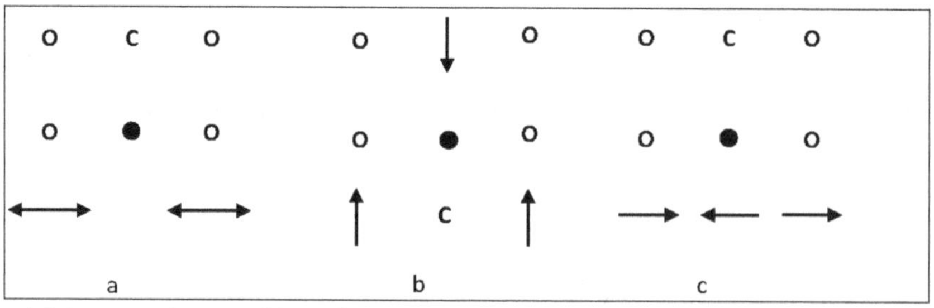

(a) Symmetric stretching, (b) Bending mode, (c) Antisymmetric stretching.

In the first approximation these three modes are described as independent harmonic oscillators, and hence the excited vibrational states are completely described by the number of excited quanta in each mode. These states are represented by a set of three vibrational quantum number v_1, v_2^1, v_3 where subscript 1, 2 and 3 refer to the symmetric bending and antisymmetric modes, respectively. The superscript 1 gives the angular momentum of this vibration about the axis of molecule in units of $h/2\pi$. For example, $0_2\,{}^{\circ}0$ indicates that the two vibrations combine to give an angular momentum $l = 0$.

The Figure depicts the energy level diagram for the lowest levels. Each of the vibrational levels has superimposed upon it a set of rotational levels. The rotational levels are spaced by energies that are more than three orders of magnitude smaller than the vibrational energies.

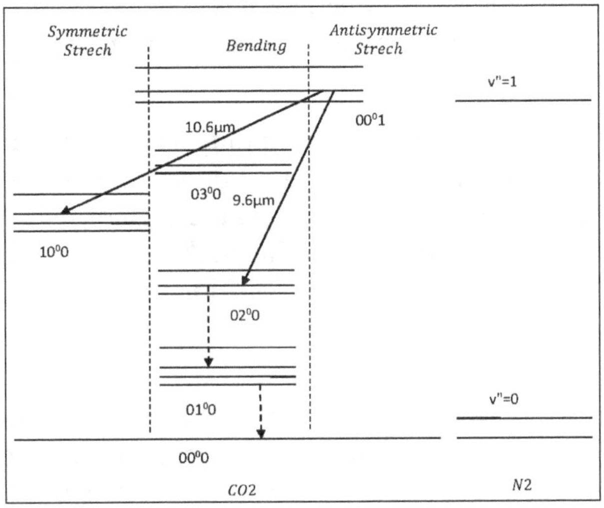

The CO_2 laser uses a mixture of CO_2, N_2 and He or water vapor. The active centres are the CO_2 molecules lasing on the transition between the rotational levels of the vibrational bands of the electronic ground state. The figure also shows the vibrational energy level for the ground state of N_2 molecule.

Nitrogen being a homonuclear diatomic molecule does not posses a permanent electric dipole moment and hence radiation decay from v" = 1 to v" = 0 is forbidden. The lifetime of the level v" = 1 is therefore quite long and ~0.1 s at one torr. The vibrational level 00 °1 of CO_2 is in near coincidence ($\Delta E = 18$ cm^{-1}) with v"= 1 level of nitrogen. The broken lines in the figure show non-radiative transitions.

When a discharge is passed in a tube containing mixture of CO_2, N_2 and He, electron collisions excite the molecules to higher electronic vibrational rotational states. The electronic collision cross section for the excitation to the level 00 °1 is very large. The electron collision excite the N_2 molecules by,

$$N_2 + e^- \rightarrow N_2^- \rightarrow N_2^* + e^-$$

The cross section for this process is quite large and produce significant amount of metastable N_2 in the discharge. These nitrogen molecules transfer their internal energy to CO_2 anymmetric stretching modes through near resonance collision.

$$CO_2 + N_2^* \rightarrow CO_2^* + N_2 + \Delta E$$

The cross section for this process is also large therefore, the excitation of CO_2 00 °1 level is unusually efficient. Further, the discharge containing the CO_2 produces dissociation of CO_2 into CO and oxygen. As Co vibrational frequency is not grossly different from that of the CO_2 asymmetric mode, it will rapidly transfer its internal energy to the CO_2 system through near resonant collision in the same manner as N_2. The high density of population arises principally from three factors:

- The Long Vibrational Lifetime of the Asymmetric Mode of CO_2.
- The Near Resonant N_2– CO_2 Energy Transfer.
- The Large Cross Section for the Production of the Vibrationally Excited Nitrogen through Inelastic Collision with Electrons population Gain from Other Levels of the Asymmetric Mode.

The 10 °0 and 02 °0 levels are in resonance. The coupling between these levels is sufficiently strong so that 10 °0 and 02 °0 level system behaves like a single level. The separation between these levels ~100cm^{-1} that is significantly less than the thermal energy of motion k_BT ~ 210 cm^{-1} at room temperature. The similarity of vibrational motion in combination with the near degeneracy of the levels causes the rate of transfer between them by collision process, corresponding to,

$$10^{\circ}0 + 00^{\circ}0 \rightarrow 02^{\circ}0 + 00^{\circ}0 + \Delta E$$

to be extremely rapid. Thus 10 °0 and 02 °0 levels of CO_2 reach thermal equilibrium in a very short time. For population inversion, between upper 00 °1 level and lower 10 °0 and 02 °0 levels, it is necessary that population of lower levels should decay very fast. The presence of He has a considerable influence on the population decrease of lower levels. The He atoms collide with CO_2 and increase the rate of relaxation of 01 °0 levels.

The rate is approximately twenty times in collision of the type,

$$01^{\circ}0 + He \rightarrow 00^{\circ}0 + He + \Delta E$$ that for the corresponding

$$01^{\circ}0 + 00^{\circ}0 \rightarrow 00^{\circ}0 + 00^{\circ}0 + \Delta E$$

process where two CO_2 molecules collide. This reduces the bending mode population that in turn reflects itself in reduction of lower level population, and hence increases the gain.

Another function of He is to keep CO_2 cold. The excited CO_2 can decay spontaneously and directly to the ground state, liberating energy as heat. It is necessary to keep temperature of CO_2 as low to avoid population of lower level by thermal excitation. Helium has high thermal conductivity and hence helps to conduct heat away to the walls keeping CO_2 cold. Thus while N_2 helps to increase the population of the upper level, helium helps to depopulate the lower level.

The laser oscillation is at a wavelength corresponding to the transition from the upper vibrational level 00 °1 to the lower vibrational level 10 °0. With each vibrational level, rotational levels are also associated. Thus a number of laser lines are possible. The population of J' = 21 of 00 °1 has the maximum population and transition corresponding to J' = 21 to J'' = 22 or P(22) is strongest and this corresponds to the wavelength of 10.6μm. Another laser line of λ = 9.6 μm from transition between 00 °1 and 02 °0 is also prominent.

A Schematic of CO_2 laser is shown in the figure. The discharge is produced in tube having diameter of 2.5 cm and length 5m. Brewster windows are used at the end. These windows are made of NaC1, CaC1, BaFC1, ZnSe or KC1 as these materials are transparent to 10.6μm radiation. As glass strongly absorbs most of the frequencies in infrared region therefore, near confocal Si mirrors coated with A1 are used.

He, N_2 and CO_2 are filled in the ratio 8:1:1 with pressure of He, N_2 and CO_2 as 0.7 torr, 1.2 torr and 0.33 torr, respectively. To remove the dissociation products such as Co and O, that would contaminate the laser, the continuous flow of the gas mixture is maintained in the tube. Grating can be used to separate out different laser lines. CO_2 laser can be made to operate in pulsed or CW mode. Pulse energy ~2KJ and pulse width is of few μs. For CW operation, power > 50 W is easily available.

Applications of CO_2 Laser

The advantage of using this laser is based on the fact that it provides very intense heating source over a very small area. It has applications in:

- Material processing: Cutting, drilling, material removal, welding, cladding, alloying, hardening, melting etc.
- In medicine it is used for cutting and cauterizing.

Nitrogen Laser

N_2 is a homonuclear molecule. The electronic ground state of N_2 is $X^1\Sigma_g^+$. Out of the many electronic state possible, three triplet excited states, that is, $A^3\Sigma_u^-$, $B^3\Pi_g$ and $A^3\Sigma^+_u$ along with ground state are shown in figure and the equilibrium internuclear distances r_e for various states are as:

State	r_e(nm)
$C^3\Pi_u$	0.11487
$B^3\Pi_g$	0.12126
$A^3\Sigma^+u$	0.12866
$X^1\Sigma^+g$	0.10977

Energy Levels of N_2 Molecule.

The values of r_e indicate that the minimum of potential for C state lies almost vertically above that of the X state. On the other hand the minima of A and B shifted to high r. The electron collides with the molecules and excites them to various levels, when the electric discharge is passed through the gas, According to Franck Condon principle the probability of transition from X to C is much larger than that of X to A or X to B. Thus, population inversion occurs at level C (v = 0 level) relative to level B (v = 0 level). A lasing action is observed between C and B levels (corresponding to 0-0 and 0-1 vibrational levels). As the lifetime of state C is about 40 ns and that of B is about 100μs, therefore, the molecule remains in C state for a short duration and after transition accumulates in level B. Hence the population difference between B and C decreases very rapidly and this renders laser action impossible.

The population inversion can therefore be achieved only for a short time and after this the laser action is self-terminating. Due to high gain of this self-terminating transition, oscillations take place in the form of amplified spontaneous emissions. Thus the laser can be operated without mirror. The mirror placed at one end of the tube reduces the threshold power and also provides unidirectional output. This also reduces the beam divergence. The pulse width for nitrogen laser is normally 10ns and emission wavelength is at 337.1 nm. Peak power up to 1MW can be achieve.

A pulsed high voltage of about 20kV, triggered by a thyratron or spark gap is applied across the tube through which N_2 at a pressure of about 100 torr, is flowing. On one end of the tube a mirror is used while a window used on the other end has a very high transmission. The length of the tube is between 0.1 – 1.0m.

Titanium Sapphire Laser

In 1960 T. H. Maiman demonstrated the operation of the first optical maser, or laser, a crystal of synthetically grown ruby, or sapphire (Al_2O_3) doped with a small amount of chromium was used in the experiments. P. F. Moulton demonstrated a widely tunable laser at Lincoln Laboratory, twenty-two years later, where titanium instead of chromium was incorporated as an impurity into sapphire. Titanium-doped sapphire or Ti:Al_2O_3, has the largest tuning range of any laser that is, from 660 to 1180 nm.

The Ti:Al_2O_3 crystals used in the initial experiments exhibited significant scattering and an absorption at the laser wavelength and these losses affected the efficiency of the laser, and only pulsed operation was possible. Further with the advances in the development of higher-quality laser crystals of Ti:Al_2O_3, (with significantly smaller losses).room-temperature continuous-wave operation became a reality.

Sapphire is an ideal host crystal in both the ruby and Ti:Al_2O_3 laser. It is transparent from ultraviolet to infrared and also, it is non-hygroscopic and very hard, that is required for producing good optical-quality surfaces. The most common tunable lasers are the organic dye lasers. A dye laser consists of an organic dye (such as polymethine, xanthenes, or coumarin dye) in a liquid solvent or host. A typical dye laser such as Rhodamine 6G can be tuned from 570 to 610 nm; other dyes have laser bandwidths that cover portions of the spectrum from the ultraviolet to the infrared.

In a Ti:Al_2O_3 laser the laser transition is between two electronic levels of a single Ti^{3+} ion, and in dye lasers the laser transition is from one molecular electronic level to another. The broadening of the absorption and luminescence bands is caused by the multitude of rotational levels associated with each molecular electronic level. However there are undesirable characteristics of dye lasers as toxicity of the dyes and solvents, degradation of the dyes with time, and amplitude noise in the laser output (because the dyes are flowed to reduce thermal loads).

Another class of tunable solid state lasers is the alkali-halide color-center lasers. A color center is a crystal defect in which an electron becomes trapped. For example, an F color-center consists of an electron trapped at an anion vacancy of the crystal lattice. The excitations of the trapped electron strongly couple to the phonons of the lattice and give rise to broad absorption and emission bands. Color-center lasers can span the wavelength range from 800 nm to 4 μm where organic dye lasers are of limited usefulness. Color-center lasers have many drawbacks:

- In many alkali-halide crystals the color-centers are not stable at room temperature.
- Degrade on the time scale of a day.

In a tunable solid state laser, the interaction between the ion and host crystal is such that lattice vibrations or phonons usually accompany the emission or absorption of photons. As a result the absorption and emission spectra become broadened in each laser. These vibronic transitions can provide gain over the large bandwidth required for tunable lasers.

Spectroscopy

In titanium-doped sapphire the titanium ions substitute for the aluminum ions and exist in only 3+ charge states. The energy levels of the titanium ions are simple where a single d electron is in the outermost shell while the remaining 18 electrons have a filled-shell configuration of a neutral argon atom. When the titanium ions are placed in a host crystal, the electrostatic field of neighbouring atoms, or the crystal field, remove the five-fold angular momentum degeneracy of the single d electron. In Ti:Al_2O_3 the 3d electron electrostatically interacts with the electronic charges of six surrounding oxygen ions that are positioned at the corners of an octahedron.

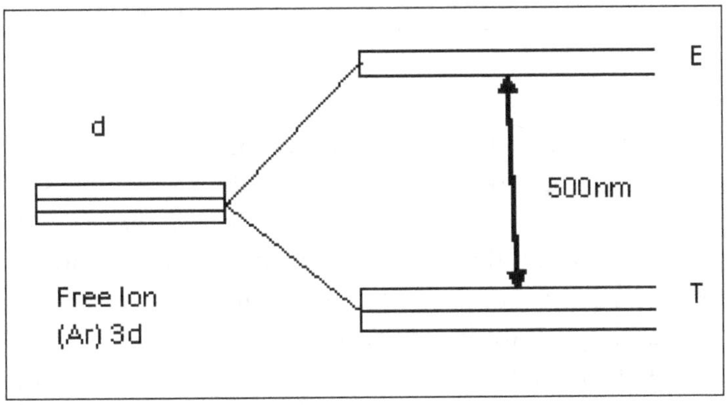

Energy-level diagram of Ti^{3+}.

The electronic configuration of the free ion is that of an argon shell plus a single 3d electron. The electronic energy levels of the Ti^{3+} ions in Ti:Al_2O_3 are further perturbed by the sapphire host lattice. When the Ti^{3+} ion is in the excited state, the overall energy of the system gets lowered if the position of the Ti^{3+} ion displaces itself with respect to the surrounding oxygen atoms understood to be due to the Jahn-Teller effect. This displacement removes the degeneracy of the two excited angular momentum states that leads to a splitting of the absorption band. Also, as the Ti^{3+} ion moves to its new equilibrium position, it kicks the surrounding lattice and excites vibrations (or phonons); that is why the Ti:Al_2O_3 laser is called a vibronic laser.

The coupling of the electronic energy levels of the Ti^{3+} ions with the vibrational energy levels of the surrounding sapphire lattice is essential for Ti: Al_2O_3 to operate as a laser. The energy-level diagram resembles that of a large polyatomic molecule such as that of an organic dye molecule. When the Ti^{3+} ion either absorbs or emits a photon, the 3d electron rearranges it's orbital more quickly than the heavier Ti^{3+} nucleus can move understood due to the Franck-Condon principle. The Gaussian-shaped curves at points A and C in the figure represent the probability of finding the Ti^{3+} at a particular position in the lowest vibrational state of the T and E levels, respectively. Figure shows absorption of light as the transition from point A to B. The transition is to either of the Jahn-Teller split upper states, and it results in a broad absorption. The Jahn-Teller splitting is not totally resolved and is manifested as a main peak with a shoulder. At point B, the ion displaces itself and lowers its energy by emitting phonons.

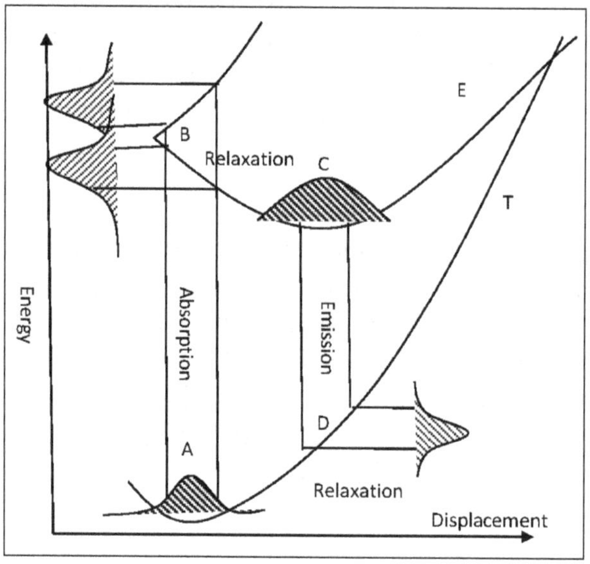

The transition from point C to point D shows the emission of light and the ion relaxes quickly to the ground state by emitting phonons. Two important results are to be noted; first, the emission of light following the absorption of light is at a longer wavelength (Stokes shifted). A population inversion in the red emission band, necessary for amplification, is more easily achieved because the emission terminates on high vibrational levels of the ground state that are unpopulated because of the fast vibrational relaxation rate. Second, a large emission bandwidth required for broad tenability results because the spread in probability of the Ti^{3+} ion position at the bottom of the E potential can connect via vertical transitions to a large spread of vibrational levels of the T potential. Further excited levels of the 3d electron lie far above the E levels and hence the Ti^{3+} exhibits no excited state absorption (ESA).

Laser Structure

- Continuous wave Titanium Sapphire laser.
- Mode Locked Titanium Sapphire laser.

Pumping is done either by argon ion laser (for cw operation) or a doubled Nd:YAG or Nd:YLF lasers for pulsed operations is used as pumping sources.

Applications

- Titanium Sapphire Lasers are used in Infrared spectroscopy of semiconductors and in laser radar, rangefinders, and remote sensing.
- They are used in medical applications such as photodynamic therapy.
- They are used to produce short pulses of X-rays by focusing the mode locked pulses onto solid target from which high density and high temperature radiating plasmas are produced, plasmas that in turn emit large fluxes of X-rays.

Free Electron Laser (FEL)

More than 50 years ago the first coherent red light has been generated by a ruby laser and thereafter more than 20 years ago the first coherent infrared light has been generated at the Stanford University (USA), and this laser was named Free Electron Laser (FEL).

Principle of Free Electron Laser (FEL)

This new type of coherent radiation source appeared from the very beginning to be quite promising. The reason lies on the peculiarity of its operating principle. In a FEL the active medium is not made of atoms or molecules, it consists of a beam of free electrons, propagating at relativistic velocities in a spatially periodic magnet (undulator), where the electrons experience the Lorentz force, execute transverse oscillations and emit synchrotron radiation in the forward direction as has been shown in figure. The wavelength λ of the emitted radiation depends on the electron energy E, on the period of the undulator magnet λ_u and on its magnetic field B,

$$\lambda = \left(\frac{1}{2\lambda_u}\right)\left(1+K^2\right)\left(\frac{mc^2}{E}\right)^2$$

Here, $K = \frac{eB\lambda_u}{2\pi mc^2}$ = Pitch Parameter

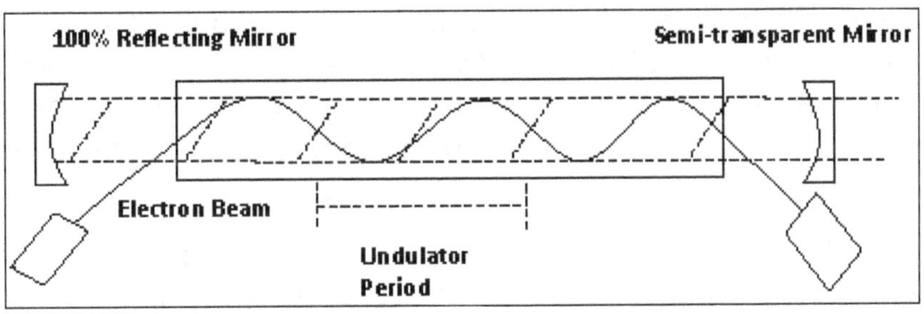

Synchrotron emission in Undulator Magnet.

The above relationship can be easily derived by taking into account the synchronism condition, i.e. considering that the electron, that is oscillating with a spatial period λ_u and it is moving with at a longitudinal speed v, maintains, oscillation after oscillation, the same phase relation with respect to the emitted electromagnetic wave of wavelength λ, running along the same direction at the speed c. Thus,

$$v = c\lambda_u / (\lambda_u + \lambda)$$

Taking into account the relationship between electron energy and relativistic velocities, the process described so far can be viewed as a kind of spontaneous emission by electrons propagating in an undulator magnet. By spontaneous mean that the emission occurs in absence of a field quasi resonant with the emitted radiation and co-propagating with the electrons. The possibility of realizing a stimulated process is quite natural if the photons emitted along the passage through the undulator are trapped in an optical cavity and then interact again with a fresh electron beam as is shown in the figure. These electrons are forced to emit stimulated synchrotron radiation.

One can associate to this process a gain and if the gain is larger than the cavity losses, the onset of the laser action is achieved that is in full analogy with conventional lasers, when saturation is reached, highly coherent (transverse and longitudinal) radiation can be extracted from the optical cavity. A quantity of much importance is the small signal gain coefficient that characterize the gain of the device for its operation in low or high gain regime,

$$g = \left(\frac{16\pi}{\gamma}\right)\lambda[m]N^2L_u[m]j\left[\frac{A}{m^2}\right]\xi f_b(\xi)^2 / 1.7\times10^4$$

Here, $\gamma = E/mc^2$ - relativistic factor,

N = No of undulator periods,

L_u = Undulator Length,

j = Electron current density,

$$\xi = \frac{\left(\frac{K}{2}\right)^2}{1+\frac{K^2}{2}}$$

$f_b(\xi) = J_0(\xi) - J_1(\xi)$ bessel function

This suggests that the gain decreases drastically at short wavelength. The gain decreases indeed with the wavelength and, in the meantime, with the increase of the electron energy, that is required in order to go to shorter wavelengths. In this region (~up to the VUV-10 to 200 nm) the operation of FEL oscillator requires very high brilliant electron beams (high j) that can be obtained in high quality linac accelerators or in Storage Rings.

The figure gives the layout of FEL (single passage). In the spectral region where the mirror reflectivity is too small (typically in the VUV region) a different operation scheme, based on single passage self-amplified spontaneous emission (SASE), is possible if the electron beam brightness is good enough to have very high gain. In this operation mode, the electron beam interacts with its own spontaneous radiation and an unstable wave is excited. The radiation intensity I grow exponentially along the undulator.

The most interesting features of FELs are:

- λ can be tuned (tenability) by varying electron energy, period and amplitude of magnetic field; in addition harmonic generation can provide even shorter λ.
- Time structure: depending on the e.beam time structure it ranges from c.w. to ultrashort pulsed regime (fraction of ps).
- Coherence: Transverse and longitudinal for oscillators and coherent seed amplifiers, only transverse for SASE ·
- Brilliance: The FEL brilliance can be larger, in some spectral regions (in particular in VUV-X), by many order of magnitude than the brilliance of the existing sources (lasers and synchrotron radiation).

There are many FEL devices in operation around the world ranging from millimeter waves to UV.

X-ray Laser

After T. Maiman demonstrated the first laser in 1960, the optical community immediately began seeking shorter and shorter wavelengths, through the visible and into the ultraviolet. By the early 1970s, lasers had reached the vacuum ultraviolet. However, the push toward shorter wavelengths stalled around 110 nm as problems emerged with short-wavelength laser physics.

As the energy of a laser transition increases, the excited state lifetime decreases, and the pump energy required to produce a population inversion increases. In fact, the two effects combine, so pump energy must be concentrated in a pump pulse that can be deposited very quickly in the laser material. A further complication is the lack of materials able to reflect or transmit light efficiently in the deep or extreme ultraviolet, making conventional mirrors and windows impractical.

With these problems in mind, scientists shifted their approach and instead of trying to make a laser oscillator, they sought to highly excite cylindrical plasmas that could generate intense X-ray pulses by amplifying stimulated emission along their lengths. The obvious way to excite the plasma was by firing intense optical laser pulses onto a target, and fortunately, high-power lasers became possible.

Shorter Wavelengths and Smaller Lasers

The selenium experiments marked a major milestone for Livermore, but had two important limitations. The 20-nm wavelength is relatively long. Biological imaging researchers wanted wavelengths shorter than 4.5 nm, and some definitions of X-rays include only wavelengths shorter than 10 nm. The logical way to shorter wavelengths was by using neon-like ions of heavier elements. Molybdenum brought Livermore to 13.3 nm, but Nova didn't have the power to reach 4.5 nm with heavier neon-like ions. Reaching that target required shifting to nickel-like ions, heavy elements stripped of all but their 28 innermost electrons. Europium was first, at 7.1 nm, and Livermore later reached 4.316 nm in tungsten. Hagelstein, in 1986, devised a dual-pulse approach, with the first pulse ionizing the plasma and the second exciting the ions to produce a population inversion in nickel-like niobium.

Lacking mirrors suitable for X-ray oscillators, all those X-ray laser demonstrations generated only amplified spontaneous emission with limited coherence. Early this year, Rocca's group reported an important new twist— aiming high harmonics from a titanium-sapphire laser into the excited plasma. Where plasma gain matched a harmonic frequency, the plasma amplified the harmonic seed pulse, generating gain-saturated output at 18.9 nm from nickel-like molybdenum and 13.9 nm from nickel-like silver. Those experiments are a big step toward a practical coherent table-top X-ray source, and the shorter wavelength is in a band that is attractive for future generations of semiconductor photolithography.

Methods and Applications of Laser Spectroscopy

Laser Absorption Spectroscopy

A frequently used method is laser absorption spectroscopy, where a tunable narrow-linewidth laser (frequently a single-frequency laser) is tuned through some wavelength range, and the light absorption in some sample (i.e., a reduction of optical power of the probe beam) is measured as a function of that wavelength. Obviously, the spectral resolution is limited by the laser linewidth, which is therefore often minimized with suitable laser designs. Extremely high precision is required and achieved in the area of optical frequency metrology, e.g. for realizing extremely precise optical clocks.

Absorption features are not always directly investigated by measuring wavelength-dependent absorption. Instead, one may exploit more subtle effects on modulated laser beams (frequency modulation spectroscopy) or detect sound waves induced by laser pulses. There are many variants of laser absorption spectroscopy, often developed with the goal to further improve the sensitivity.

Frequency Comb Spectroscopy

Some techniques of modern laser spectroscopy employ frequency combs as generated with mode-locked lasers. As such a frequency comb contains some number of exactly equidistant lines in the spectrum, all of its frequency components are known (apart from some noise) if only two parameters are fixed (possibly stabilized with some feedback techniques): the comb spacing, which is related to the pulse repetition rate and the carrier–envelope offset frequency. Therefore, frequency combs can serve for extremely precise frequency measurements in wide wavelength ranges, if they are produced with a high optical bandwidth (sometimes more than octave-spanning) and with proper frequency stabilization. There are various important applications of frequency comb spectroscopy in frequency metrology (in particular, ultraprecise optical clocks) and in other fields.

Raman Spectroscopy

In Raman spectroscopy, one exploits the phenomenon that a medium irradiated with narrow-band continuous-wave laser light does not only scatter light at the same optical frequency (Rayleigh scattering), but partly with slightly reduced frequencies (Raman scattering). The corresponding loss of photon energy stays in the medium; it leads to the excitation of vibrational or rotational modes. By analyzing the optical spectrum of the weak Raman-shifted light (after suppressing Rayleigh-scattered light with a notch filter), one can retrieve information on the vibrational or rotational modes of the medium, also on its temperature. Raman spectroscopy has a wide range of applications for example in biology and medicine, for distributed temperature sensing in optical fibers, the analysis of artwork and the detection of explosives.

Photoacoustic Spectroscopy

The wavelength-dependent absorption of light in a sample is usually measured through the reduction of optical power of a light beam going through the sample. Photoacoustic spectroscopy, however, is based on a different method: one detects sound waves which are excited by the absorption of intense laser pulses. Such pulses can heat a sample gas or solid material, for example, causing a week pressure wave which can be detected with a microphone. Processing the microphone signal with a lock-in amplifier, one can achieve a high sensitivity.

Ultrafast Laser Spectroscopy

Although probably most techniques of laser spectroscopy are based on continuous-wave lasers, there are also various methods where mode-locked lasers are used, which produce trains of ultrashort pulses. Some examples:

- There are pump–probe measurements which allow one to measure the temporal evolution of systems on ultrashort time scales. In simple cases, no spectral information is obtained. However, it is also possible to get time-resolved spectroscopic information when using broadband probe pulses. This is called transient-absorption spectroscopy.
- There are methods of time-resolved photoelectron spectroscopy, where a first laser pulse excites the sample under investigation, and a second laser pulse ionizes the material. One then measures the distribution of kinetic energies of photoelectrons from this process for different time delays between the pump and probe pulses.

Laser-induced Breakdown Spectroscopy

Intense laser pulses can vaporize and ionize materials, involving the effect of laser-induced breakdown. The light flash emerging from the vaporized material can be analyzed with a spectrograph, and spectroscopic fingerprints can then be used to identify certain atoms. For this kind of spectroscopic method, the wavelength of the laser pulses is not particularly important, as the laser light is used only for concentrated delivery of energy. Wavelength resolution is obtained only on the side of photodetection.

Applications of Laser Spectroscopy

Methods of laser spectroscopy are often used for detecting the composition of materials, often including quantitative measurements of concentrations. Some examples:

- Concentrations of trace gases in the atmosphere are measured e.g. with laser radar (LIDAR) methods in the context of environmental monitoring. Similarly, pollutants can be detected in water, and concentrations of medically active substances can be measured.

Spectroscopic devices based on the analysis of laser-induced fluorescence are partly used even in space, e.g. for monitoring vegetation and optimizing agriculture.

- Biology, medicine and chemistry can profit from precise and rapid methods of material analysis, which are partly non-destructive.
- Typical security applications include the detection of explosives and drugs.
- The detailed composition of pieces of artwork like paintings can be analyzed e.g. for revealing their age and checking their authenticity.
- There are numerous applications in industrial process monitoring and quality control, and in fundamental research, e.g. the diagnostics of combustion processes.
- In various contexts, spectroscopy can be used for temperature measurements based on the temperature dependence of level populations.

References

- Laser-spectroscopy: rp-photonics.com, Retrieved 01, March 2020
- Mode-locking: rp-photonics.com, Retrieved 29, July 2020
- About-semiconductor-laser-diode: fiberlabs.com, Retrieved 13, April 2020
- Ruby-laser-definition-construction-working: physics-and-radio-electronics.com Retrieved 10, August 2020

Permissions

All chapters in this book are published with permission under the Creative Commons Attribution Share Alike License or equivalent. Every chapter published in this book has been scrutinized by our experts. Their significance has been extensively debated. The topics covered herein carry significant information for a comprehensive understanding. They may even be implemented as practical applications or may be referred to as a beginning point for further studies.

We would like to thank the editorial team for lending their expertise to make the book truly unique. They have played a crucial role in the development of this book. Without their invaluable contributions this book wouldn't have been possible. They have made vital efforts to compile up to date information on the varied aspects of this subject to make this book a valuable addition to the collection of many professionals and students.

This book was conceptualized with the vision of imparting up-to-date and integrated information in this field. To ensure the same, a matchless editorial board was set up. Every individual on the board went through rigorous rounds of assessment to prove their worth. After which they invested a large part of their time researching and compiling the most relevant data for our readers.

The editorial board has been involved in producing this book since its inception. They have spent rigorous hours researching and exploring the diverse topics which have resulted in the successful publishing of this book. They have passed on their knowledge of decades through this book. To expedite this challenging task, the publisher supported the team at every step. A small team of assistant editors was also appointed to further simplify the editing procedure and attain best results for the readers.

Apart from the editorial board, the designing team has also invested a significant amount of their time in understanding the subject and creating the most relevant covers. They scrutinized every image to scout for the most suitable representation of the subject and create an appropriate cover for the book.

The publishing team has been an ardent support to the editorial, designing and production team. Their endless efforts to recruit the best for this project, has resulted in the accomplishment of this book. They are a veteran in the field of academics and their pool of knowledge is as vast as their experience in printing. Their expertise and guidance has proved useful at every step. Their uncompromising quality standards have made this book an exceptional effort. Their encouragement from time to time has been an inspiration for everyone.

The publisher and the editorial board hope that this book will prove to be a valuable piece of knowledge for students, practitioners and scholars across the globe.

Index

www.ingramcontent.com/pod-product-compliance
Lightning Source LLC
LaVergne TN
LVHW080721210926
848485LV00003B/502